OBJECT ORIENTED TECHNIQUES

(WITH JAVA)

Lalit Kishore Arora
M.Tech (C.S.), MCA, B.Sc. (Comp. Sc.)
Asstt. Professor, MCA Department
Ajay Kumar Garg Engineering College,
Ghaziabad

Dr. Raj Kumar Bhatia
Ph.D (C.S.)
Asstt. Professor, Deptt of Computer Science
Gurukal Kangri Vishwavidyalaya, Haridwar

S.K. KATARIA & SONS®
Publisher of Engineering Books
4885/109, Prakash Mahal, Dr. Subhash Bhargav Lane,
Opposite Delhi Medical Association, Daryaganj, New Delhi–110002 (INDIA)
Phone: +91-11-43551243; *Mobile:* +91-9871775858, 7678425209
e-mail: katariabook@gmail.com, katariabooks@yahoo.com
Website: www.skkatariaandsons.com
(e-books, subscriptions and bulk buying also available on our website)

Published by:

S.K. Kataria & Sons®
4885/109, Prakash Mahal, Dr. Subhash Bhargav Lane,
Opposite Delhi Medical Association, Daryaganj, New Delhi-110002 (INDIA)
Phone: +91-11-43551243; *Mobile:* +91-9871775858, 7678425209
e-mail: katariabook@gmail.com, katariabooks@yahoo.com
Website: www.skkatariaandsons.com

First Edition: 2010-11; Second Edition: July, 2011, Reprint: 2026

Price: ₹ 595/-

Laser Typesetting by : Plus Computers, Delhi *Printed at :* New A.S. Printing Press, Delhi

PREFACE

This book Object Oriented Technique with JAVA is written to serve as a text book for B.Tech Vth Semester students of Gautam Budha Technical University. There are no prerequisites beyond a general familiarity with the ideas of computers and programs. There is enough material for a full semester of college-level programming.

This book, has been divided into two parts. Part-I is deals with Object-Oriented Techniques. The Object-Oriented Approach of software development based on modeling object from the real world. In this part we have also described implementation with modeling and design.

Part-II dedicated to programming in JAVA, which covers all features of object oriented programming such as object, classes, inheritance and polymorphism. In the second revised edition we try to cover the feature of Java5. Therefore we have covered new feature of Java5 such as Boxing, enchanced for loop, object class, instance of operator and etc.

This book also cover the advance features of Java such as Java applet, Java Beans, Java Swings and JDBC. In each chapter, sufficient worked examples have been given. At the end of each chapter, solved problems and programs have been given for the practice.

Wherever necessary, concepts are given in boxes to facilitate easy grasping and better understanding.

We have also given Two appendices in last, where you can find the questions related to interview and your will also find the previous year papers and some model papers. The author will feel highly obliged to all readers for their constructive suggestions healthy criticism of the book which will go a long way in the improvement of the text.

— Authors

ACKNOWLEDGEMENT

It is a pleasure to be able to thank all the people who helped us turn this third edition of Object Oriented Techniques with Java from a list of good ideas into a real book.

We are pleased to express our thanks to the following Academician for allowing us to work under their valuable guidance:

- Dr. Vinod Kumar Sharma, Professor, Gurukul Kangri Vishwavidyalaya, Haridwar
- Dr. Karamjeet Bhatia, Professor, Gurukul Kangri Vishwavidyalaya, Haridwar
- Dr. Atul Kumar Jain, Chairman, Ideal Institute of Technology, Ghaziabad.

My special thanks goes to Mr. Ankit Gupta, Sr. Executive, RelSol Pvt. Ltd., Delhi, Mr. Deepak Gupta, Tech. Lead, Portech India Info. Service Pvt. Ltd., Noida, who played a major role in scripting this book.

We are greatly intended to Mr. Sanjeev Kataria and other publishing professionals at **S.K. Kataria and Sons** for their active involvement throughout the production process and for bringing out the book in its present from in record time.

— Authors

Dedication

To my Parents and Gurus for their love and support to achieve any goal.

—Lalit Kishore Arora

To my Parents and friends with whose love and support made this goal possible.

—Dr. Raj Kumar Bhatia

CONTENTS

Chapter 1

INTRODUCTION

1.1. INTRODUCTION

Programming languages may be thought to be belonging to different generations, depending on the way a program is structured and the way data and program are connected.

First-Generation Languages (1954–1958). These first-generation languages (Fortan I, ALGOL 58) have the following features (Fig. 1.1):

1. Subprograms were seen as more labour-saving devices.
2. Data were globally defined.

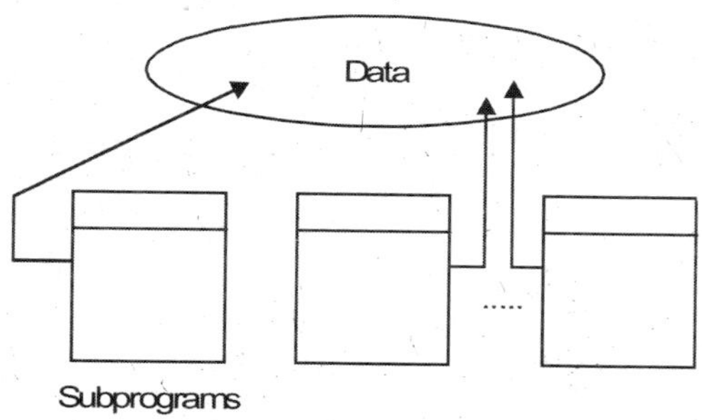

Fig. 1.1. First Generation Languages

Second-Generation Languages (1959–1961). To this generation belong such languages as Fortran II, ALGOL 60, COBOL and LISP. They have the following features (Fig. 1.2):

A. Nesting of subprograms was allowed.
B. Various methods were used for passing parameters from are subprogram to another.

CHAPTER OUTLINE

C. Structured programming constructs were used.

Third-Generation Languages (1962–1970). The languages belonging to this generation are PL/1, ALGOL 68, PASCAL, and Simula. The features of these languages are as under (Fig. 1.3):

- Programming-in-the-large
- Separately compiled modules
- Data types were introduced

The Generation Gap (1970–1990). An excess of languages developed during the seventies.

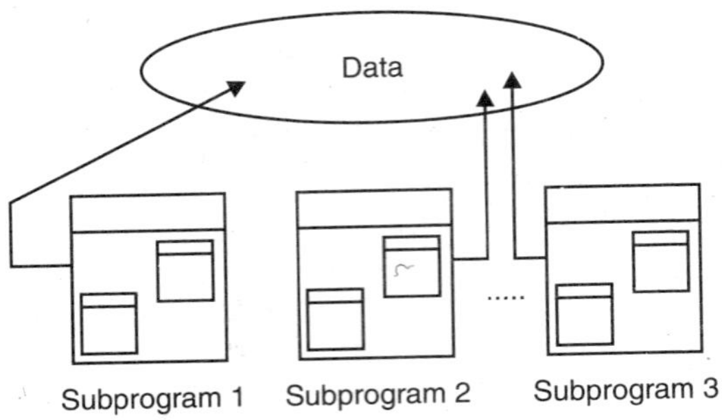

Fig. 1.2. Second Generation Languages

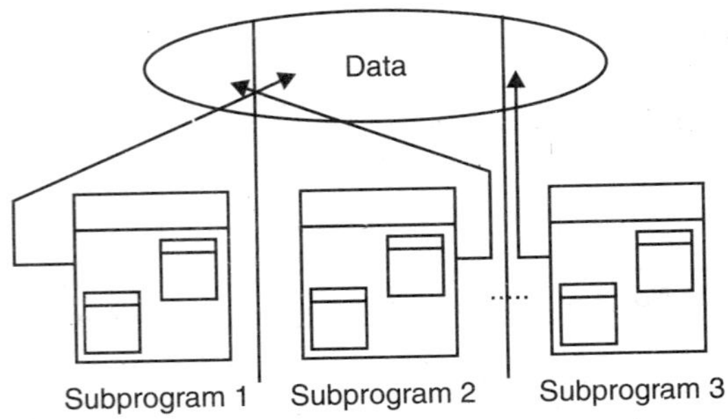

Fig. 1.3. Third Generation Languages

Object-Based and Object-Oriented Programming Languages (1990–Onwards). These languages (Ada, Smalltalk, C++, Object PASCAL, Eiffel, CLOS, etc.) have the following features (Fig. 1.4):

1. Data-driven design methods.
2. Theory of data typing emerged.
3. Little or no global data.
4. Physical structure of an application appears like a graph, rather than a tree.

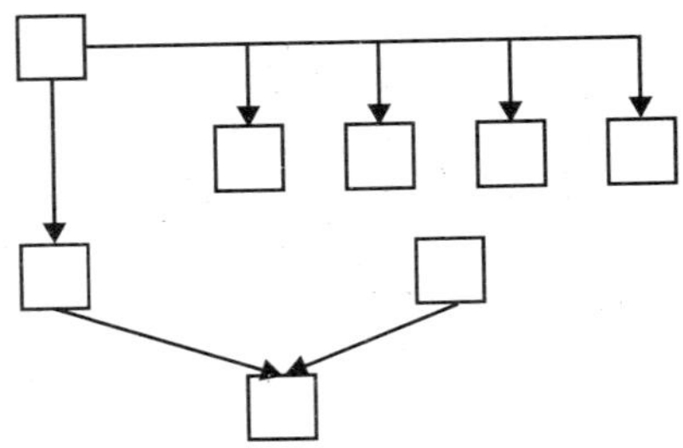

Fig. 1.4. Object-Oriented Languages

Following table gives the evolution of generation of languages.

Table 1.1. Evolution of Generation of Languages

1st Generation	2nd Generation	3rd Generation	Generation Gap (1970–1990)	Object Based and Object Oriented Generation 1990 - on wards
Fortan I	Fortan II	PL/1		
ALGOL 58	ALGOL 60	ALGOL 68		Ada (Contribn. from Alphard (CLU))
	COBOL	PASCAL		Object PASCAL, Eiffel Smalltalk
	LISP	SIMULA		C ++ (Contribn. from C) CLOS (Contribution from LOOPS + and Flavours)

One way to distinguish a procedure-oriented language from a object-oriented language is that the former is organized around procedures and functions (verbs) whereas the latter is organized around pieces of data (nouns). Thus, in a procedure-oriented program based design, a module represents a major function, such as 'Read a Master Record', whereas in an object-oriented program based software design, 'Master Record' is a module.

Simula-80 had the fundamental ideas of classes and objects. Alphard, CLU, Euclid, Gypsy, Mesa and Modula supported the idea of data abstraction. Use of object-oriented concepts led the development of C to C++; Pascal to Object Pascal, Eiffel, and Ada; LISP to Flavours, LOOPS, and to Common LISP Object System (CLOS).

The concepts of object orientation came from many computer scientists working in different areas of computer science. We give below, almost chronologically, a list of prominent scientists whose contributions were significant (Table 1.2).

Table 1.2. Object Oriented Languages and their Scientist

S.No.	*Computer Scientist*	*Contributions*
1.	Larry Constantine	Gave the idea of coupling and cohesion in 1960s that provided the principles of modular design of programs.
2.	D. J. Dahl and K. Nygaard	Introduced the concept of 'class' in the language Simula in 1966.
3.	Allan Kay, Adele Goblberg, and others	Developed, in 1969, the first incarnation of Smalltalk - the purest form of object orientation where they introduced the concepts of inheritance, message, and dynamic binding.
4.	Edsger Dijkstra	Gave the idea of semantically separated layers of abstraction during software building, which is the central concept of encapsulation.
5.	Barbara Liskov	Developed the theory of Abstract Data Type (ADT) and also developed, in 1970s, the CLU Language that supported the notion of hidden internal data representation.
6.	David Parnas	Forwarded the principle of information hiding in 1972.
7.	Jean Ichbiah and others	Developed ADA that had, for the first time, the features of genericity and package.
8.	Bjarne Stroustrup	Grafted object orientation on C, in 1991, to develop C++ that is portable across many machines and operating systems due to its foundation on C.
9.	Bertrand Meyer	Combined the best idea of computer science to the best idea of object orientation, in 1995, to develop Eiffel.
10.	Grady Booch, Ivar Jacobson, and James Rumbaugh	Developed, in the late 1990s, the Unified Modeling Language (UML) that has the power of graphically depicting the object-oriented concepts.

1.2. 'OBJECT': AN OVERVIEW

According to New Webster's Dictionary (1981), an object is:

- some visible or tangible thing;
- that toward which the mind is directed in any of its states or activities;
- that to which efforts are directed.

Thus an object refers to a thing, such as a chair, a customer, a university, a painting, a plan, or a mathematical model. The first four of these examples are real-world objects, while the last two are conceptual or abstract objects. Software engineers build abstract objects that represent real-world objects that are of interest to a user.

> **Concept**
>
> *An object is anything, real or abstract*

In the context of object-oriented methodologies, the second dictionary definition is more appropriate:

An object is any thing, real or abstract, which is characterized by the state it occupies and by the *activities* defined on that object that can bring about change in the state.

In a later section we shall expand the idea of the state and the activities. Suffice it is to state at this point that the state of an object indicates the information the object stores within itself at any point of time, and that the activities are the operations and can change the information content or the state of the object.

Two other definitions are worth mentioning:

1. An object is anything, real or abstract, about which we store data and those methods that manipulate the data. (Martin and Odell, 1992).
2. A system built with object-oriented methods is one whose components are encapsulated chunks of data and function, which can inherit attributes and behaviour from other such components, and whose components communicate via messages with one another. (Yourdon, 1994).

Central Concepts Underlying Object Orientation

Various authors have suggested various concepts that, they think, are central to object orientation. We give below some of the oft-repeated concepts:

- Encapsulation
- State retention
- Information hiding
- Object identity
- Message
- Classes
- Inheritance
- Polymorphism
- Genericity

1.3. ENCAPSULATION

Encapsulation means enclosing related components within a capsule. The capsule can be referred to by a single name. In the object-oriented methodology, the components within this capsule are (1) the attributes and (2) the operations.

Attributes store information about the object. The operations can change the values of the attributes and help accessing them.

1.4. STATE RETENTION

The idea of encapsulation is not unique to object orientation. Subroutines in early high-level languages had already used the idea of encapsulation. Modules in structured design also represent encapsulation. There is however a difference between encapsulation represented in modules and that in objects. After a module completes its tasks, the module returns back to its original state. In contrast, after an operation is performed on an object, the object does not return to its original state; instead it continues to retain its final state till it is changed when another operation is performed on it.

1.5. INFORMATION HIDING

One result of encapsulation is that details of what take place when an operation is performed on an object are suppressed from public view. Only the operations that can be performed on an object are visible to an outsider. It has two major benefits:

1. It localizes design decisions. Private design decisions (within an object) can be made and changed with minimal impact upon the system as a whole.
2. It decouples the content of information from its form.

Once again the idea of information hiding is not new. This idea was forwarded by Parnas (1972) and was used in the modular design of programs in structured design.

1.6. OBJECT IDENTITY

Every object is unique and is identified by an object reference or an object handle. A programmer can refer to the object with the help of such a reference (or handle) and can manipulate it. Thus a program statement

```
Customer cust_rec1 = new Customer ();
```

defines a variable cust-rec1 and causes this variable to hold the handle of the object Customer, created newly. This object belongs to the class customer. The assignment operator (=) directs the class Customer to create an instance of its own.

1.7. MESSAGES

An object *obj1* requests another object *obj2*, via a message, to carry out an activity using one of the operations of *obj2*. Thus *obj1* should

1. Store the handle of obj2 in one of its variable.
2. Know the operation of obj2 that it wishes to execute.
3. Pass any supplementary information, in the form of arguments, that may be required by obj2 to carry out the operation.

Further, *obj2* may pass back the result of the operation to *obj1*.

Thus, the structure of a message is defined as under:

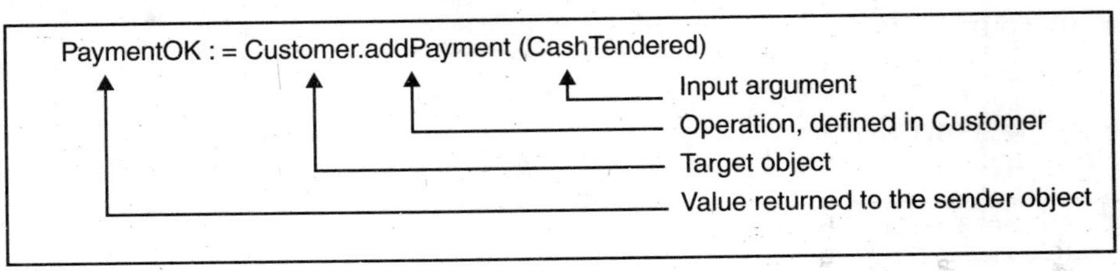

The UML representation of the message is as under:

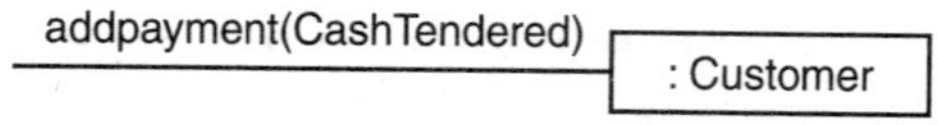

The input arguments are generally parameter values defined in (or variable at) *obj1*. But they can also be other objects as well. In fact, in Smalltalk, there is no need for any data. Objects point to other objects (via variables) and communicate with one another by passing back and forth handles of other objects.

Messages can be of three types:

1. Informative (past-oriented, update, forward, or push)
2. Interrogative (present-oriented, real, backward, or pull)
3. Imperative (further-oriented, force or action)

An *informative messgae* provides the target object information on what has taken place elsewhere in order to update itself:

```
employee.updateAddress (address:Address)
```

An interrogative message requests the target object for some current information about itself:

```
inventory.getStatus
```

An *imperative message* asks the object to take some action in the immediate future on itself, another object, or even on the environment around the system.

```
payment.computeAmount (quantity, price)
```

1.8. CLASSES

A class is stencil from which objects are created (instantied); that is, instances of a class are objects. Thus customer 1, customer 2 and so on, objects of the class customer, and product 1, product 2, ... are objects of the class product.

The UML definition of a *class* if a "a description of a set of objects that share the same attributes, operations, methods, relationships, and semantics". It does not include concrete software implementation such as a Java class; thus it includes all specifications that precede implementation. In the UML, an implemented software class is called an *implementation class.*

Oftentimes a term *type* is used to describe a set of objects with the same attributes and objects. Its difference from a class is that it may not include any methods. Recall that a *method* is the implementation of an operation, specifying the operation's algorithm or procedure.

Although objects of a class are structurally identical, each object:

1. has a separate handle or reference and
2. can be in different states.

Normally, operations and attributes are defined at the object level, but they can be defined at the level of a class. Thus, creating a new customer is a class-level operation :

```
new Customer();
```

Customer( ) is a class operation that creates a new customer (constructor).

Similarly, *noOfCustomersCreated* that keeps a count of the number of customer objects created by the class customer is a class-level attribute:

```
noOfCustomersCreated:Integer
```

noOfCustomersCreated is a class attribute that is incremented by 1 each time the operation New is executed.

The UML notation of a class, an instance of a class, and an instance of a class with a specific name are as under:

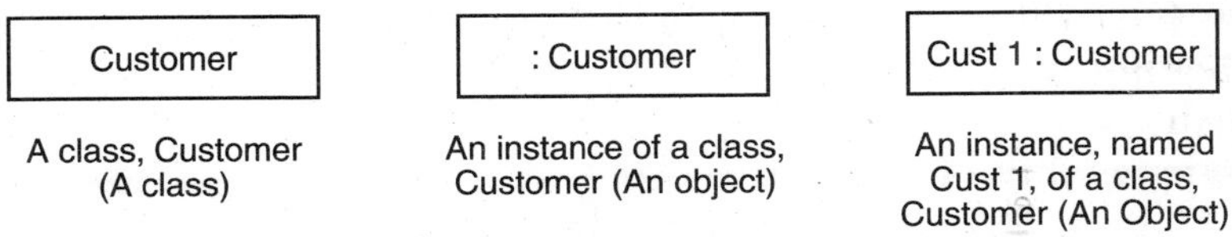

1.9. INHERITANCE

Inheritance (by D from C) is a facility by which a subtype D implicitly defines upon it all the attributes and operations of a supertype C, as if those attributes and operations had been defined upon D itself.

Note that we have used the terms subtypes and supertypes instead of the terms subclasses and superclasses (although the latter two terms are popularly used in this context) because we talk of only operations and attributes, and not methods.

The classes Manager and Worker are both Employee. So we define attributes such as Name, Address, and EmployeeNo, and define operations such as transfer, promote, and retire in the supertype Employee. These attributes and operations are valid for, and can be used by, the subtypes, Manager and Worker, without separately defining them for these subtypes. In addition, these subtypes can define attributes and operations that are local to them. For example, an attribute OfficeRoom and operation attachOfficeRoom can be defined on the Manager, and an attribute DailyWage and an operation computeDailyWage can be defined on Worker.

Inheritance is best depicted in the form of a Gen-Spec (Generalization-Specialization) diagram. The example of Manager and Worker inheriting from employee is depicted below in the form of a Gen-Spec diagram Fig. 1.6.

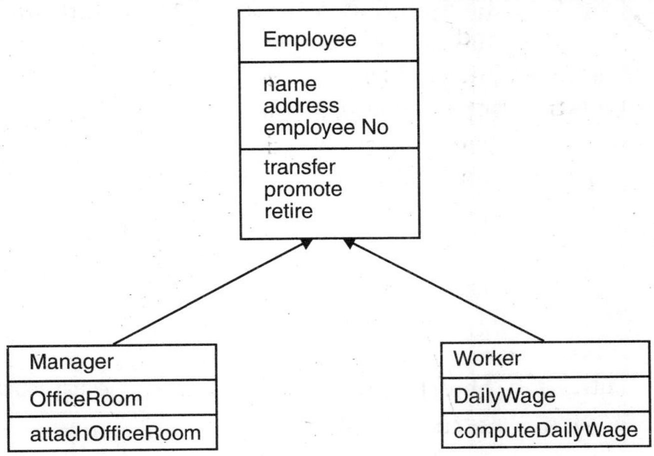

Fig. 1.6. Gen-Spec Diagram

Here, Employee is a generalized class and Manager and Worker are specialized classes.

An alternative UML notation is given in Fig. 1.7.

To define a correct subtype, two rules are to be satisfied:

1. The 100% Rule. The subtype conforms to 100% of the super-type's attributes and operations.
2. The Is-a Rule. The subtype is a member of the super-type.

The Gen-spec diagram is often called an 'Is-a' diagram.

Often a subtype can inherit attributes and operations from two super-types. Thus a Manager can be both an Employee and a Shareholder of a company. This is a case of multiple inheritance (Fig. 1.8).

While languages such as C++ and Eiffel support this features, Java and Smalltalk do not. Multiple inheritance leads to problems of:

1. Name-clash
2. Incomprehensibility of structures

More about there problems will be cited later in the text.

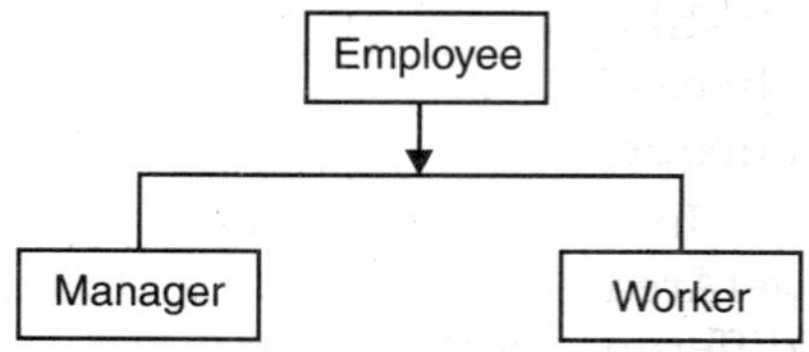

Fig. 1.7. Alternative Form of Gen-Spec Diagram

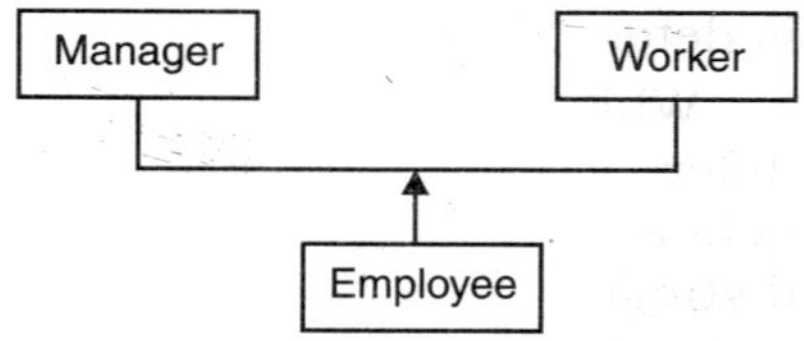

Fig. 1.8. Multiple Inheritance

1.10. POLYMORPHISM

Polymorphism is a Greek word, with poly meaning 'many' and morph meaning 'form'. Polymorphism allows the same name to be given to services in different objects, when the services are similar or related. Usually, different object types are related in a hierarchy with a common supertype, but this is not necessary (especially in dynamic binding languages, such as Smalltalk, or languages that support interface, such as Java). Two examples are shown in Fig. 1.9 and Fig. 1.10 to illustrate the use of polymorphism.

In the first example, getArea is an operation in the supertype Hexagon that specifies a general method of calculating the area of a Polygon. The subtype Hexagon inherits this operation, and therefore the method of calculating its area. But if the polygon happens to be a Triangle or a Rectangle, the same operation getArea would mean calculating area by simpler methods such as ½ (product of base and the height), or product of two adjacent sides, respectively.

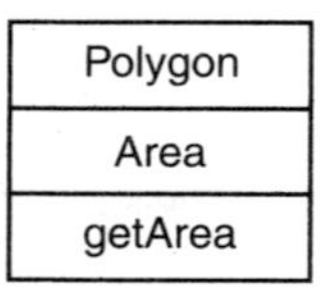

Fig. 1.9.

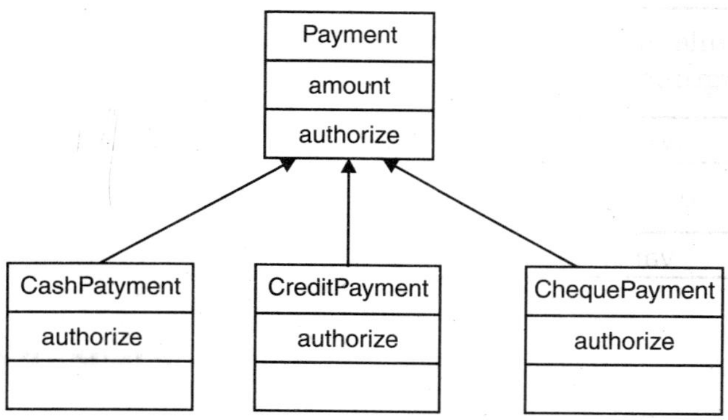

Fig. 1.10. A Second Example of Polymorphism

In the second example, Payment types are different—cash, credit, or cheque. The same operation authorize is implemented differently in different Payment types. In

CashPayment, authorize looks for counterfeit paper currency; in CreditPayment, it checks for credit worthiness; and in ChequePayment, it examines the validity of the cheque.

In these two examples, the concept of overriding has been used. The operations getArea and authorize defined on the supertype are overridden in the subtypes, where different methods are used.

Polymorphism is often implemented through dynamic binding. Also called run-time binding or late binding, it is a technique by which the exact piece of code to be executed is determined only at run-time (as opposed to compile-time), when the message is sent.

While polymorphism allows the same operation name to be defined differently across different classes, a concept called overloading allows the same operation name to be defined differently several times within the same class. Such overloaded operations are distinguished by the signature of the message, *i.e.,* by the number and/or class of the arguments. For example, two operations, one without an argument and the other with an argument, may invoke different pieces of code:

```
giveDiscount
giveDiscount (percentage)
```

The first operation invokes a general discounting scheme allowing a standard discount percentage, while the second operation allows a percentage discount that is specified in the argument of the operation.

1.11. GENERICITY

Genericity allows defining a class such that one or more of the classes that it uses internally is supplied only at run time, at the time an object of this class is instantiated. Such a class is known as a parameterized class. In C++ it is known as a *template class.* To use this facility, one has to define parameterized class argument while defining the class. In run time, when we desire to instantiate a particular class of items, we have to pass the required argument value. Thus, for example, we may define a parameterized class:

```
class Product < Product type>;
```

while instantiating a new object of this class, we supply a real class name as an argument:

```
var Product 1: Product: = Product. New <Gear>;
```

or

```
var Product 2: Product: = Product. New <Pump>;
```

SUMMARY

- Object oriented methodology provides a better way to design and develop software.
- According to Object oriented methodology, An object is any thing, real or abstract, which is characterized by the state it occupies and by the activities defined on that object that can bring about change in the state.

- Encapsulation means enclosing related components within a capsule, such as member data and member methods.
- Information hiding means the hiding some important data or members from outer world. That means the members made hidden can't be accessed globally and directly out of class.
- Classes and Objects are the base of the object oriented methodology.
- The main features of object oriented programming are encapsulation, information hiding, inheritance and polymorphism.

SOLVED PROBLEMS

Q. 1.1. Discuss the generation of Computer Languages.

Ans. 1. **First-Generation Languages (1954–1958).**

Languages: Fortran I, ALGOL 58.

2. **Second-Generation Languages (1959–1961).**

Languages: Fortran II, ALGOL 60, COBOL and LISP.

3. **Third-Generation Languages (1962–1970).**

Languages: PL/1, ALGOL 68, PASCAL, and Simula.

4. **Forth Generation (1990–Onwards).**

Languages: Object-Based and Object-Oriented Programming Languages such as Ada, Smalltalk, C++, Object PASCAL, Eiffel, CLOS.

Q. 1.2. What are the main concepts of object-oriented programming?

Ans. 1. Encapsulation
2. State retention
3. Inheritance
4. Information hiding
5. Polymorphism
6. Genericity
7. Message

Q. 1.3. What is mean of Object-Identity?

Ans. Every object is unique and is identified by an object reference or an object handle. It means all objects defined in the system has their own identity.

Q. 1.4 What is the role of messages in Object Oriented Programming Languages?

Ans. An object *obj1* requests another object *obj2,* via a message, to carry out an activity using one of the operations of *obj2.*

Messages can be of three types:

1. Informative (past-oriented, update, forward, or push)
2. Interrogative (present-oriented, real, backward, or pull)
3. Imperative (further-oriented, force, or action)

Q. 1.5. What is Encapsulation?

Ans. Binding the member data (attributes) and member functions (operations) in a single unit is called encapsulation.

Q. 1.6. What is the Inheritance?

Ans. Sharing the members (may be data or functions) among the classes is a feature of Inheritance.

Q. 1.7. What do you mean by Data hiding?

Ans. Insulation of data from direct access by the program is called data hiding or information hiding.

Q. 1.8. What do you mean by Reusability?

Ans. The reusability means, using an existing part of program or whole program in another new program. We can add additional features to an existing class without modifying. This is possible by deriving a new class from the existing one. The new class will have the combined features of both the classes. The feature of reusability can be achieved by inheritance.

Q. 1.9. What do you mean by object-based programming? *(Important)*

Ans. The object-oriented programming is the super-set of object-based programming.

Object-based programming is the style of programming that primarily supports encapsulation and object-identity. Major features that are required for object-based programming are:

- Data abstraction
- Data hiding and access mechanisms
- Automatic initialization and clear-up of objects
- Operator over-loading.

Languages that support programming with objects are said to be object-based programming languages.

Object-oriented programming incorporates all of object-based programming features along with two additional features-inheritance and dynamic binding.

Object Orient Programming ⇒

- Object-based features

+

- Inheritance

+

- Dynamic binding.

EXERCISES

Q. 1.1. What is the need of object oriented programming?

Q. 1.2. List out the names of programming languages, which refer as object oriented languages.

Q. 1.3. What are the main features of object oriented programming?

Q. 1.4. What do you understand by Object and Object-Identity?

Q. 1.5. What are the main advantages of Classes?

Q. 1.6. How a class can fulfill the features of object oriented programming?

Q. 1.7. What do you mean by reusability?

Q. 1.8. What do you mean by Polymorphism?

Q. 1.9. What are the Computer language generations?

Q. 1.10. What is the difference between object-based and object oriented programming language?

Q. 1.11. What is the benefit of information hiding?

Q. 1.12. Justify: "object oriented programs" are easy to maintain.

❑❑❑

Chapter 2

Modeling

CHAPTER OUTLINE

2.1. INTRODUCTION

The requirements for software, as delivered by typical business analysts, is not sufficiently clear, insightful, or well understood to develop software systems that meet the needs of business users. Improving the quality of software requirements will have a net positive effect on the quality, reliability, applicability, usability, and value of custom software as perceived by the business users who use it.

Modeling is a central part of all the activities that lead up to the deployment of good software. We build models to communicate the desired structure and behavior of our system. We build models to visualize and control the system's architecture. We build models to better understand the system we are building, often exposing opportunities for simplification and reuse. We build models to manage risk.

2.2. IMPORTANCE OF MODELING

A model is a representation in a certain medium of something in the same or another medium. The model captures the important aspects of the thing being modeled from a certain point of view and simplifies or omits the rest. Engineering, architecture, and many other creative fields use models.

A model is expressed in a medium that is convenient for working. Models of buildings may be drawings on paper, 3-D figures made of cardboards, or finite-element equations in a computer. A construction model of a building shows the appearance of the building but can also be used to make engineering and cost calculations.

> **Concept**
>
> *Models are building to better understand the system before development.*

A model of a software system is made in a modeling language, such as UML. The model has both semantics and notation and can take various forms that include both pictures and text. The model is intended to be easier to use for certain purposes than the final system.

An architect uses models on paper, on a computer, or as 3-D constructs to visualize and experiment with possible designs. The simplicity of creating and modifying small models permits creative thought and innovation at little cost.

A model of a software system helps developers explore several architectures and design solutions easily before writing code. A good modeling language allows the designer to get the overall architecture right before detailed design begins.

Modeling is basically used to achieve four basic goals:

1. Models help us to visualize a system as it is or as we want it to be.
2. Models permit us to specify the structure or behavior of a system.
3. Models give us a template that guides us in constructing a system.
4. Models document the decisions we have made.

Design models are created as early as when we weigh alternative solutions. This step is the detailing of the selected solution's model. From the system engineer's perspective the model also serves for managing the entire lifecycle of the system, in addition to being used to specify the system design.

Models are useful for several purposes:

- To capture and precisely state requirements and domain knowledge so that all stakeholders may understand and agree on them.
- To think about the design of a system.
- To capture design decisions in a mutable form separate from the requirements.
- To generate usable work products.
- To organize, find, filter, retrieve, examine, and edit information about large systems.
- To explore multiple solutions economically.
- To master complex systems: A model of a large software system permits dealing with complexity that is too difficult to deal with directly.

2.3. PRINCIPLES OF MODELING

The use of modeling has a wide history in all the branches of engineering. The James Rumbaugh et.al, suggests four basic principles of modeling:

1. **The choice of models has a profound influence on how a problem is attacked and how the solution is shaped.**

In other words, you must choose a right model. The right models will brilliantly clarify the most wicked development problems, offering insight that you simply could not gain otherwise; the wrong models will mislead you, causing you to focus on irrelevant issues.

2. **Every model may be expressed at different levels of abstraction (precision).**

   In other words, the best kinds of models are those that let you choose your degree of detail, depending on who is doing the viewing and why they need to view it. An analyst or an end user will want to focus on issues of what; a developer will want to focus on issues of how. Both of these stakeholders will want to visualize a system at different levels of detail at different times.

3. **The effective models are connected to reality.**

   In other words, the weak point of structured analysis techniques is the fact that there is a basic disconnect between its analysis model and the system's design model. In object oriented systems, it is possible to connect all the nearly independent views of a system into one semantic whole.

4. **No single model is sufficient. Nontrivial systems are best described a set of independent but related models.**

   In other words, having models that can be built and studied separately but that are still interrelated. As in the case of a building, you can study electrical plans in isolation, but you can also see their mapping to the floor plan and perhaps even their interaction with the routing of pipes in the plumbing plan.

Depending on the nature of the system, some models may be more important than others. For example, in data-intensive systems, models addressing static design views will dominate. In GUI intensive systems, static and dynamic use case views are quite important. In hard real time systems, dynamic process views tend to be more important. Finally, in distributed systems, such as one finds in Web-intensive applications, implementation and deployment models are the most important.

2.4. OBJECT ORIENTED MODELING

All organizations of software development need to design their software easily upgradation, easily extendable, easily usable. Before the object oriented approach all software companies were using structured programming for development of software.

In software, there are several ways to approach a model. There are two most common ways:

1. Algorithmic Approach, which can be implemented as Structured Programming.
2. Object Oriented Approach.

2.4.1. Structured Programming Approach

The structured programming was the best way to solve any problem in last decays. Every programmer knows the technique of structure programming to develop the software. Basically it focuses on the functions. In the structural programming most of the programs have a fixed well-defined structure and format. A general program can access or retrieve many functions or program can access or retrieve many functions or programs as they require for a particular problem. It is also known as "procedure-oriented programming."

> **Concept**
>
> *Mainly structured programming is based on global, local data items and the functions.*

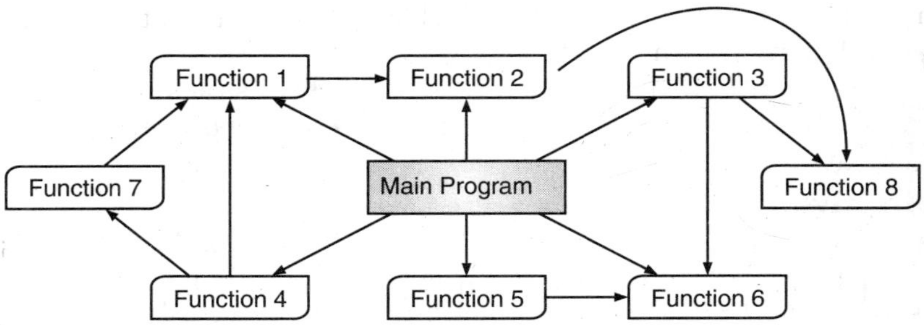

Fig. 2.1.

In the above example of a format, main program can access or execute any function and any function can execute or call any function as their requirement. This is not typical to design any system using this structure programming.

Drawbacks:

But when we use this type of programming for large programs or softwares, it is required to define some data items as global data (easily accessed by any function in program) and some data item as local data (accessed by only same function in which they are declared).

— It will produce confusion, what data are used by which function and which are global data and which are local data.

— If any software or system, which is developed, using structured programming, is very difficult for maintenance, upgrading or modification.

— It is also difficult to understand the designing of the structure of any program.

— And we cannot simulate the problem with the real world problem using structure programming.

2.4.2. Object Oriented Approach

In the object-oriented approach, users can develop their software using a new way and thinking about models organized around real-world concepts. The foundation is the object of this approach, which combines both data structure and behavior in a single entity.

The object-oriented techniques allows users to fully understand the environment they are attempting to model.

To understand the object oriented approach, see this example of an organization, which contains three departments:

- Accounts
- Management
- Development

Using the structured programming we can create the program using procedure or functions for every department according to their requirement. After creating software the structure will look like Fig. 1.2.

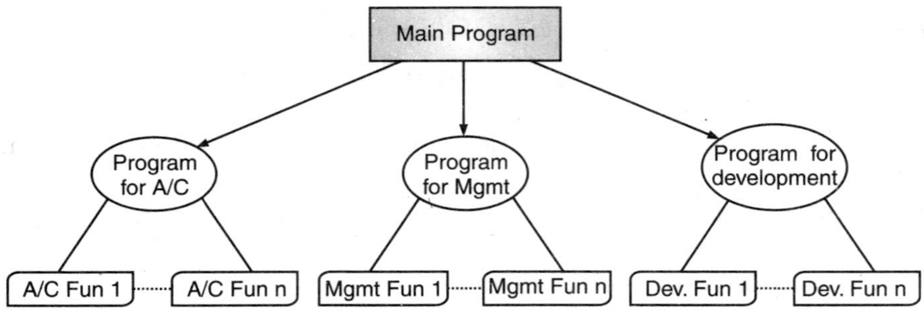

Fig. 1.2.

But using object-oriented approach, we will create an entity for each department with their data and function, called class (will discuss later). The organization of object-oriented approach for above program is as follows:

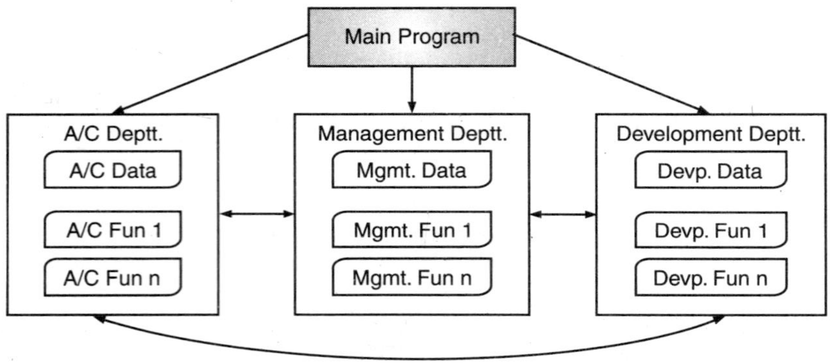

Fig. 1.3.

In this approach each department exist as single entity having their own data and own function. The data of each entity define its data structure and function defines its behavior.

Advantages

— Each entity could be easily modified, easily reused and easily upgraded to alter their data or functions easily.

— You can extend any entity (or department) according to your requirement or future plan.

— It also has the scope of future enhancement.

— You can easily relate one entity with another. It means you can build the relation between more than one entities as you required.

— You can add new function or behavior easily in any of the entity.

> **Concept**
> *Attributes and behavior bind in single unit and use as objects.*

— You can add new entity also, as extension of new department, or growth of an organization.

2.5. OBJECT ORIENTED METHODOLOGIES

Many methodologies have been proposed for object oriented software development. A methodology usually includes:

- **Notation:** Graphical representation of classes and their relationships and interactions.
- **Process:** Suggested set of steps to carry out for transforming requirements into a working system.
- **Tool:** Software for drawings and documentation.

Some popular methodologies include:

- Object Modeling Technique (OMT) by James Rumbaugh
- The Booch Method by Grady Booch
- Object-Oriented Software Engineering (OOSE) by Ivar Jacobson
- Object-Oriented Analysis (OOA) by Peter Coad & Ed Yourdon
- CRC (Classes-Responsibilities-Collaborations) Method by Rebecca Wrifs-Brock

The number of Object-oriented methodologies increased from less than 10 to more than 50 between 1989 and 1994. Methodologists battled for market share (the "method wars"), while many developers had trouble finding complete satisfaction with any single method.

SUMMARY

- Modeling is a central part of all the activities that lead up to the deployment of good software.
- We build models to communicate the desired structure and behavior of our system.
- We build models to visualize and control the system's architecture.
- We build models to better understand the system we are building, often exposing opportunities for simplification and reuse.
- We build models to manage risk.
- James Rumbaugh and Grady Booch has done great efforts to shine object oriented methodology.

SOLVED PROBLEMS

Q. 2.1. What are the popular object oriented methodologies?

Ans. Some popular methodologies are:

- Object Modeling Technique (OMT) by James Rumbaugh
- The Booch Method by Grady Booch
- Object-Oriented Software Engineering (OOSE) by Ivar Jacobson
- Object-Oriented Analysis (OOA) by Peter Coad & Ed Yourdon
- CRC (Classes-Responsibilities-Collaborations) Method by Rebecca Wrifs-Brock

Q. 2.2. What are the basic principles of Modeling?

Ans. There are four basic principles of modeling:

1. The choice of models has a profound influence on how a problem is attacked and how the solution is shaped.
2. Every model may be expressed at different levels of abstraction (precision).

3. The effective models are connected to reality.
4. No single model is sufficient. Nontrivial systems are best described a set of independent but related models.

Q.2.3. What are the main characteristics of structural programming?

Ans. Characteristics of Structural Programming:

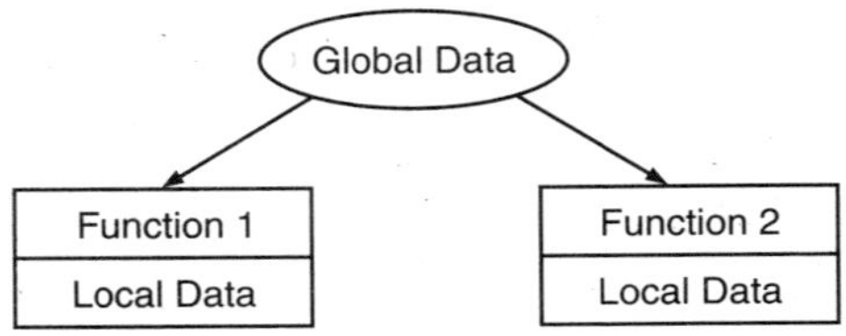

1. Large programs are divided into smaller programs known as functions.
2. Most of the functions share global data.
3. Emphasis is on doing things (algorithms).
4. Data move openly around the system from function to functions.
5. Functions transforms data from one form to another.
6. They exhibit "top-down" approach in program design.

Q. 2.4. What are the main characteristics of object-oriented programming?

(MCA 2006-07, MCA)

Ans. Characteristics of object-oriented Programming:

1. It does not allow data to move freely around the system. It ties data more closely to the functions that operate on it.
2. It protects data from accidental modifications from outside functions.
3. It emphasises on data rather than procedures.
4. It decomposes a problem into a number of entities called objects and then builds data and functions around these entities.
5. Data is hidden and cannot be accessed by external functions.
6. Objects may communicate with each other through functions.
7. It follows "buttom-up" approach in program design.
8. Objects is considered to be partitioned area of complex memory that stores data and set of operations that can access that data.

Q. 2.5. Given some name of programming languages for structured programming and object-oriented programming.

Ans. Structured Programming Languages (Procedural Oriented):

1. C
2. Pascal
3. Fortran
4. Cobol

Object-oriented Programming Languages:

1. C++
2. Java
3. Simula
4. Smalltalk
5. Ada
6. Eiffel

Q. 2.6. What is Dynamic Binding?

Ans. Dynamic binding means that the code associated with a given procedure is not known until the time of the call at run time. It is also known as **late binding**.

Q. 2.7. What do you mean by model?

Ans. A model is an abstraction of something for the purpose of understanding it before building it. Models are abstractions built to understand a problem before implementing a solution. All abstractions are subsets of reality selected for a particular purpose.

The word model has two dimensions:

1. A view of a system (object model, dynamic or functional model).
2. Stage of development (analysis, design or implementation).

Q. 2.8. What is the OMT methodology?

Ans. OMT methodology is a methodology of software engineering, to develope the software using the steps of software development life cycle.

Q. 2.9. What are the advantages of modeling?

Ans. Advantages of Modeling:

1. Testing a physical entity before building original system.
2. Communication with customers to show and understand.
3. Visualization to show the idea of designer.
4. Reduction of complexity.

EXERCISES

Q. 2.1. What is Model? Explain in brief.

Q. 2.2. Why modeling is important? Explain in detail.

Q. 2.3. What are the basic principles of Modeling? Explain in detail.

Q. 2.4. What are the benefits of Modeling?

Q. 2.5. Is there any limitation of Modeling? If yes, Explain them.

❑❑❑

Chapter 3

UML

3.1. INTRODUCTION

The Unified Modeling Language (UML), had originally been developed to create models for software systems. The word **unified** stands for the claim that the language can be used for software systems of a large number of different domains—from business and economics systems to the development of standard software products to technical systems, such as an airbag control. The language uses controlled extension mechanisms (stereotypes) to let you adapt it to your domain, so that UML actually lives up to that claim.

The Unified Modeling Language (UML) is a general-purpose visual modeling language that is used to specify, visualize, construct, and document the artifacts of a software system. It captures decisions and understanding about systems that must be constructed. It is used to understand, design, browse, configure, maintain, and control information about such systems.

The examples of artifacts are requirements, architecture, design, source code, test cases, prototypes, etc.

The UML is intended for use with all development methods, lifecycle stages, application domains, and media. The modeling language is intended to unify past experience about modeling techniques and to incorporate current software best practices into a standard approach. UML includes semantic concepts, notation, and

CHAPTER OUTLINE

guidelines. It has static, dynamic, environmental, and organizational parts. It is intended to be supported by interactive visual modeling tools that have code generators and report writers.

> **Concept**
> *UML is a language for visualizing, specifying, constructing, and documenting the artifacts of software intensive systems.*

The UML specification does not define a standard process but is intended to be useful with an iterative development process. It is intended to support most existing object-oriented development processes.

The UML is suitable for modeling various kinds of system such as enterprise information systems, distributed web-based applications, real-time embedded systems, etc.

3.2. UML LANGUAGE SPECIFICATIONS

The UML is appropriate for modeling systems ranging from enterprise information systems to distributed Web-based applications and even to hard real time embedded systems. It is a very expressive language, addressing all the views needed to develop and then deploy such systems.

Even though it is expressive, the UML is not difficult to understand and to use. Learning to apply the UML effectively starts with forming a conceptual model of the language, which requires learning three major elements: the UML's basic building blocks, the rules that dictate how these building blocks may be put together, and some common mechanisms that apply throughout the language.

The Unified Modeling Language (UML) provides views for development and deployment. The Unified Modeling Language (UML) is process independent.

The UML is process independent, although optimally it should be used in a process that is:

- Use-case driven
- Architecture – centric
- Iterative
- Incremental

3.3. GOALS OF UML

The Basic goals of the Unified Modeling Language (UML) are as follows:

- Provide modellers with a ready to use expressive and visual modeling language to develop and exchange meaningful models.
- Provide extensibility and specialization mechanisms to extend core concepts.
- Provide a formal basis for understanding the specification language.
- Encourage/growth of the object tools market.
- Supports higher level of development with concepts such as components frameworks or patterns.

3.4. CONCEPTUAL MODEL OF THE UML

To understand the UML, you must to design a conceptual model of the language, and this requires learning three major elements:

1. The UML's basic building blocks,
2. **The Rules:** that state how those building blocks may be put together,
3. Some common Mechanisms: that applies throughout the UML.

Once you have grasped these ideas, you will be able to read UML models and create some basic ones. As you gain more experience in applying the UML, you can build on this conceptual model, using more advanced features of the language.

3.5. ARCHITECTURE

The UML language architecture was provided by OMG to align UML with other OMG's technologies.

The UML architecture follows the meta-modeling architecture of OMG's Meta-Object Facility (MOF).

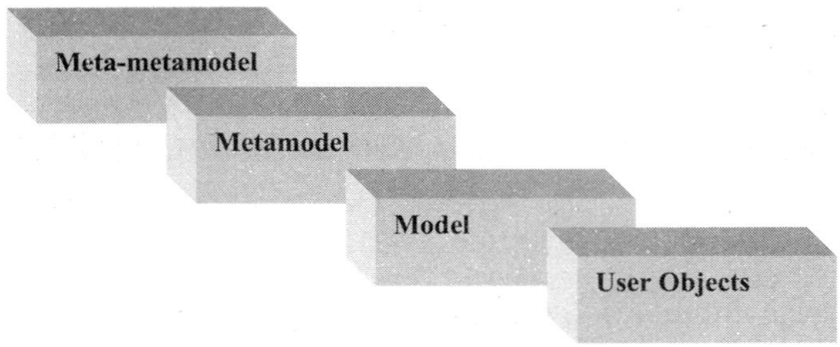

The UML architecture contains following Four layers:

1. Meta-metamodel layer
2. Metamodel layer
3. Model
4. User Objects

1. **Meta-metamodel layer.**
   The infrastructure for a metamodeling architecture. Defines the language for specifying metamodels.
   Examples: MetaClass, MetaAttribute, MetaOperation.
2. **Metamodel layer.**
   An instance of a metamodel. Defines the language for specifying a model.
   Examples: Class, Attribute, Operations, Component.
3. **Model.**
   An instance of a metamodel. Defines a language to define an information domain.
   Examples: Agency, AgencyCode, numberUnits, staffStrength.
4. **User Objects.**
   An instance of model. Defines a specific information model.
   Example: <CPSP, 5, 100>.

3.6. BASIC BUILDING BLOCKS

There are three basic building blocks in the UML:

1. **Elements (or Things):** main "component" of the model.
2. **Relationships:** relationships that tie elements together.
3. **Diagrams:** mechanism to group interesting collections of elements and relationships.

These building blocks will be used to represent large and small complex structures.

3.7. ELEMENTS

There are Four basic types of elements:

(*i*) structural

(*ii*) behavioural
(*iii*) grouping
(*iv*) annotation

They will be used to specify well-formed models.

3.7.1. Structural Elements

Structural Elements are the nouns of UML models. These are the mostly static parts of a model, representing elements that are either conceptual or physical. In all, there are seven kinds of structural elements:

(*a*) Class
(*b*) Interface
(*c*) Collaboration
(*d*) Active Class
(*e*) Use Case
(*f*) Component
(*g*) Node

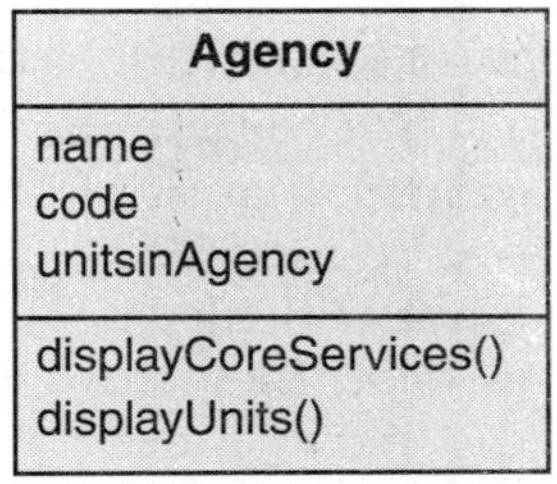

Fig. 3.1. Class

(*a*) **Class** is a description of a set of objects that share the same attributes, operations, relationships, and semantics. A class implements one or more interfaces. Graphically, a class is rendered as a rectangle, usually including its name, attributes, and operations, as in given Fig. 3.1.

Fig. 3.2. Interface

(*b*) **Interface** is a collection of operations that specify a service of a class or component. An interface therefore describes the externally visible behavior of that element. An interface might represent the complete behavior of a class or component or only a part of that behavior. An interface defines a set of operation specifications (that is, their signatures) but never a set of operation implementations. Graphically, an interface is rendered as a circle together with its name. An interface rarely stands alone. Rather, it is typically attached to the class or component that realizes the interface, as in given Fig. 3.2.

(*c*) **Collaboration** defines an interaction and is a society of roles and other elements that work together to provide some cooperative behavior that's bigger than the sum of all the elements. Therefore, collaborations have structural, as well as behavioral, dimensions. A given class might participate in several collaborations. These collaborations therefore represent the implementation of patterns that make up a system. Graphically, a collaboration is rendered as an ellipse with dashed lines, usually including only its name, as in given Fig. 3.3.

Fig. 3.3. Collaboration

(*d*) **Use case** is a description of set of sequence of actions that a system performs that yields an observable result of value to a particular actor. A use case is used to structure the behavioral things in a model. A use case is realized by a collaboration. Graphically, a use case is rendered as an ellipse with solid lines, usually including only its name, as in given Fig. 3.4.

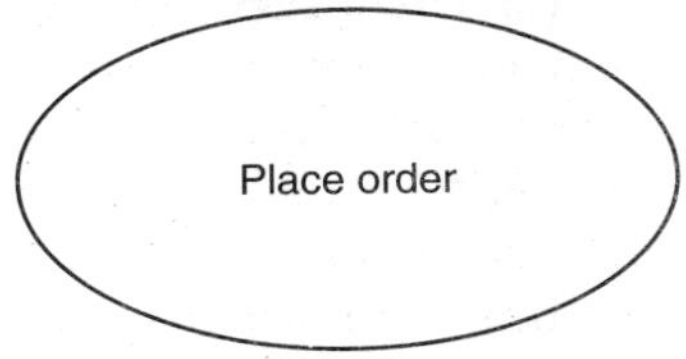

Fig. 3.4. Use Case

The remaining three things—active classes, components, and nodes—are all class-like, meaning they also describe a set of objects that share the same attributes, operations, relationships, and semantics. However, these three are different enough and are necessary for modeling certain aspects of an object-oriented system, and so they warrant special treatment.

(*e*) **Active class** is a class whose objects own one or more processes or threads and therefore can initiate control activity. An active class is just like a class except that its objects represent elements whose behavior is concurrent with other elements. Graphically, an active class is rendered just like a class, but with heavy lines, usually including its name, attributes, and operations, as in given Fig. 3.5.

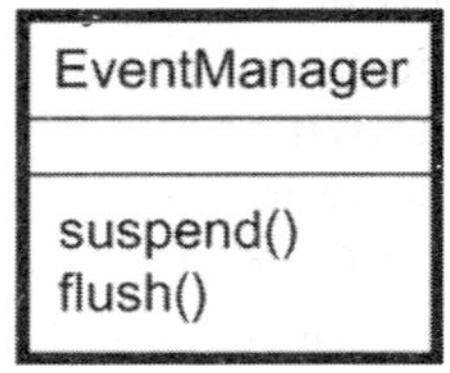

Fig. 3.5. Active Class

The remaining two elements—component, and nodes—are also different. They represent physical things, whereas the previous five things represent conceptual or logical things.

(*f*) **Component** is a physical and replaceable part of a system that conforms to and provides the realization of a set of interfaces. In a system, you'll encounter different kinds of deployment components, such as COM+ components or Java Beans, as well as components that are artifacts of the development process, such as source code files. A component typically represents the physical packaging of otherwise logical elements, such as classes, interfaces, and collaborations.

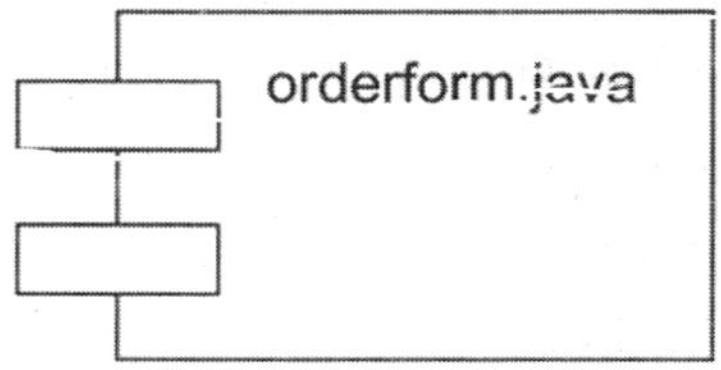

Fig. 3.6. Component

Graphically, a component is rendered as a rectangle with tabs, usually including only its name, as in given Fig. 3.6.

(*g*) **Node** is a physical element that exists at run time and represents a computational resource, generally having at least some memory and, often, processing capability. A set of components may reside on a node and may also migrate from node to node. Graphically, a node is rendered as a cube, usually including only its name, as in given Fig. 3.7.

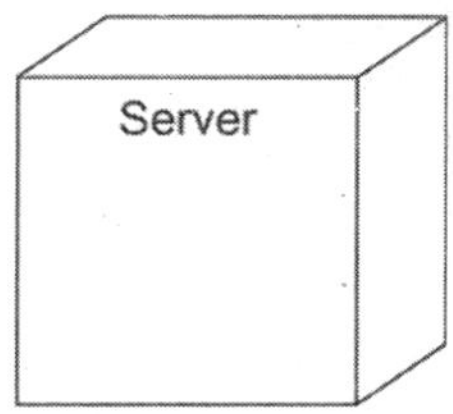

Fig. 3.7. Node

All above seven elements—classes, interfaces, collaborations, use cases, active classes, components, and nodes—are the basic structural things that you may include in a UML model. There are also variations on these seven, such as actors, signals, and utilities (kinds of classes), processes and threads (kinds of active classes), and applications, documents, files, libraries, pages, and tables (kinds of components).

3.7.2. Behavioral Elements

Behavioral Elements are the dynamic parts of UML models. These are the verbs of a model, representing behavior over time and space. In all, there are two primary kinds of behavioral Elements.

(*a*) Interaction; (*b*) State Machine

(*a*) **Interaction** is a behavior that comprises a set of messages exchanged among a set of objects within a particular context to accomplish a specific purpose. The behavior of a society of objects or of an individual operation may be specified with an interaction. An interaction involves a number of other elements, including messages, action sequences (the behavior invoked by a message), and links (the connection between objects). Graphically, a message is rendered as a directed line, almost always including the name of its operation, as in given Fig. 3.8.

Fig. 3.8. Messages

(*b*) **State machine** is a behavior that specifies the sequences of states an object or an interaction goes through during its lifetime in response to events, together with its responses to those events. The behavior of an individual class or a collaboration of classes may be specified with a state machine. A state machine involves a number of other elements, including states, transitions (the flow from state to state), events (things that trigger a transition), and activities (the response to a transition). Graphically, a state is rendered as a rounded rectangle, usually including its name and its substates, if any, as in given Fig. 3.9.

Fig. 3.9. States

Both elements-interactions and state machines—are the basic behavioral things that you may include in a UML model. Semantically, these elements are usually connected to various structural elements, primarily classes, collaborations, and objects.

3.7.3. Grouping Elements

Grouping things are the organizational parts of UML models. These are the boxes into which a model can be decomposed. In all, there is one primary kind of grouping thing, namely, packages.

Package is a general-purpose mechanism for organizing elements into groups. Structural things, behavioral things, and even other grouping things may be placed in a package. Unlike components (which exist at run time), a package is purely conceptual (meaning that it exists only at development time). Graphically, a package is rendered as a tabbed folder, usually including only its name and, sometimes, its contents, as in given Fig. 3.10.

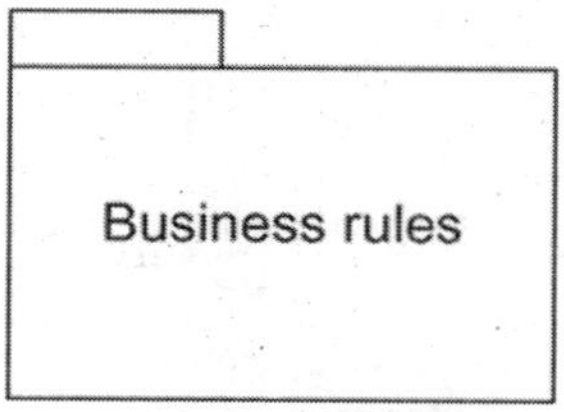

Fig. 3.10. Packages

Packages are the basic grouping things with which you may organize a UML model. There are also variations, such as frameworks, models, and subsystems (kinds of packages).

3.7.4. Annotational Elements

Annotational Elements are the explanatory parts of UML models. These are the comments you may apply to describe, illuminate, and remark about any element in a model. There is one primary kind of annotational Element, called a note.

A Note is simply a symbol for rendering constraints and comments attached to an element or a collection of elements. Graphically, a note is rendered as a rectangle with a dog-eared corner, together with a textual or graphical comment, as in given Fig. 3.11.

Fig. 3.11. Notes

This element is the one basic annotational thing you may include in a UML model. You'll typically use notes to adorn your diagrams with constraints or comments that are best expressed in informal or formal text. There are also variations on this element, such as requirements (which specify some desired behavior from the perspective of outside the model).

3.8. RELATIONSHIPS

There are four kinds of relationships in the UML:

1. Dependency
2. Association
3. Generalization
4. Realization

These relationships are the basic relational building blocks of the UML. You use them to write well-formed models. See chapter 4 for more details.

3.9. DIAGRAMS

A diagram is the graphical presentation of a set of elements, most often rendered as a connected graph of vertices (things) and arcs (relationships). Diagram is the graphical presentation of a set of elements and relationships where:

(*a*) **Nodes** are elements (*b*) **Edges** are relationship

Diagrams are used to visualize a system from different perspectives, so a diagram is a projection into a system. For all but the most trivial systems, a diagram represents an elided view of the elements that make up a system. The same element may appear in all diagrams, only a few diagrams (the most common case), or in no diagrams at all (a very rare case). In theory, a diagram may contain any combination of things and relationships. In practice, however, a small number of common combinations arise, which are consistent with the five most useful views that comprise the architecture of a software-intensive system.

The UML is characterized by nine major diagrams:

1. Class diagram
2. Object diagram
3. Use case diagram
4. Sequence diagram
5. Collaboration diagram
6. Statechart diagram
7. Activity diagram
8. Component diagram
9. Deployment diagram

Details of diagrams are discussed in chapter 4.

3.10. RULES

The UML's building blocks can't simply be thrown together in a random fashion. Like any language, the UML has a number of rules that specify what a well-formed model should look like. A well-formed model is one that is semantically self-consistent and in harmony with all its related models.

The UML has semantic rules for

• Names	For identification of elements, relationships, and diagrams.
• Scope	The context that gives specific meaning to a name.
• Visibility	How those names can be seen and used by others.
• Integrity	How things properly and consistently relate to one another.
• Execution	What it means to run or simulates a dyamic model.

Models built during the development of a software-intensive system tend to evolve and may be viewed by many stakeholders in different ways and at different times. For this reason, it is common for the development team to not only build models that are well-formed, but also to build models that are:

• Elided	Certain elements are hidden to simplify the view.
• Incomplete	Certain elements may be missing.
• Inconsistent	The integrity of the model is not guaranteed.

These less-than-well-formed models are unavoidable as the details of a system unfold and churn during the software development life cycle. The rules of the UML encourage you—but do not force you—to address the most important analysis, design, and implementation questions that push such models to become well-formed over time.

3.11. COMMON MECHANISMS

A building is made simpler and more harmonious by the conformance to a pattern of common features. A house may be built in the Victorian or French country style largely by using certain architectural patterns that define those styles. The same is true of the UML. It is made simpler by the presence of four common mechanisms that apply consistently throughout the language.

1. Specifications
2. Adornments
3. Common divisions
4. Extensibility mechanisms

Read Chapter 4 for more details.

SUMMARY

- The Unified Modeling Language (UML), had originally been developed to create models for software systems.
- The word unified stands for the claim that the language can be used for software systems of a large number of different domains—from business and economics systems to the development of standard software products to technical systems, such as an airbag control.
- The Unified Modeling Language (UML) is a general-purpose visual modeling language that is used to specify, visualize, construct, and document the artifacts of a software system.
- The UML is intended for use with all development methods, lifecycle stages, application domains, and media.
- The modeling language is intended to unify past experience about modeling techniques and to incorporate current software best practices into a standard approach.
- UML includes semantic concepts, notation, and guidelines. It has static, dynamic, environmental, and organizational parts.
- UML is intended to be supported by interactive visual modeling tools that have code generators and report writers.
- To understand the UML, you must learn three major elements: (1) UML's basic building blocks, (2)UML Rules, and (3) Some common Mechanisms.

SOLVED PROBLEMS

Q. 3.1. What is the history of UML?

Ans:
- Booch and Rumbaugh started working towards a Unified Modeling Language (UML) in 1994 under the auspices of Rational Inc.
- UML only offers a model notation, not a methodology for how to do good modeling.
- UML is used by the development method Objectory (Jacobson at Rational).
- UML was proposed by Rational Inc. and by Hewlett-Packard as a standard for object-oriented analysis and design and was adopted by the OMG.
- Vendors modify their CASE tools to make them consistent with UML.

Q. 3.2. What are the four basic type elements in UML?

Ans: These are the basic types of elements in UML:

(*i*) Structural elements
(*ii*) Behavioral elements
(*iii*) Grouping elements
(*iv*) Annotation elements

Q. 3.3 List out the all Structural Elements in UML.

Ans: There are seven kinds of structural elements:

1. Class
2. Interface
3. Collaboration
4. Active Class
5. Use Case
6. Component
7. Node

Q. 3.4 List out the all Behavioral Elements in UML.

Ans: There are two primary kinds of behavioral Elements.

1. Interaction
2. State Machine

Q. 3.5 Explain the Grouping Elements in UML.

Ans: *Grouping elements* are the organizational parts of UML models. These are the boxes into which a model can be decomposed. There is only one primary kind of grouping elements which is packages.

Package is a general-purpose mechanism for organizing elements into groups. Structural things, behavioral things, and even other grouping things may be placed in a package. Unlike components (which exist at run time), a package is purely conceptual (meaning that it exists only at development time).

Q. 3.6. Explain the Annotational Elements in UML.

Ans: Annotational Elements are the explanatory parts of UML models. These are the comments you may apply to describe, illuminate, and remark about any element in a model. There is one primary kind of annotational Element, called a note.

Q. 3.7. How many Relationships exists in UML?

Ans: There are four kinds of relationships in the UML:

1. Dependency
2. Association
3. Generalization
4. Realization

Q. 3.8. How many diagrams are available in UML?

Ans: The UML is characterized by nine major diagrams:

1. Class diagram
2. Object diagram
3. Use case diagram
4. Sequence diagram
5. Collaboration diagram
6. Statechart diagram
7. Activity diagram
8. Component diagram
9. Deployment diagram

EXERCISES

Q. 3.1. What UML stands for and what this is?

Q. 3.2. In which company UML has been initially developed?

Q. 3.3. Why UML required?

Q. 3.4. What are the basic architecture of UML?

Q. 3.5. What are the main goals of UML?

Q. 3.6. What do you mean by visual modeling language?

Q. 3.7. Explain all the structural elements of UML.

Q. 3.8. Explain all the behavioral elements of UML.

Q. 3.9. What is the difference between class and interface according to UML?

Q. 3.10. What do you mean by Active class?

Q. 3.11. What is the role of use cases?

Q. 3.12. "Nodes are the most important elements in the UML" Explain.

Q. 3.13. What are the diagrams available in the UML?

Q. 3.14. Explain the UML Rules to form a model.

Chapter 4

Basic Structural Modeling

4.1. CLASSES

Classes are the most important building block of any object-oriented system. A class is a description of a set of objects that share the same attributes, operations, relationships, and semantics.

A class is stencil from which objects are created (instantiated); that is, instances of a class are objects. Thus customer1, customer2, and so on, are objects of the class customer; and product1, product2, ... are objects of the class product.

The UML definition of a class is "a description of a set of objects that share the same attributes, operations, methods, relationships, and semantics". It does not include concrete software implementation such as a Java class; thus it includes all specifications that precede implementation. In the UML, an implemented software class is called an implementation class.

Oftentimes a term type is used to describe a set of objects with the same attributes and objects. Its difference from a class is that it may not include any methods. Recall that a method is the implementation of an operation, specifying the operation's algorithm or procedure.

4.2. RELATIONSHIPS

There are four kinds of relationships in the UML:

CHAPTER OUTLINE

1. Dependency
2. Association
3. Generalization
4. Realization

These relationships are the basic relational building blocks of the UML. You use them to write well-formed models.

4.2.1. Dependency

A dependency is a semantic relationship between two things in which a change to one thing (the independent thing) may affect the semantics of the other thing (the dependent thing). Graphically, a dependency is rendered as a dashed line, possibly directed, and occasionally including a label, as in given Fig. 4.1.

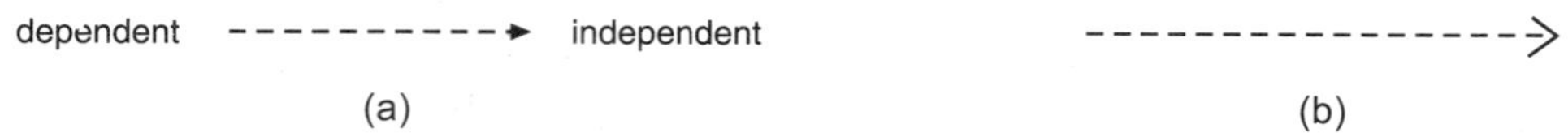

Fig. 4.1. Dependencies

4.2.2. Associations

An association is a structural relationship that describes a set of links, a link being a connection among objects. Aggregation is a special kind of association, representing a structural relationship between a whole and its parts. Graphically, an association is rendered as a solid line, possibly directed, occasionally including a label, and often containing other adornments, such as multiplicity and role names, as in given Fig. 4.2.

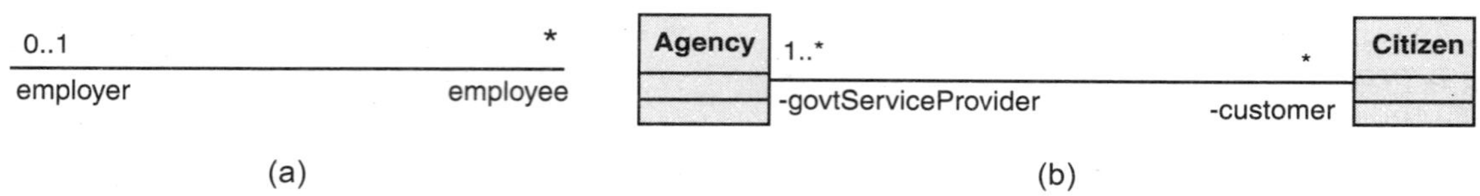

Fig. 4.2. Associations

4.2.3. Generalizations

A generalization is a specialization/generalization relationship in which objects of the specialized element (the child) are substitutable for objects of the generalized element (the parent). In this way, the child shares the structure and the behavior of the parent. Graphically, a generalization relationship is rendered as a solid line with a hollow arrowhead pointing to the parent, as in given Fig. 4.3.

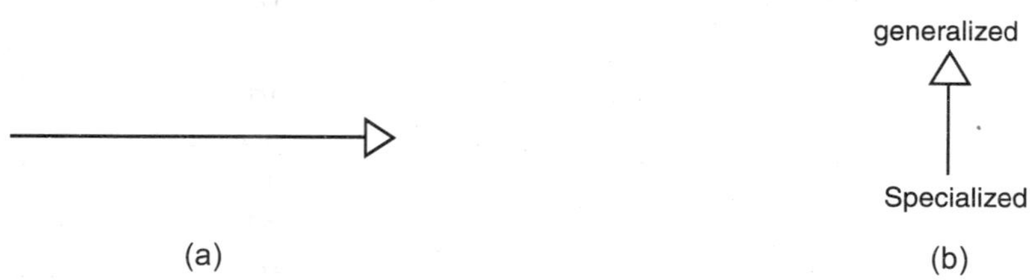

Fig. 4.3. Generalizations

4.2.4. Realizations

A realization is a semantic relationship between classifiers, wherein one classifier specifies a contract that another classifier guarantees to carry out. You'll encounter realization relationships in two places: between interfaces and the classes or components that realize them, and between use cases and the collaborations that realize them. Graphically, a realization relationship is rendered as a cross between a generalization and a dependency relationship, as in given Fig. 4.4.

(a) (b)

Fig. 4.4: Realization

These four elements are the basic relational things you may include in a UML model. There are also variations on these four, such as refinement, trace, include, and extend (for dependencies).

4.3. DIAGRAMS

A diagram is the graphical presentation of a set of elements, most often rendered as a connected graph of vertices (things) and arcs (relationships). You can diagram is the graphical presentation of a set of elements and relationships where:

(*a*) Nodes are elements (*b*) Edges are relationship

You draw diagrams to visualize a system from different perspectives, so a diagram is a projection into a system. For all but the most trivial systems, a diagram represents an elided view of the elements that make up a system. The same element may appear in all diagrams, only a few diagrams (the most common case), or in no diagrams at all (a very rare case). In theory, a diagram may contain any combination of things and relationships. In practice, however, a small number of common combinations arise, which are consistent with the five most useful views that comprise the architecture of a software-intensive system.

The UML is characterized by nine major diagrams:

1. Class diagram
2. Object diagram
3. Use case diagram
4. Sequence diagram
5. Collaboration diagram
6. State-chart diagram
7. Activity diagram
8. Component diagram
9. Deployment diagram

A ***class diagram*** shows a set of classes, interfaces, and collaborations and their relationships. These diagrams are the most common diagram found in modeling object-oriented systems. Class diagrams address the static design view of a system. Class diagrams that include active classes address the static process view of a system.

An ***object diagram*** shows a set of objects and their relationships. Object diagrams represent static snapshots of instances of the things found in class diagrams. These diagrams address the static design view or static process view of a system as do class diagrams, but from the perspective of real or prototypical cases.

A ***use case diagram*** shows a set of use cases and actors (a special kind of class) and their relationships. Use case diagrams address the static use case view of a system. These diagrams are especially important in organizing and modeling the behaviors of a system.

Both ***sequence diagrams*** and ***collaboration diagrams*** are kinds of interaction diagrams. An shows an interaction, consisting of a set of objects and their relationships, including the messages that may be dispatched among them. Interaction diagrams address the dynamic view of a system. A ***sequence diagram*** is an interaction diagram that emphasizes the time-ordering of messages; a ***collaboration diagram*** is an interaction diagram that emphasizes the structural organization of the objects that send and receive messages. Sequence diagrams and collaboration diagrams are isomorphic, meaning that you can take one and transform it into the other.

A ***state-chart diagram*** shows a state machine, consisting of states, transitions, events, and activities. State-chart diagrams address the dynamic view of a system. They are especially important in modeling the behavior of an interface, class, or collaboration and emphasize the event-ordered behavior of an object, which is especially useful in modeling reactive systems. Details are discussed in chapter 5.

An ***activity diagram*** is a special kind of a state-chart diagram that shows the flow from activity to activity within a system. Activity diagrams address the dynamic view of a system. They are especially important in modeling the function of a system and emphasize the flow of control among objects. Details are discussed in chapter 5.

A ***component diagram*** shows the organizations and dependencies among a set of components.

Component diagrams address the static implementation view of a system. They are related to class diagrams in that a component typically maps to one or more classes, interfaces, or collaborations. Details are discussed in chapter 6.

A ***deployment diagram*** shows the configuration of run-time processing nodes and the components that live on them. Deployment diagrams address the static deployment view of an architecture. They are related to component diagrams in that a node typically encloses one or more components. Details are discussed in chapter 6.

This is not a closed list of diagrams. Tools may use the UML to provide other kinds of diagrams, although these nine are by far the most common you will encounter in practice.

4.4. COMMON MECHANISMS

A building is made simpler and more harmonious by the conformance to a pattern of common features. A house may be built in the Victorian or French country style largely by using certain architectural patterns that define those styles. The same is true of the UML. It is made simpler by the presence of four common mechanisms that apply consistently throughout the language.

1. Specifications
2. Adornments
3. Common divisions
4. Extensibility mechanisms

4.4.1. Specifications

The UML is more than just a graphical language. Rather, behind every part of its graphical notation there is a specification that provides a textual statement of the syntax and semantics of that building block. For example, behind a class icon is a specification that provides the full set of attributes, operations (including their full signatures), and behaviors that the class embodies; visually, that class icon might only show a small part of this specification. Furthermore, there might be another view of that class that presents a completely different set of parts yet is still consistent with the class's underlying specification. You use the UML's graphical notation to visualize a system; you use the UML's specification to state the system's details. Given this split, it's possible to build up a model incrementally by drawing diagrams and then adding semantics to the model's specifications, or directly by creating a specification, perhaps by reverse engineering an existing system, and then creating diagrams that are projections into those specifications. The UML's specifications provide a semantic backplane that contains all the parts of all the models of a system, each part related to one another in a consistent fashion. The UML's diagrams are thus simply visual projections into that backplane, each diagram revealing a specific interesting aspect of the system.

Notes and other adornments are discussed in later Chapter.

4.4.2. Adornments

Most elements in the UML have a unique and direct graphical notation that provides a visual representation of the most important aspects of the element. For example, the notation for a class is intentionally designed to be easy to draw, because classes are the most common element found in modeling object-oriented systems. The class notation also exposes the most important aspects of a class, namely its name, attributes, and operations.

A class's specification may include other details, such as whether it is abstract or the visibility of its attributes and operations. Many of these details can be rendered as graphical or textual adornments to the class's basic rectangular notation. For example, Fig. 4.5. shows a class, adorned to indicate that it is an abstract class with two public, one protected, and one private operation.

Every element in the UML's notation starts with a basic symbol, to which can be added a variety of adornments specific to that symbol.

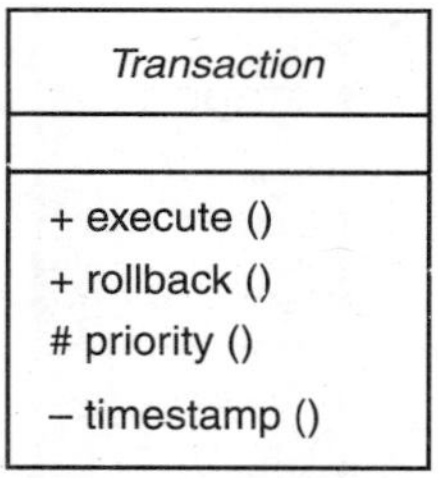

Fig. 4.5. Adornments

4.4.3. Common Divisions

In modeling object-oriented systems, the world often gets divided in at least a couple of ways.

First, there is the division of class and object. A class is an abstraction; an object is one concrete manifestation of that abstraction. In the UML, you can model classes as well as objects, as shown in Fig. 4.6.

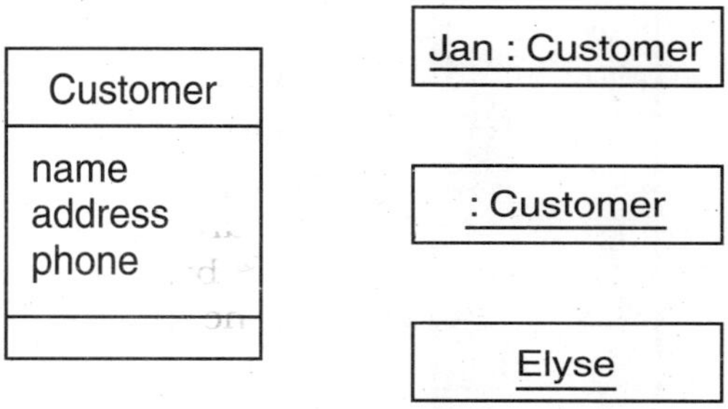

Fig. 4.6. Classes and Objects

In this figure, there is one class, named Customer, together with three objects: Jan (which is marked explicitly as being a Customer object),:Customer (an anonymous Customer object), and Elyse (which in its specification is marked as being a kind of Customer object, although it's not shown explicitly here).

Almost every building block in the UML has this same kind of class/object dichotomy. For example, you can have use cases and use case instances, components and component instances, nodes and node instances, and so on. Graphically, the UML distinguishes an object by using the same symbol as its class and then simply underlying the object's name. Objects are discussed in Chapter 4 in details.

Second, there is the separation of interface and implementation. An interface declares a contract, and an implementation represents one concrete realization of that contract, responsible for faithfully carrying out the interface's complete semantics. In the UML, you can model both interfaces and their implementations, as shown in Fig. 4.7. Interfaces are discussed in Chapter 4.

In this figure, there is one component named spellingwizard.dll that implements two interfaces, IUnknown and ISpelling.

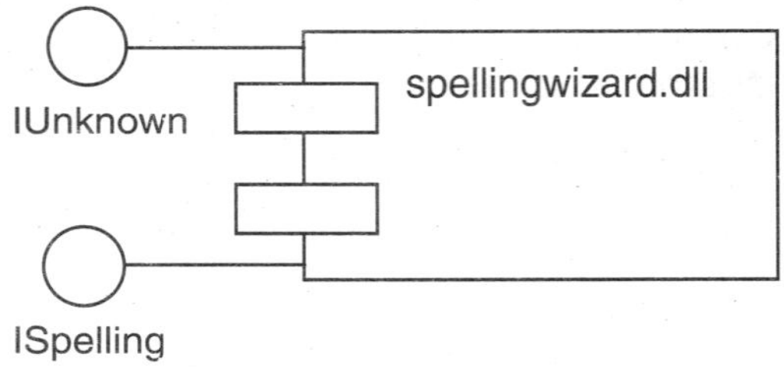

Fig. 4.7. Interfaces And Implementations

Almost every building block in the UML has this same kind of interface/implementation dichotomy. For example, you can have use cases and the collaborations that realize them, as well as operations and the methods that implement them.

4.4.4. Extensibility Mechanisms

The UML provides a standard language for writing software blueprints, but it is not possible for one closed language to ever be sufficient to express all possible nuances of all models across all domains across all time. For this reason, the UML is opened-ended, making it possible for you to extend the language in controlled ways. The UML's extensibility mechanisms include:

- Stereotypes
- Constraints
- Tagged values

A stereotype extends the vocabulary of the UML, allowing you to create new kinds of building blocks that are derived from existing ones but that are specific to your problem. For example, if you are working in a programming language, such as Java or C++, you will often want to model exceptions. In these languages, exceptions are just classes, although they are treated in very special ways. Typically, you only want to allow them to be thrown and caught, nothing else. You can make exceptions first class citizens in your modelso meaning that they are treated like basic building blockso by marking them with an appropriate stereotype, as for the class Overflow in Fig. 4.8.

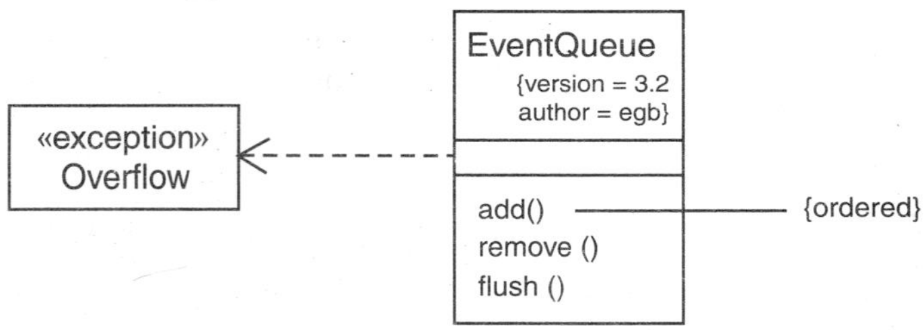

Fig. 4.8. Extensibility Mechanisms

4.5. CLASS AND OBJECT DIAGRAMS

The class diagrams are used to model the static design view of a system. Class diagrams are also the foundation for a couple of related diagrams: component diagrams and deployment diagrams.

The Object diagrams are used to model the instances of things contained in class diagrams. An object diagram shows a set of objects and their relationships at a point in time. The object diagram is also used to model the static design view or static process view of a system.

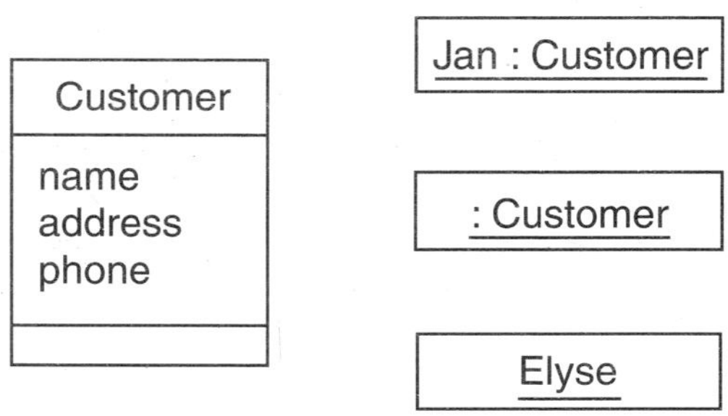

Fig. 4.9. Class and Object Diagram

In this figure, there is one class, named Customer, together with three objects: Jan (which is marked explicitly as being a Customer object), :Customer (an anonymous

Customer object), and Elyse (which in its specification is marked as being a kind of Customer object, although it's not shown explicitly here).

Class and Object diagrams are not only important for visualizing, specifying, and documenting structural models, but also for constructing the static aspects of systems through forward and reverse engineering.

4.5.1. Basic Terms and Concepts

There are some basic terms are used to distinguish the all parts of classes. See Fig. 4.10 here we shown some terms outside the box to indicate the all parts of class diagram.

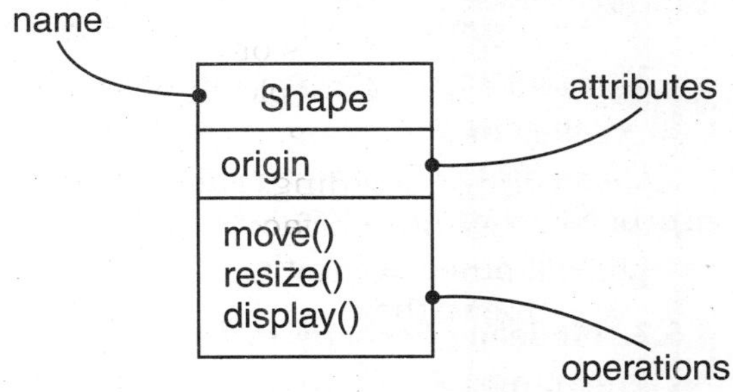

Fig. 4.10. Class Diagram

There are three basic terms are used to explain the class diagram:

1. **Name:** A class name must be unique within its enclosing package. Every class must have a name that distinguishes it from other classes.
2. **Attributes:** Attributes are related to the semantics of aggregation. A class may have any number of attributes or no attributes at all. An attribute represents some property of the thing you are modeling that is shared by all objects of that class.
3. **Operations:** An operation is the implementation of a service that can be requested from any object of the class to affect behavior.

Class diagrams commonly contain the following things:

- Classes
- Interfaces
- Collaborations
- Dependency, generalization, and association relationships

Object diagrams also having some basic term for use to call each element. See Fig. 4.11, here we shown some terms using rounded arcs to indicate the all parts of object diagram.

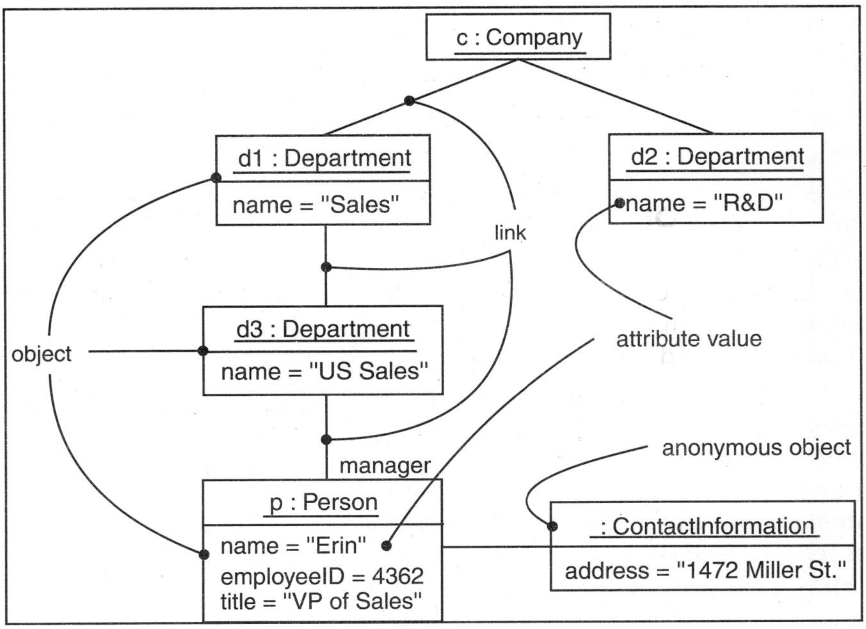

Fig. 4.11. Object Diagram

Following basic terms are used to explain the Object diagram:

1. **Link:** Links are used to bind one object to others.
2. **Attribute Value:** Each object contains their own attribute values represents property of the thing you are modeling.
3. **Anonymous Object:** The object without name.

The Object diagrams commonly contain:

- Objects
- Links

Class and Object diagrams may also contain packages or subsystems, both of which are used to group elements of your model into larger chunks.

Like all other diagrams, class and object diagrams may contain notes and constraints.

4.5.2. Modeling Techniques for Class and Object Diagrams

The James Rumbaugh and et. al., suggested following techniques for modeling Class and Object diagrams:

(*a*) Modeling Simple Collaborations
(*b*) Modeling a Logical Database Schema
(*c*) Modeling Object Structures
(*d*) Forward and Reverse Engineering

(*a*) Modeling Simple Collaborations

Classes rarely stand alone. Rather, when you build models, you will typically focus on groups of classes that interact with one another. In the UML, these societies of classes form collaborations and are usually visualized in class diagrams.

When you create a class diagram, you just model a part of the elements and relationships that make up your system's design view. For this reason, each class diagram should focus on one collaboration at a time.

To model a collaboration:

- Identify the mechanism you'd like to model. A mechanism represents some function or behavior of the part of the system you are modeling that results from the interaction of a society of classes, interfaces, and other things.
- For each mechanism, identify the classes, interfaces, and other collaborations that participate in this collaboration. Identify the relationships among these things, as well.
- Use scenarios to walk through these things. Along the way, you'll discover parts of your model that were missing and parts that were just plain semantically wrong.
- Be sure to populate these elements with their contents. For classes, start with getting a good balance of responsibilities. Then, over time, turn these into concrete attributes and operations.

(*b*) Modeling a Logical Database Schema

- Many of the systems you'll model will have persistent objects, which means that they can be stored in a database for later retrieval. Most often, you'll use a relational database, an object oriented database, or a hybrid object/relational database for persistent storage. The UML is well suited to modeling logical database schemas, as well as physical databases themselves.
- The UML's class diagrams are a superset of entity-relationship (E-R) diagrams, a common modeling tool for logical database design. Whereas classical E-R diagrams

focus only on data, class diagrams go a step further by permitting the modeling of behavior, as well. In the physical database, these logical operations are generally turned into triggers or stored procedures.

To model a schema:

- Identify those classes in your model whose state must transcend the lifetime of their applications.
- Create a class diagram that contains these classes and mark them as persistent (a standard tagged value). You can define your own set of tagged values to address database-specific details.
- Expand the structural details of these classes. In general, this means specifying the details of their attributes and focusing on the associations and their cardinalities that structure these classes.
- Watch for common patterns that complicate physical database design, such as cyclic associations, one-to-one associations, and n-ary associations. Where necessary, create intermediate abstractions to simplify your logical structure.
- Consider also the behavior of these classes by expanding operations that are important for data access and data integrity. In general, to provide a better separation of concerns, business rules concerned with the manipulation of sets of these objects should be encapsulated in a layer above these persistent classes.
- Where possible, use tools to help you transform your logical design into a physical design.

(c) Modeling Object Structures

When you model your system's design view, a set of class diagrams can be used to completely specify the semantics of your abstractions and their relationships. With object diagrams, however, you cannot completely specify the object structure of your system. For an individual class, there may be a multitude of possible instances, and for a set of classes in relationship to one another, there may be many times more possible configurations of these objects. Therefore, when you use object diagrams, you can only meaningfully expose interesting sets of concrete or prototypical objects. This is what it means to model an object structure of an object diagram shows one set of objects in relation to one another at one moment in time.

To model an object structure:

- Identify the mechanism you'd like to model. A mechanism represents some function or behavior of the part of the system you are modeling that results from the interaction of a society of classes, interfaces, and other things.
- For each mechanism, identify the classes, interfaces, and other elements that participate in this collaboration; identify the relationships among these things, as well.
- Consider one scenario that walks through this mechanism. Freeze that scenario at a moment in time, and render each object that participates in the mechanism.
- Expose the state and attribute values of each such object, as necessary, to understand the scenario.
- Similarly, expose the links among these objects, representing instances of associations among them.

(d) Forward and Reverse Engineering

Modeling is important, but you have to remember that the primary product of a development team is software, not diagrams. Of course, the reason you create models is to predictably deliver at the right time the right software that satisfies the evolving goals of its users and the business. For this reason, it's important that the models you create and the implementations you deploy map to one another and do so in a way that minimizes or even eliminates the cost of keeping your models and your implementation in sync with one another.

Forward engineering is the process of transforming a model into code through a mapping to an implementation language. Forward engineering results in a loss of information, because models written in the UML are semantically richer than any current object-oriented programming language. In fact, this is a major reason why you need models in addition to code. Structural features, such as collaborations, and behavioral features, such as interactions, can be visualized clearly in the UML, but not so clearly from raw code.

To forward engineer a class diagram:

- Identify the rules for mapping to your implementation language or languages of choice.
- This is something you'll want to do for your project or your organization as a whole.
- Depending on the semantics of the languages you choose, you may have to constrain your use of certain UML features. For example, the UML permits you to model multiple inheritance, but Smalltalk permits only single inheritance. You can either choose to prohibit developers from modeling with multiple inheritance (which makes your models language-dependent) or develop idioms that transform these richer features into the implementation language (which makes the mapping more complex).
- Use tagged values to specify your target language. You can do this at the level of individual classes if you need precise control. You can also do so at a higher level, such as with collaborations or packages.
- Use tools to forward engineer your models.

To reverse engineer a class diagram:

- Identify the rules for mapping from your implementation language or languages of choice. This is something you'll want to do for your project or your organization as a whole.
- Using a tool, point to the code you'd like to reverse engineer. Use your tool to generate a new model or modify an existing one that was previously forward engineered.
- Using your tool, create a class diagram by querying the model. For example, you might start with one or more classes, then expand the diagram by following specific relationships or other neighbouring classes. Expose or hide details of the contents of this class diagram as necessary to communicate your intent.

Forward engineering (the creation of code from a model) an object diagram is theoretically possible but pragmatically of limited value. In an object-oriented system, instances are things that are created and destroyed by the application during run time. Therefore, you can't exactly instantiate these objects from the outside.

Reverse engineering (the creation of a model from code) an object diagram is a very different thing. In fact, while you are debugging your system, this is something that you or your tools will do all the time. For example, if you are chasing down a dangling link, you'll want to literally or mentally draw an object diagram of the affected objects to see where, at a given moment in time, an object's state or its relationship to other objects is broken.

To reverse engineer an object diagram:

- Chose the target you want to reverse engineer. Typically, you'll set your context inside an operation or relative to an instance of one particular class.
- Using a tool or simply walking through a scenario, stop execution at a certain moment in time.
- Identify the set of interesting objects that collaborate in that context and render them in an object diagram.
- As necessary to understand their semantics, expose these object's states.
- As necessary to understand their semantics, identify the links that exist among these objects.
- If your diagram ends up overly complicated, prune it by eliminating objects that are not germane to the questions about the scenario you need answered. If your diagram is too simplistic, expand the neighbours of certain interesting objects and expose each object's state more deeply.

4.6. COLLABORATION DIAGRAMS

A collaboration diagram is an interaction diagram that emphasizes the structural organization of the objects that send and receive messages. A collaboration diagram shows a set of objects, links among those objects, and messages sent and received by those objects. The objects are typically named or anonymous instances of classes, but may also represent instances of other things, such as collaborations, components, and nodes. You use collaboration diagrams to illustrate the dynamic view of a system.

A collaboration diagram emphasizes the organization of the objects that participate in an interaction. As Fig. shows, you form a collaboration diagram by first placing the objects that participate in the interaction as the vertices in a graph. Next, you render the links that connect these objects as the arcs of this graph. Finally, you adorn these links with the messages that objects send and receive. This gives the reader a clear visual cue to the flow of control in the context of the structural organization of objects that collaborate.

4.6.1. Terms and Concepts

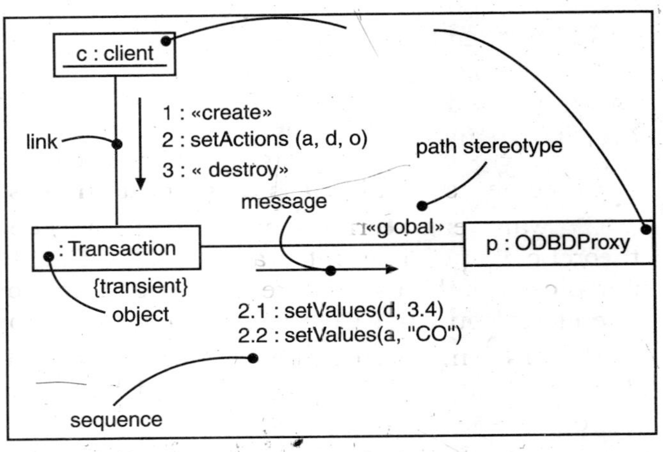

There some basic terms which are used to refer the parts of the collaboration diagram, such as:

1. **Links:** Links are used to bind one object to others.
2. **Objects:** Objects are real instances of the classes.
3. **Path Stereotype:** Path Stereotype used to indicate how one object is linked to another such as <local> or <global>.
4. **Message:** Message is communication between

objects, prefixed by the sequence number. Both the transmission of a signal and the call of an operation are represented by a message.

5. **Sequence:** Sequence number used to indicate the time order of a message.

4.6.2. Depicting a Message

A message is a communication between two lifelines. It can be synchronous or asynchronous, and it can invoke an operation, transport a signal, or create an object.

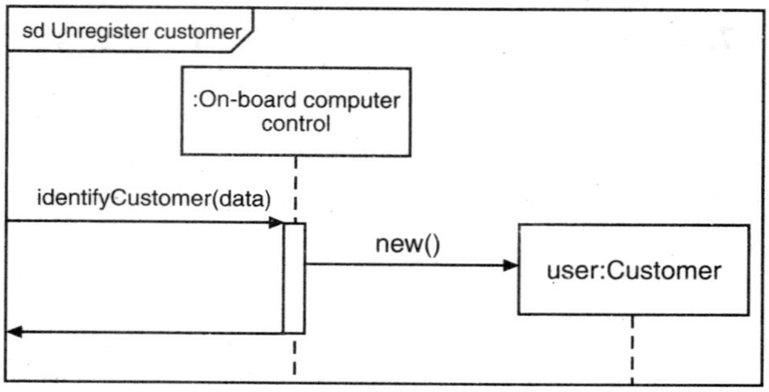

Fig. 4.12.

Parts of the system send messages to each other to exchange information and jointly execute a behavior. In the sense of UML and particularly of SysML, this is not to be understood as mere operation calls in the context of a programming language.

For example, if you model a pumping system that pumps a liquid from A to B , then this transmission can be seen as a message between A and B (send signal). Both the transmission of a signal and the call of an operation are represented by a message. The notation is an arrow from the sender's lifeline to the receiver's lifeline. Different arrow shapes mark different characteristics of a message. Figure 4.12 describes a message to create objects. The arrow is dashed. For all other forms of messages the arrow is solid, but with different arrowheads.

4.7. SEQUENCE DIAGRAMS

A sequence diagram emphasizes the time ordering of messages. As Fig. shows, you form a sequence diagram by first placing the objects that participate in the interaction at the top of your diagram, across the X-axis. Typically, you place the object that initiates the interaction at the left, and increasingly more subordinate objects to the right. Next, you place the messages that these objects send and receive along the Y-axis, in order of increasing time from top to bottom. This gives the reader a clear visual cue to the flow of control over time.

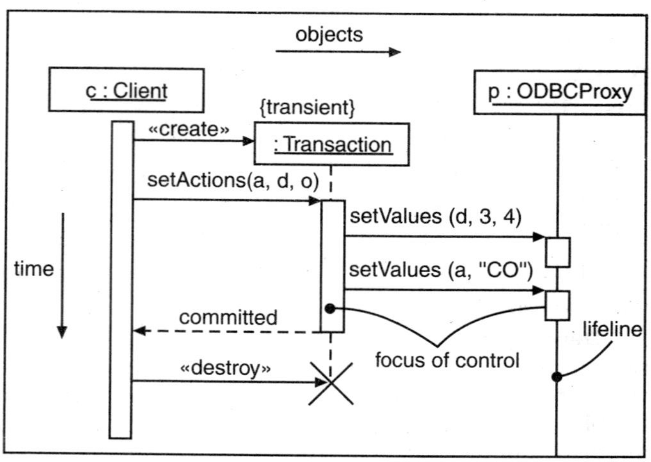

Fig. 4.13.

4.7.1. Terms and Concepts

There some basic terms which are used to refer the parts of the sequence diagram, such as:

1. **Objects:** Links are used to bind one object to others.
2. **Time:** Represents total time duration.
3. **Lifeline:** Represents the existence of an object over a period of time.

4. **Focus of Control:** The focus of control is a tall, thin rectangle that shows the period of time during which an object is performing an action, either directly or through a subordinate procedure. The top of the rectangle is aligned with the start of the action; the bottom is aligned with its completion (and can be marked by a return message).

4.7.2. Collaboration Vs. Sequence Diagrams

Collaboration diagrams and sequence diagrams both show interactions, but they emphasize different aspects. Sequence diagrams show time sequences clearly but do not show object relationships explicitly. Collaboration diagrams show object relationships clearly, but time sequences must be obtained from sequence numbers. Sequence diagrams are often most useful for showing scenarios; collaboration diagrams are often more useful for showing detailed design of procedures.

SUMMARY

- Classes are the most important building block of any object-oriented system. A class is a description of a set of objects that share the same attributes, operations, relationships, and semantics.
- There are four kinds of relationships in the UML: Dependency, Association, Generalization and Realization.
- The UML is characterized by nine major diagrams: Class diagram, Object diagram, Use case diagram, Sequence diagram, Collaboration diagram, State-chart diagram, Activity diagram, Component diagram, Deployment diagram.
- There are four common mechanisms that apply consistently throughout the language, which are: Specifications, Adornments, Common divisions, Extensibility mechanisms.
- The class diagrams are used to model the static design view of a system. Class diagrams are also the foundation for a couple of related diagrams: component diagrams and deployment diagrams.
- The Object diagrams are used to model the instances of things contained in class diagrams. An object diagram shows a set of objects and their relationships at a point in time.

SOLVED PROBLEMS

Q. 4.1. Explain all the relationships of UML with notations.

Ans: There are four kinds of relationships in the UML:

1. Dependency
2. Association
3. Generalization
4. Realization

These relationships are the basic relational building blocks of the UML. You use them to write well-formed models.

Dependency: A dependency is a semantic relationship between two things in which a change to one thing (the independent thing) may affect the semantics of the other thing (the dependent thing). Graphically, a dependency is rendered as a dashed line, possibly directed, and occasionally including a label, as in given Fig. 4.14.

Fig. 4.14.

Associations: An association is a structural relationship that describes a set of links, a link being a connection among objects. Aggregation is a special kind of association, representing a structural relationship between a whole and its parts. Graphically, an association is rendered as a solid line, possibly directed, occasionally including a label, and often containing other adornments, such as multiplicity and role names, as in given Fig. 4.15.

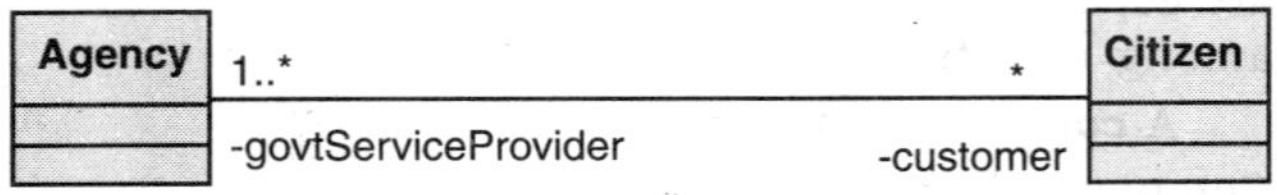

Fig. 4.15.

Generalizations: A generalization is a specialization/generalization relationship in which objects of the specialized element (the child) are substitutable for objects of the generalized element (the parent). In this way, the child shares the structure and the behavior of the parent. Graphically, a generalization relationship is rendered as a solid line with a hollow arrowhead pointing to the parent, as in given Fig. 4.16.

Fig. 4.16.

Realizations: A realization is a semantic relationship between classifiers, wherein one classifier specifies a contract that another classifier guarantees to carry out. You'll encounter realization relationships in two places: between interfaces and the classes or components that realize them, and between use cases and the collaborations that realize them. Graphically, a realization relationship is rendered as a cross between a generalization and a dependency relationship, as in given Fig. 4.17.

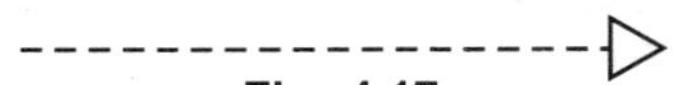

Fig. 4.17.

These four elements are the basic relational things you may include in a UML model. There are also variations on these four, such as refinement, trace, include, and extend (for dependencies).

Q. 4.2. Explain the role of all type of diagrams.

Ans: The UML is characterized by nine major diagrams:

1. Class diagram
2. Object diagram
3. Use case diagram
4. Sequence diagram
5. Collaboration diagram
6. State-chart diagram
7. Activity diagram
8. Component diagram
9. Deployment diagram

A ***class diagram*** shows a set of classes, interfaces, and collaborations and their relationships. These diagrams are the most common diagram found in modeling object-oriented systems. Class diagrams address the static design view of a system. Class diagrams that include active classes address the static process view of a system.

An ***object diagram*** shows a set of objects and their relationships. Object diagrams represent static snapshots of instances of the things found in class diagrams. These diagrams address the static design view or static process view of a system as do class diagrams, but from the perspective of real or prototypical cases.

A ***use case diagram*** shows a set of use cases and actors (a special kind of class) and their relationships. Use case diagrams address the static use case view of a system. These diagrams are especially important in organizing and modeling the behaviors of a system.

Both ***sequence diagrams*** and ***collaboration diagrams*** are kinds of interaction diagrams. An shows an interaction, consisting of a set of objects and their relationships, including the messages that may be dispatched among them. Interaction diagrams address the dynamic view of a system. A ***sequence diagram*** is an interaction diagram that emphasizes the time-ordering of messages; a ***collaboration diagram*** is an interaction diagram that emphasizes the structural organization of the objects that send and receive messages. Sequence diagrams and collaboration diagrams are isomorphic, meaning that you can take one and transform it into the other.

A ***state-chart diagram*** shows a state machine, consisting of states, transitions, events, and activities. State-chart diagrams address the dynamic view of a system. They are especially important in modeling the behavior of an interface, class, or collaboration and emphasize the event-ordered behavior of an object, which is especially useful in modeling reactive systems.

An ***activity diagram*** is a special kind of a state-chart diagram that shows the flow from activity to activity within a system. Activity diagrams address the dynamic view of a system. They are especially important in modeling the function of a system and emphasize the flow of control among objects.

A ***component diagram*** shows the organizations and dependencies among a set of components. Component diagrams address the static implementation view of a system. They are related to class diagrams in that a component typically maps to one or more classes, interfaces, or collaborations.

A ***deployment diagram*** shows the configuration of run-time processing nodes and the components that live on them. Deployment diagrams address the static deployment view of an architecture. They are related to component diagrams in that a node typically encloses one or more components.

This is not a closed list of diagrams. Tools may use the UML to provide other kinds of diagrams, although these nine are by far the most common you will encounter in practice.

Q. 4.3. Draw the notations for class and object diagrams.

Ans: The class diagrams are used to model the static design view of a system. Class diagrams are also the foundation for a couple of related diagrams: component diagrams and deployment diagrams. See Fig. 4.18. for a Customer class diagram.

The ***Object diagrams*** are used to model the instances of things contained in class diagrams. An object diagram shows a set of objects and their relationships at a point in time. The object diagram is also used to model the static design view or static process view of a system.

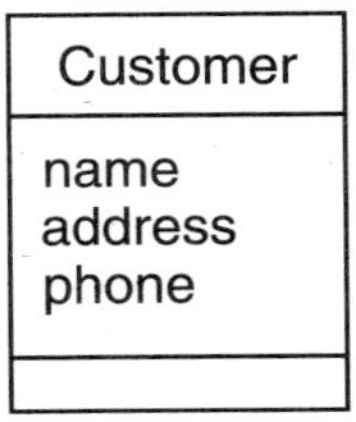

Fig. 4.18.

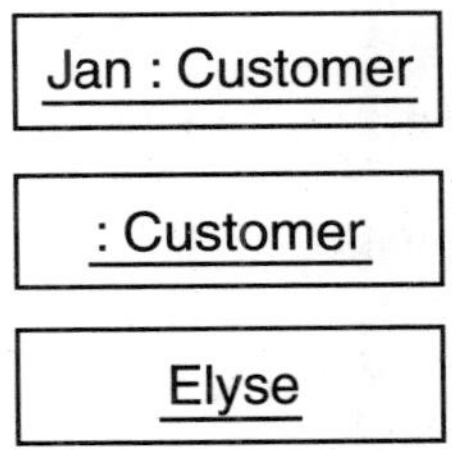

Fig. 4.19.

In this figure, there are three different type of object diagrams of Customer class, Jan (which is marked explicitly as being a Customer object),:Customer (an anonymous Customer object), and Elyse (which in its specification is marked as being a kind of Customer object, although it's not shown explicitly here).

Q. 4.4. What is Collaboration Diagram?

Ans: A collaboration diagram is an interaction diagram that emphasizes the structural organization of the objects that send and receive messages.

A collaboration diagram shows a set of objects, links among those objects, and messages sent and received by those objects. The objects are typically named or anonymous instances of classes, but may also represent instances of other things, such as collaborations, components, and nodes. You use collaboration diagrams to illustrate the dynamic view of a system.

Q. 4.5 What is Sequence Diagram?

Ans: A sequence diagram emphasizes the time ordering of messages. Sequence diagrams are based on the interaction model that describes the interaction between elements of the system. It answers the question: "When does who call whom and how?" In contrast, the activity model answers the question: "What happens in which sequence?", and the state model: "How does an object respond to events in a specific state?"

Q. 4.6. What is the difference between object and instance? *(Important)*

Ans. Each class describes a possibly infinite set of individual objects. Each object is said to be an instance of its class. Each instance of the class has its own value for each attributes but shares the attribute names and operations with their instances of the class. An object contains an implicit reference to its own class; it "knows what kind of thing it is".

Q. 4.7. What is the difference between operation and methods?

Ans. An operation is a function or transformation that may be applied to or by objects in a class. Operation shows the behavior of the class.

A method is the *implementation* of an operation for a class.

For example, the class **File** may have and operation **print**. Different methods could be implemented to print ASCII files, print binary files and print picture files. All these methods logically perform the same task-printing a file, thus name of operation refer by **print.**

Q. 4.8. Draw a sequence diagram for the cooling soft-drink.

Ans:

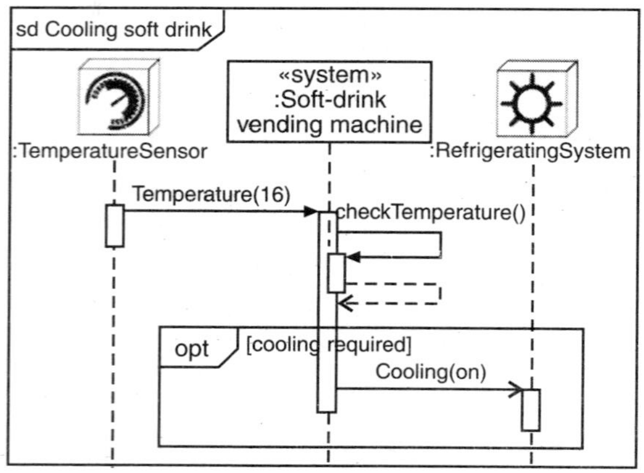

Fig. 4.20.

Q. 4.9. What is the main purpose of object modeling?

Ans. The purpose of object modeling is to describe objects of the system. Objects serve two purposes:

1. Objects promote understanding of the real world.
2. Objects provide a practical basis for computer implementation.

Q. 4.10. What do you mean by "Role names"?

Ans.

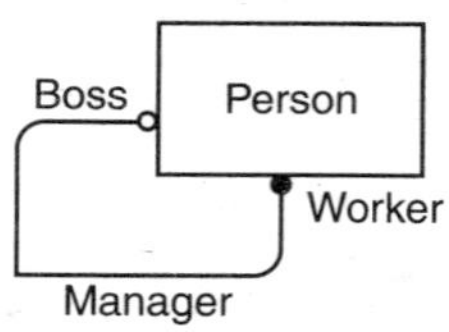

Q. 4.11. Prepare a class diagram from the direct diagram in figure. *(UPTU 2002-03)*

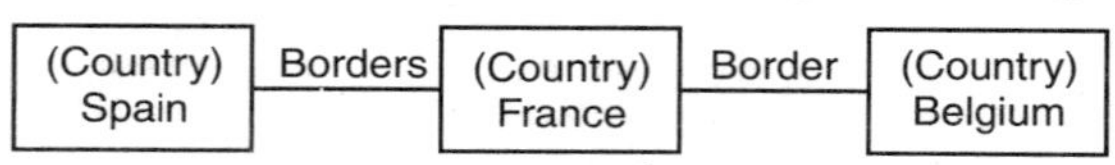

Ans.

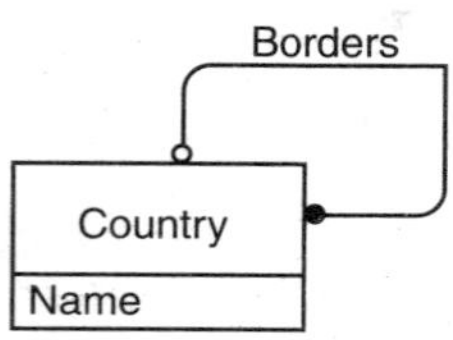

Chapter 5

Basic Behavioural Modeling

5.1. USE CASES

A use case describes a set of sequences, in which each sequence represents the interaction of the elements outside the system (its actors) with the system itself (and its key abstractions). These behaviors are in effect system-level functions that you use to visualize, specify, construct, and document the intended behavior of your system during requirements capture and analysis. A use case represents a functional requirement of your system as a whole. For example, one central use case of a bank is to process loans.

A use case describes an interaction with actors as a sequence of messages between the system and one or more actors. The term actor includes humans, as well as other computer systems and processes. Fig. 5.1 shows a use case diagram for a telephone catalog sales application. The model has been simplified as an example.

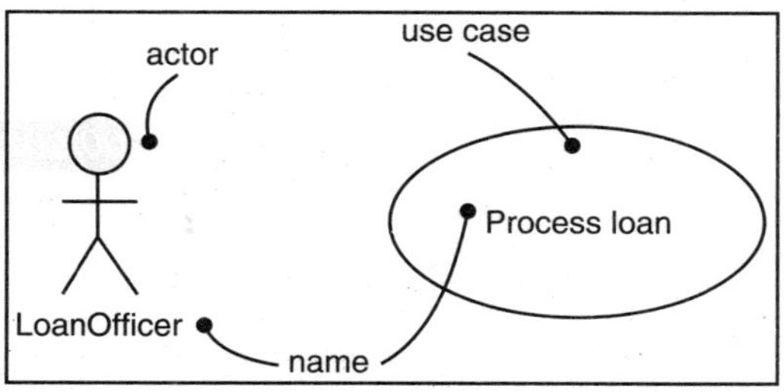

Fig. 5.1. Use Case

CHAPTER OUTLINE

5.2. USE CASE DIAGRAMS

6A use case is a coherent unit of externally visible functionality provided by a system unit and expressed by sequences of messages exchanged by the system unit and one or more actors of the system unit. The purpose of a use case is to define a piece of coherent behavior without revealing the internal structure of the system.

The definition of a use case includes all the behavior it entails—the mainline sequences, different variations on normal behavior, and all the exceptional conditions that can occur with such behavior, together with the desired response. From the user's point of view, these may be abnormal situations. From the system's point of view, they are additional variations that must be described and handled. In the model, the execution of each use case is independent of the others, although an implementation of the use cases may create implicit dependencies among them due to shared objects. Each use case represents an orthogonal piece of functionality whose execution can be mixed with the execution of other use cases.

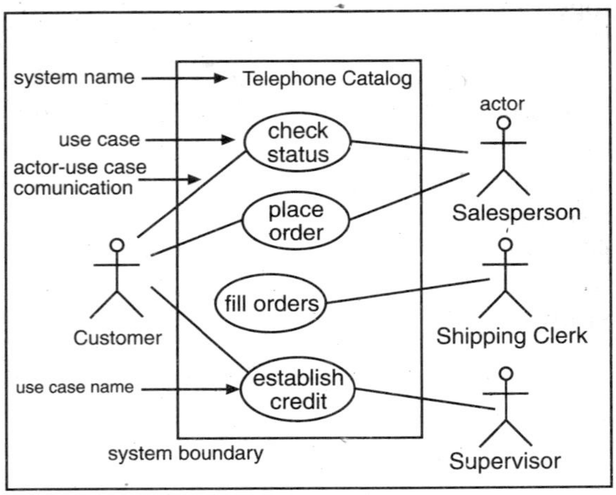

Fig. 5.2. Use Case Diagram

The dynamics of a use case may be specified by UML interactions, shown as statechart diagrams, sequence diagrams, collaboration diagrams, or informal text descriptions. When use cases are implemented, they are realized by collaborations among classes in the system. One class may participate in multiple collaborations and therefore in multiple use cases.

A use case can participate in several relationships, in addition to association with actors (Table 5.1).

Table 5.1. Kinds of Use Case Relationships

Relationship	Function	Notation
Association	The communication path between an actor and a use case that it participates in.	———
Extend	The insertion of addition behaviour into a base use case that does not know about it.	extend - - - -▸
Use case generalization	A relationship between a general use case and a more specific use case that inherites and adds features to it.	———▸
Include	The insertion of additional behaviour into a base use case that explicitly describe the insertion.	include - - - -▸

A use case is drawn as an ellipse with its name inside or below it. It is connected by solid lines to actors that communicate with it. Although each use case instance is independent, the description of a use case can be factored into other, simpler use cases. This is similar to the way the description of a class can be defined incrementally from the description of a superclass. A use case can simply incorporate the behavior of other use cases as fragments of its own behavior. This is called an include relationship. In this case, the new use case is not a special case of the original use case and cannot be substituted for it. A use case can also be defined as an incremental extension to a base use case. This is called an extend relationship.

There may be several extensions of the same base use case that may all be applied together. The extensions to a base use case add to its semantics; it is the base use case that is instantiated, not the extension use cases.

The include and extend relationships are drawn as dashed arrows with the keyword "include" or "extend". The include relationship points at the use case to be included; the extend relationship points at the use case to be extended. A use case can also be specialized into one or more child use cases. This is use case generalization. Any child use case may be used in a situation in which the parent use case is expected. Use case generalization is drawn the same as any generalization, as a line from the child use case to the parent use case with a large triangular arrowhead on the parent end. Figure 5.3 shows use case relationships in the catalog sales application.

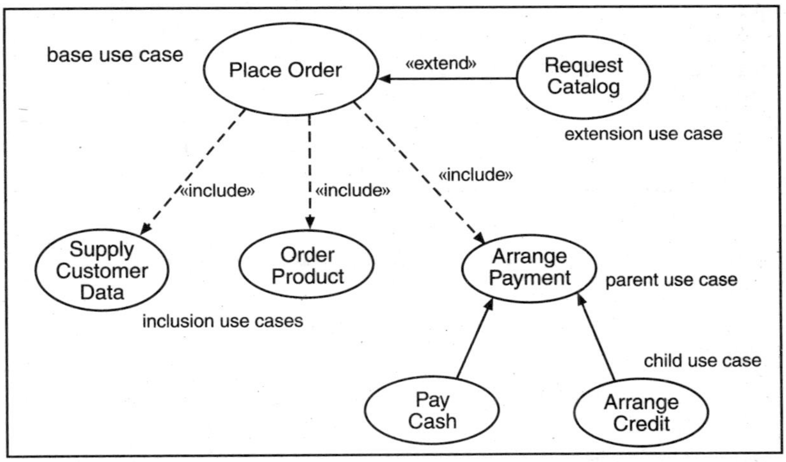

Fig. 5.3. Use case relationship

5.3. ACTIVITY DIAGRAMS

Activity diagrams are one of the five diagrams in the UML for modeling the dynamic aspects of systems. An activity diagram is essentially a flowchart, showing flow of control from activity to activity.

An activity graph contains activity states. An activity state represents the execution of a statement in a procedure or the performance of an activity in a workflow.

A completion transition in an activity diagram fires when the preceding activity is complete. Activity states usually do not have transitions with explicit events, but they may be aborted by transitions on enclosing states.

An activity graph may also contain action states, which are similar to activity states, except that they are atomic and do not permit transitions while they are active.

An activity diagram may contain branches, as well as forking of control into concurrent threads. Concurrent threads represent activities that can be performed concurrently by different objects or persons in an organization.

An activity diagram is the notation for an activity graph (Fig. 5.4). It includes some special shorthand symbols for convenience. These symbols can actually be used on any statechart diagram, although mixing notation may be ugly much of the time.

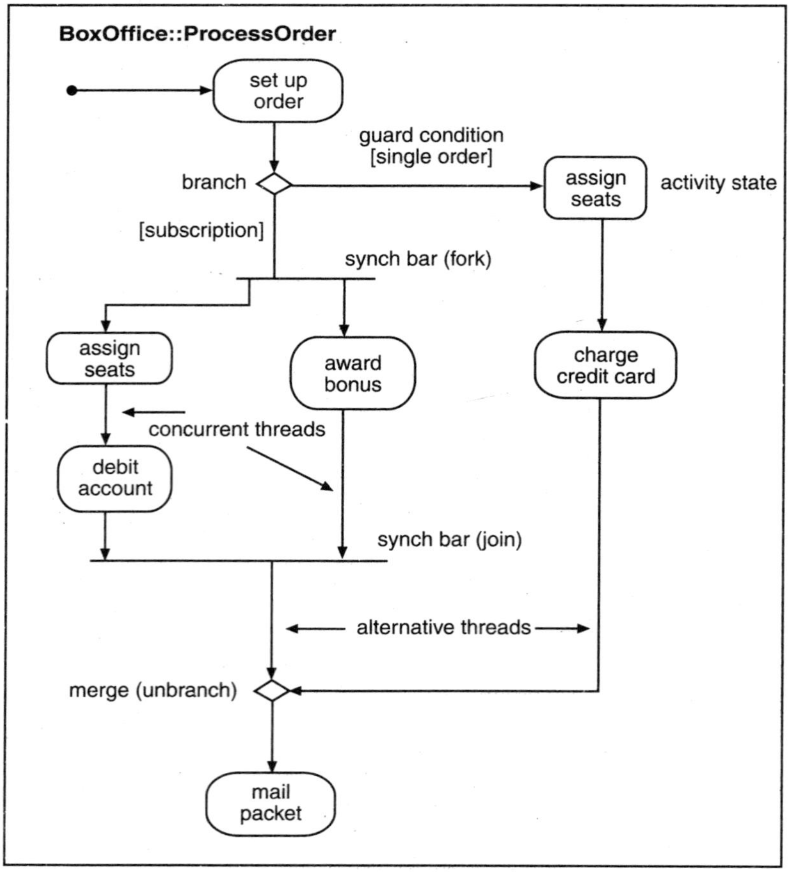

Fig. 5.4. Avtivity diagram

An activity state is shown as a box with rounded ends containing a description of the activity. (Normal state boxes have straight sides and rounded corners.) Simple completion transitions are shown as arrows. Branches are shown as guard conditions on transitions or as diamonds with multiple labelled exit arrows. Figure 5.4 shows an activity diagram for processing an order by the box office.

5.4. STATE MACHINE DIAGRAM

A system is always in a state that abstracts a combination of values given in the system. Events arriving at the system lead to reactions-depending on the state—that change values and results in new states.

A state machine diagram describes this behavior. It contains state machines with states and state transitions that are triggered by events. The semantics of the state model is defined in such great detail in the UML specification that it can also be executed.

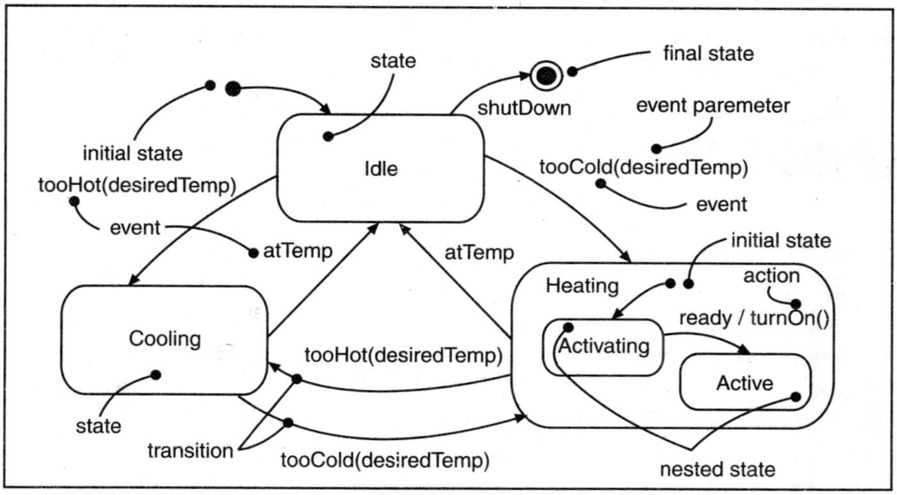

Fig. 5.5. State machine diagram

5.4.1. State Machine

A state machine describes behavior of a structure that is generally a class. With regard to the concrete execution, we often speak of a context object.

A state machine has no notation of its own. It is represented by the entire state machine diagram. This means that the diagram frame forms the border of the state machine. It is not possible to show a graphical connection between the state machine and the pertaining class in the diagram (Fig. 5.6). However, the relationship is explicitly present in the model.

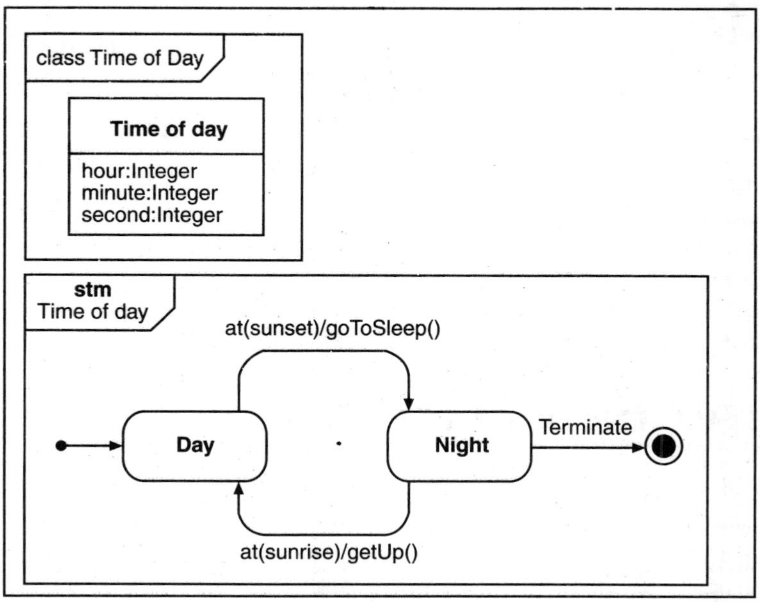

Fig. 5.6. State machine time of day

A state machine forms a semantic capsule around states, transitions, and pseudo-states. This normally includes the execution rules. A state machine has a queue for events that are consumed by states and transitions in the sequence in which they occur. This means that they lead to state transitions, or to internal actions in a state.

> **Concept**
>
> *A state machine describes the states and state transitions of a structure.*

State machines are subject to the so-called run-to-completion semantics, which says that the next event will be consumed only provided that all actions of the previous event have been completed. One exception is the state behavior that does not have to be completed before the next event can be processed. An event that cannot be consumed, for example, because there is no matching transition, is discarded.

The state machine in Fig. 5.6, can consume the two time events at ( ) and the event terminate. The latter event activates the transition that leads to the final state, so that the state machine terminates since no other region is active.

5.4.2. State

A state represents a set of value combinations, for example, values for the attributes of a class. Figure 5.7 you can see a class, Time of Day, in a class diagram and a corresponding simple state machine diagram. The state Day stands for all value combinations of the attributes hour, minute, and second, between the two events, sunrise and sunset. 16 This example shows clearly that a state is an abstraction of possible value combinations. Not every value combination is a meaningful state, for example, the state "13 : 17 : 31" (HH : MM : SS), which is not modeled here. What is meaningful and what isn't eventually depends on the domain we model, of course.

> **Concept**
>
> *A state represents a set of value combinations for the underlying element. It has a name, and may have an internal behavior that is executed based on defined events.*

States are represented by rounded rectangles (Fig. 5.7). The name of the state is within the rectangle. Another compartment shows the internal behavior, which is triggered by events, similar to transitions, but which does not cause the state to be exited.

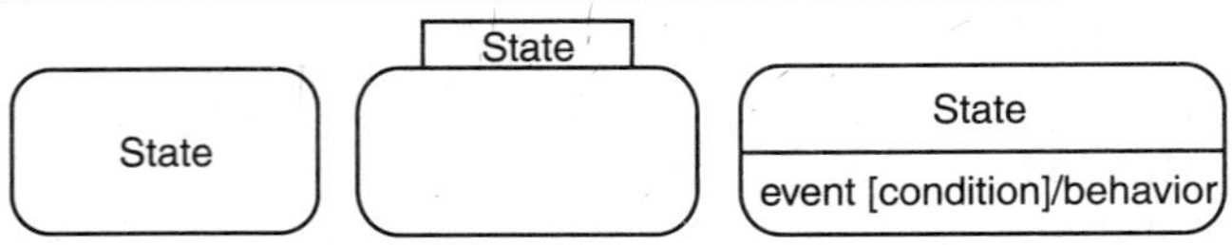

Fig. 5.7. Notation for State

In addition to the internal behavior, a state can also have three special behaviors that are triggered on the basis of predefined events:

- The entry behavior is executed immediately upon entering the state.
- The exit behavior is executed immediately before exiting the state.
- The do behavior is executed while the state is active.

In Fig. 5.8, you can see internal behavior in the state machine of a railroad crossing. This example utilizes the fact that the state behavior is interruptible. For example, the barriers are closed in the state block. The state behavior close() is interrupted by the event endPosition(down). The entry behavior and the exit behavior are not interruptible.

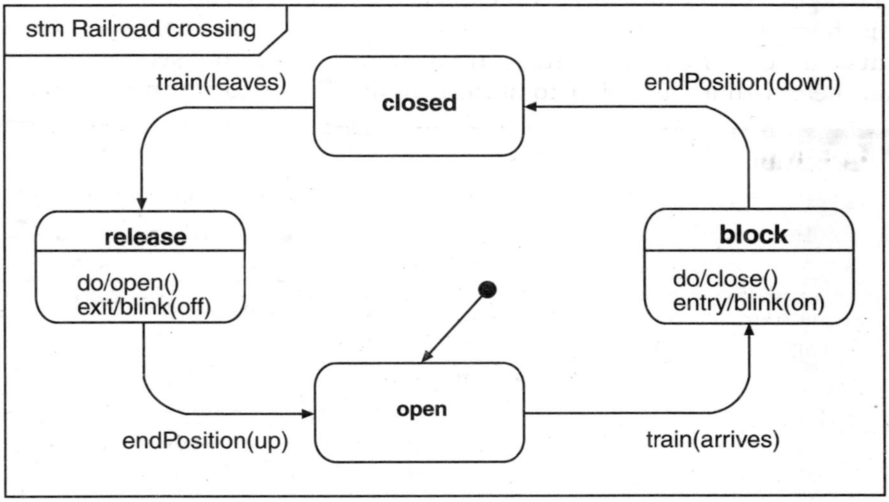

Fig. 5.8. State behaviour

An important UML extension versus the classic Finite State Machine (FSM) is the possibility to decompose states into smaller units. This allows us to first describe states on a general level and then further detailing them.

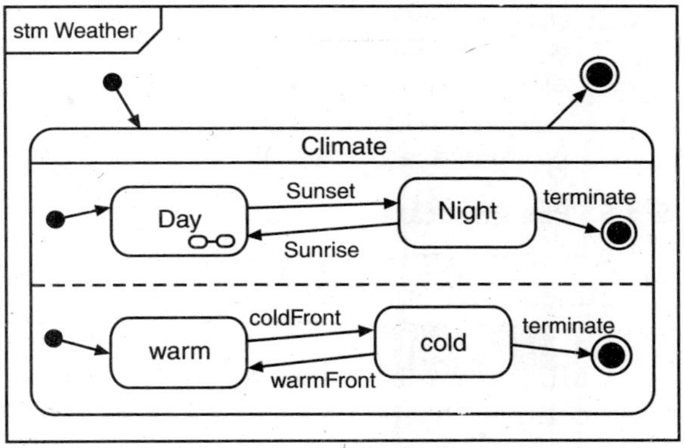

Fig. 5.9. State machine diagram with transistor

5.4.3. Transition

A transition specifies a state transition. It is a directed relationship between two states, and defines a trigger and a condition that both lead to the state transition, as well as a behavior that is executed during the transition.

In Fig. 5.9, the transition from the orthogonal state Climate to the final state has no trigger. It is activated automatically as soon as the source state has terminated. In this example, this happens as soon as the final states in both regions of the state Climate have been reached. A transition is denoted as an arrow with solid line and a description in the following syntax:

```
trigger1, trigger2, ... [condition] / behavior
```

5.4.4. Trigger and Event

> **Concept**
>
> *A trigger references exactly one event and establishes a relationship to a behavior.*
>
> *An event specifies some occurrence that can be measured with regard to location and time.*

A trigger is the connecting link between event and behavior in the model. Denoted at a transition means that the transition is activated as soon as the event occurs.

UML knows four types of events that can trigger a transition (Fig. 5.10):

1. A call event occurs when an operation is invoked.
2. A change event occurs when a Boolean expression tests to true. This is the case when a value has changed accordingly in the system.
3. A signal event occurs when a signal has been received.
4. A time event is further distinguished. A relative time event occurs when the state from which the corresponding transition originates has been active over a certain period of time (keyword after(<time>)). An absolute time event occurs when the defined absolute time occurs (keyword at(<time>)).

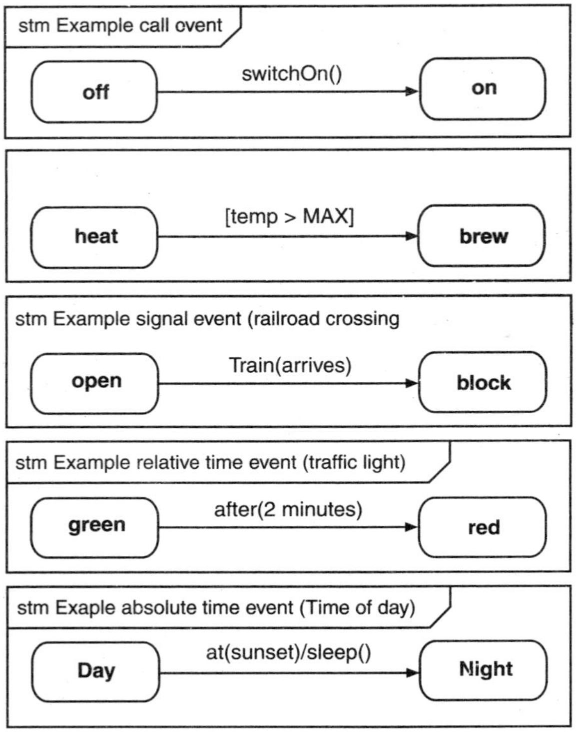

Fig. 5.10. State machine diagram for each event

5.4.5. Pseudo-state

A pseudo-state is a control element that influences the sequence of events in a state machine. It is not a real state, and a pseudo-state does not represent a value combination either.

UML knows 10 different pseudo-states. Each of them describes a special property that controls state transitions. For example, the initial state is a pseudo-state. It is responsible for marking the initial state of a state machine.

All ten pseudo-states are given below:

1. Initial State
2. Shallow history
3. Deep history
4. Join
5. Fork
6. Junction
7. Choice
8. Exit point
9. Entry point
10. Terminate

There are two types of history state: shallow history and deep history. To better understand history states, let's first look at the state machine of a hairdryer (Figure 5.11). Don't pay attention to the circle symbol with the letter "H" for now. As soon as the hairdryer is switched on, it jumps into the initial state configuration, cold and low. Unless you are particularly concerned about damaging your hair, you will probably switch the hairdryer from low to high and from cold to hot. The state hot lets you additionally select either normal or extreme. When you switch it off, the hairdryer changes to off state.

The next time you use the hairdryer it will come on in the default state configuration—low and cold. So again, you have to go through the hassle of resetting it to high and hot. After the third, fourth, or fifth time you do this, you might come to think that this hairdryer is not fit for everyday use. This situation can be improved, at least in the model, by using history states.

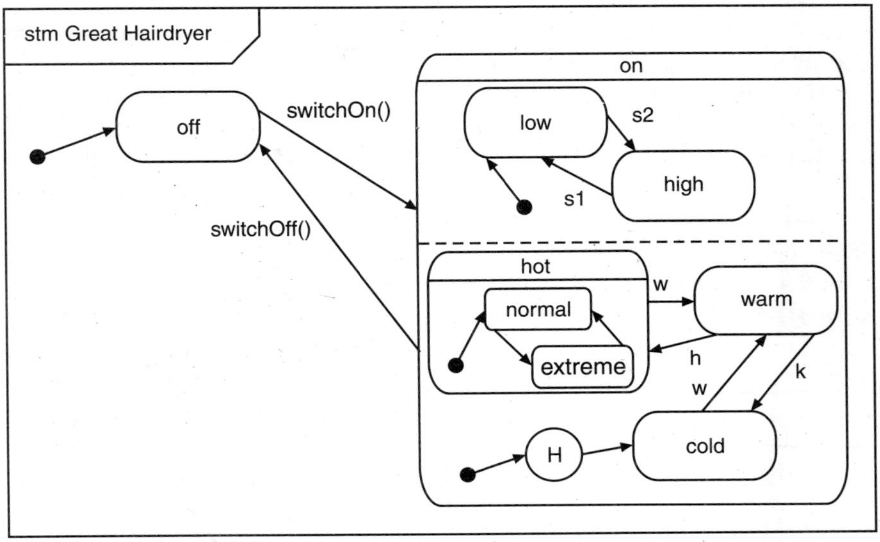

Fig. 5.11. A state machine, Hairdryer

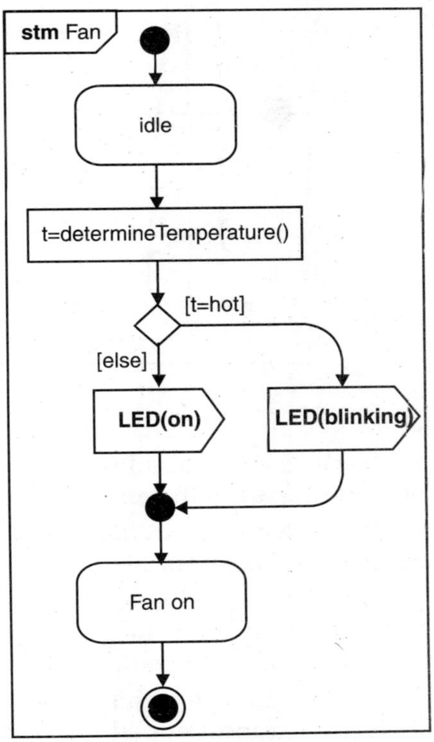

Fig. 5.12. A state machine with choice and junction

We'll now have a closer look at the circle symbol with the "H". This is the notation for shallow history. Shallow history causes the substate of the pertaining region that was last active to be saved. When you now switch the hairdryer on a second time, when traversing the shallow history, the device will read the state that was last active and activate it directly. This means that the hairdryer will immediately jump into the low and hot/normal configuration after the second time you use it.

As the name implies, deep history has a "deeper" effect than shallow history. Using shallow history, when you switch it on again, the device sets itself to the state hot, as described above. Within this state, the substate normal is activated since the initial state points to it. However, when using deep history, the device also restores the substates in the configuration that was active last. More specifically, when you switch it on again, the hairdryer jumps into the state configuration low and hot/extreme. The notation for deep history is the same as for shallow history, with an additional asterisk (*) behind the "H".

A junction connects transitions and merges them to form a path. It is possible to merge several incoming transitions, or to put several outgoing transitions into a junction. A junction has no additional semantics. Its notation is a black circle (Fig. 5.12). In contrast to an initial state, which always has the same notation, a junction has always at least one incoming transition. The merging of transitions is comparable to the merge node in activity diagrams, except that the symbol is different.

A choice has several outgoing transitions that each have a condition. In contrast to junctions, the conditions of a choice can be evaluated on the basis of values that are determined only at the time of the state transition (Fig. 5.12). It is therefore also referred to as a dynamic choice . A choice is not used for merging of transitions. The notation for choices is identical to the one for a decision in an activity diagram—a rhombus.

Fork and join are similar to the elements of the same names from the activity diagram. A fork has one incoming transition and several outgoing transitions that lead to different states, positioned orthogonally to one another. This means that all target states are activated. The outgoing transitions may have neither conditions nor triggers (Fig. 5.13).

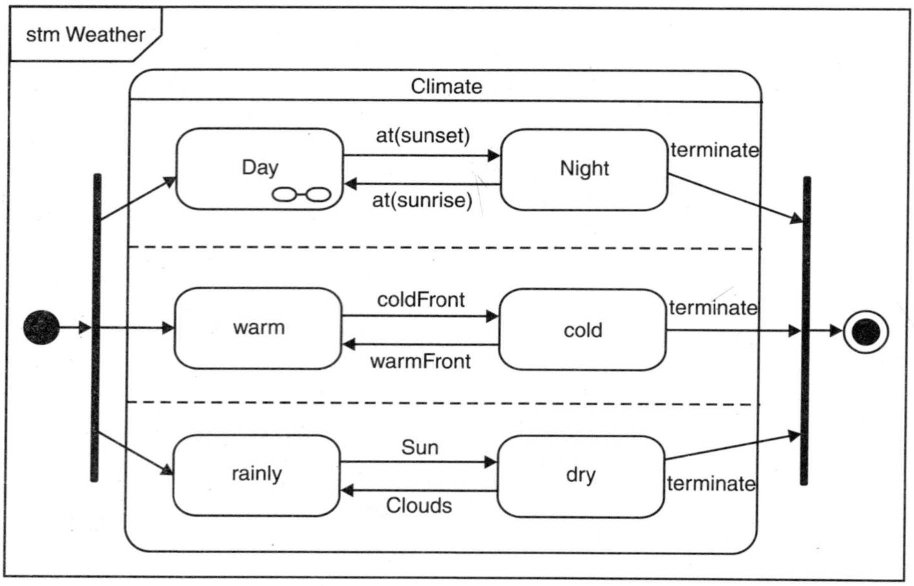

Fig. 5.13. Example for fork and join

In contrast, a join has several incoming transitions, which originate from states that are positioned orthogonally to one another, and one outgoing transition (Fig. 5.13).

Both elements are denoted as filled bars. They differ in the context of incoming and outgoing transitions.

An entry point describes a special entry into a state machine. A transition leads from the entry point to a state of the state machine, or to one state each of every region in case of an orthogonal state.

An exit point terminates a state machine. If a transition from any region in the state machine reaches an exit point, then the state machine is terminated, and the transition originating from the exit point is activated.

An entry point is denoted as a white circle, and the exit point as a circle with a cross. Viewed from the outside the circle appears on the border of the corresponding state. In the inner view, transitions terminate or start at the circle (Fig. 5.14).

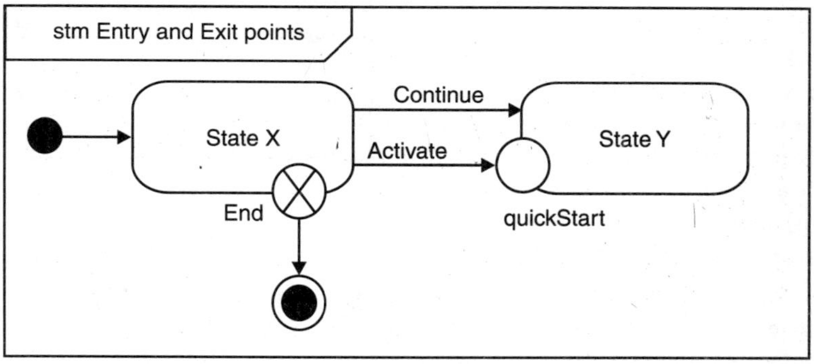

Fig. 5.14. (a) Entry and exit points

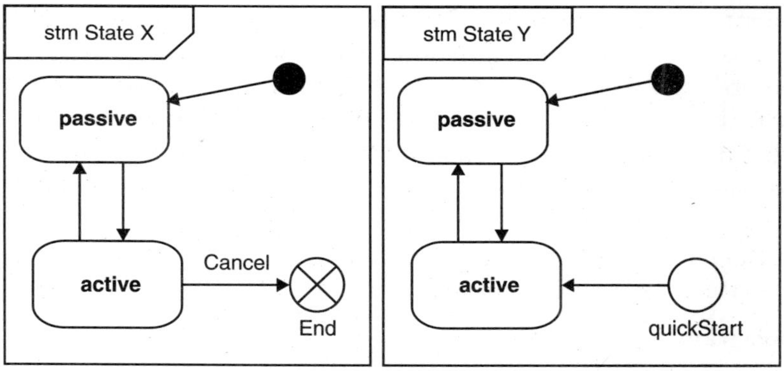

Fig. 5.14. (b) Entry and exit points

5.5. PROCESS AND THREAD

When you model a system of the real world, you must take into account its process view, which encompasses the threads and processes that form the system's concurrency and synchronization mechanisms.

In the UML, you model each independent flow of control as an active object that represents a process or thread that can initiate control activity. A process is a heavyweight flow that can execute concurrently with other processes; a thread is a lightweight flow that can execute concurrently with other threads within the same process.

5.5.1. Active Class

The UML provides a graphical representation of an active class, as Figure 5.15 shows. Active classes are kinds of classes, so have all the usual compartments for class name, attributes, and operations. Active classes often receive signals, which you typically enumerate in an extra compartment.

An active object is an object that owns a process or thread and can initiate control activity. An active class is a class whose instances are active objects. A process is a heavyweight flow that can execute concurrently with other processes. A thread is a lightweight flow that can execute concurrently with other threads within the same process. Graphically, an active class is rendered as a rectangle with thick lines. Processes and threads are rendered as stereotyped active classes (and also appear as sequences in interaction diagrams).

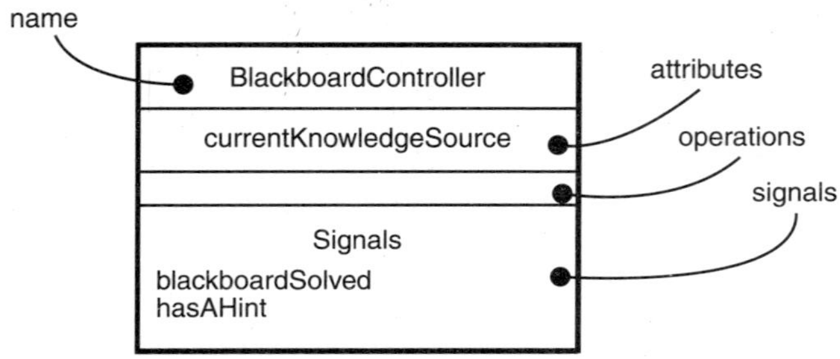

Fig. 5.15. Active Class

In a concurrent system, there is more than one flow of control – that is, more than one thing can take place at a time. In a concurrent system, there are multiple simultaneous flows of control, each rooted at the head of an independent process or a thread. If you take a snapshot of a concurrent system while it's running, you'll logically see multiple loci of execution.

In the UML, you use an active class to represent a process or thread that is the root of an independent flow of control and that is concurrent with all peer flows of control.

An active class represents an independent flow of control, whereas a plain class embodies no such flow. In contrast to active classes, plain classes are implicitly called passive because they cannot independently initiate control activity.

The UML defines two standard stereotypes that apply to active classes:

1. **process:** Specifies a heavyweight flow that can execute concurrently with other processes.
2. **thread:** Specifies a lightweight flow that can execute concurrently with other threads within the same process.

5.5.2. Synchronization

When a flow passes through an operation, we say that at a given moment, the focus of control is in the operation. If that operation is defined for some class, we can also say that at a given moment, the focus of control is in a specific instance of that class. You can have multiple flows of control in one operation (and therefore in one object), and you can have different flows of control in different operations (but still result in multiple flows of control in the one object).

The problem arises when more than one flow of control is in one object at the same time. If you are not careful, anything more than one flow will interfere with another, corrupting the state of the object. This is the classical problem of mutual exclusion. A failure to deal with it properly yields all sorts of race conditions and interference that cause concurrent systems to fail in mysterious and unrepeatable ways.

The key to solving this problem in object-oriented systems is by treating an object as a critical region. There are three alternatives to this approach, each of which involves attaching certain synchronization properties to the operations defined in a class. In the UML, you can model all three approaches.

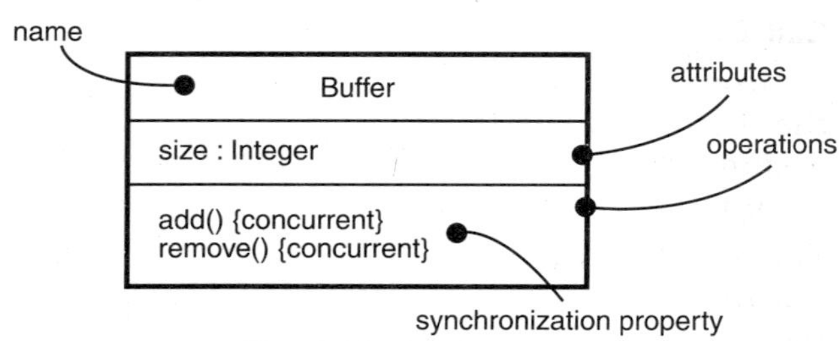

Fig. 5.16. Synchronization

1. **Sequential:** Callers must coordinate outside the object so that only one flow is in the object at a time. In the presence of multiple flows of control, the semantics and integrity of the object cannot be guaranteed.
2. **Guarded:** The semantics and integrity of the object is guaranteed in the presence of multiple flows of control by sequentializing all calls to all of the object's guarded operations. In effect, exactly one operation at a time can be invoked on the object, reducing this to sequential semantics.
3. **Concurrent:** The semantics and integrity of the object is guaranteed in the presence of multiple flows of control by treating the operation as atomic.

5.5.3. Event and Signals

An event is the specification of a significant occurrence that has a location in time and space. In the context of state machines, an event is an occurrence of a stimulus that can trigger a state transition. A signal is a kind of event that represents the specification of an asynchronous stimulus communicated between instances.

Events may be external or internal. External events are those that pass between the system and its actors. For example, the pushing of a button and an interrupt from a collision sensor are both examples of external events. Internal events are those that pass among the objects that live inside the system. An overflow exception is an example of an internal event.

In the UML, you can model four kinds of events:

- signals,
- calls,
- the passing of time,
- and a change in state

Signal

A signal represents a named object that is dispatched (thrown) asynchronously by one object and then received (caught) by another. Exceptions are supported by most contemporary programming languages and are the most common kind of internal signal that you will need to model.

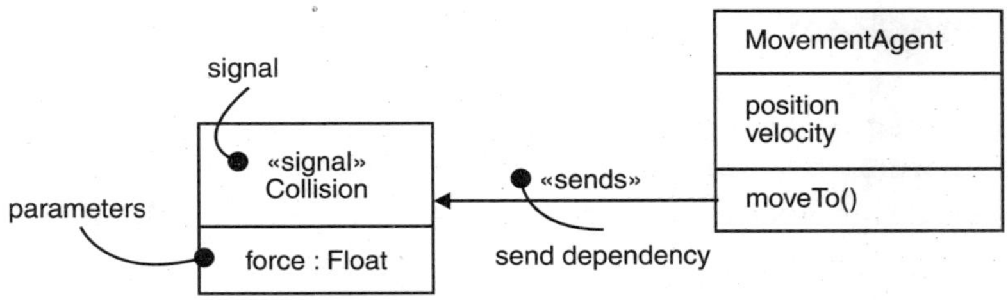

Fig. 5.17. Signal

Call Events

Just as a signal event represents the occurrence of a signal, a call event represents the dispatch of an operation. In both cases, the event may trigger a state transition in a state machine.

Whereas a signal is an asynchronous event, a call event is, in general, synchronous. This means that when an object invokes an operation on another object that has a state machine, control passes from the sender to the receiver, the transition is triggered by the event, the operation is completed, the receiver transitions to a new state, and control returns to the sender.

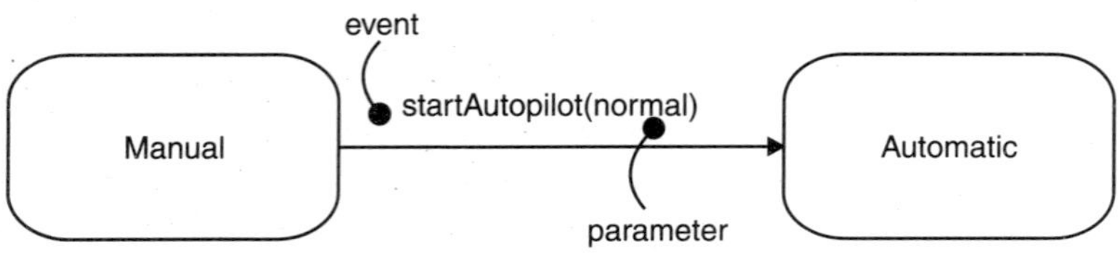

Fig. 5.18. Call Event

Time and Change Events

A time event is an event that represents the passage of time. As Fig. 5.19 shows, in the UML you model a time event by using the keyword after followed by some expression that valuates to a period of time. Such expressions can be simple (for example, after 2 seconds) or complex (for example, after 1-ms since exiting Idle). Unless you specify it explicitly, the starting time of such an expression is the time since entering the current state.

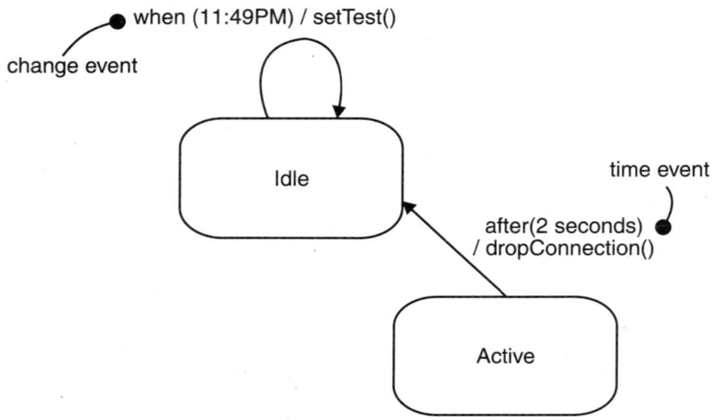

Fig. 5.19.

A change event is an event that represents a change in state or the satisfaction of some condition. As Figure 5.19 shows, in the UML you model a change event by using the keyword when followed by some Boolean expression. You can use such expressions to mark an absolute time (such as when time = 11:59) or for the continuous test of an expression (for example, when altitude < 1000).

5.6. TIME DIAGRAM

A timing mark is a denotation for the time at which an event occurs. Graphically, a timing mark is formed as an expression from the name given to the message (which is typically different from the name of the action dispatched by the message).

A time expression is an expression that evaluates to an absolute or relative value of time.

A timing constraint is a semantic statement about the relative or absolute value of time.

Graphically, a timing constraint is rendered as for any constraint—that is, a string enclosed by brackets and generally connected to an element by a dependency relationship.

Location is the placement of a component on a node. Graphically, location is rendered as a tagged value—that is, a string enclosed by brackets and placed below an element's name, or as the nesting of components inside nodes.

The passing of messages represents the dynamic aspect of any system, so when you model the time-critical nature of a system with the UML, you can give a name to each message in an interaction to be used as a timing mark. Messages in an interaction are usually not given names.

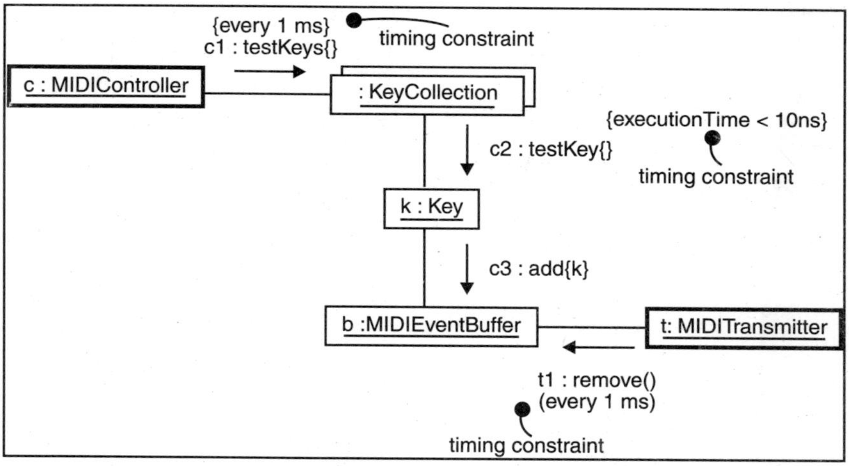

Fig. 5.20. Time Diagram

They are mainly rendered with the name of an event, such as a signal or a call. As a result, you can't use the event name to write an expression because the same event may trigger different messages. If the designated message is ambiguous, use the explicit name of the message in a timing mark to designate the message you want to mention in a time expression. A timing mark is nothing more than an expression formed from the name of a message in an interaction. Given a message name, you can refer to any of three functions of that message—that is **startTime, stopTime** and **executionTime**. You can then use these functions to specify arbitrarily complex time expressions, perhaps even using weights or offsets that are either constants or variables (as long as those variables can be bound at execution time). Finally, as shown in Fig. 5.20, you can place these time expressions in a timing constraint to specify the timing behavior of the system. As constraints, you can render them by placing them adjacent to the appropriate message, or you can explicitly attach them using dependency relationships.

5.7. INTERACTION DIAGRAM

There are two type of Interaction diagram:

- Sequence diagrams and
- Collaboration diagrams

Interaction diagram shows an interaction, consisting of a set of objects and their relationships, including the messages that may be dispatched among them. A sequence diagram is an interaction diagram that emphasizes the time ordering of messages; a collaboration diagram is an interaction diagram that emphasizes the structural organization of the objects that send and receive messages.

Interaction diagrams is used to model the dynamic aspects of a system. For the most part, this involves modeling concrete or prototypical instances of classes, interfaces, components, and nodes, along with the messages that are dispatched among them, all in the context of a scenario that illustrates a behavior. Interaction diagrams may stand alone to visualize, specify, construct, and document the dynamics of a particular society of objects, or they may be used to model one particular flow of control of a use case.

Interaction diagrams are not only important for modeling the dynamic aspects of a system, but also for constructing executable systems through forward and reverse engineering.

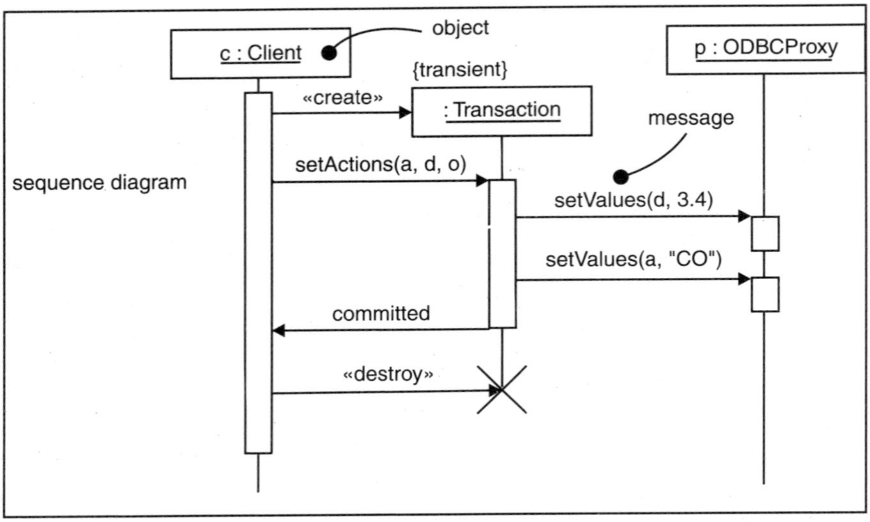

(a) Sequence diagram

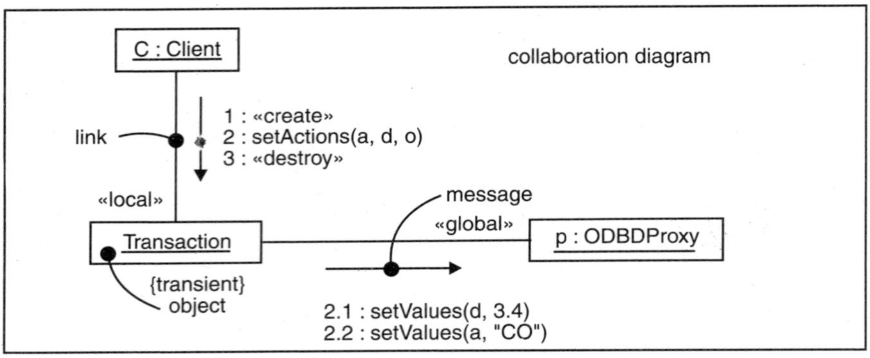

(b) Coleboration diagram

Fig. 5.21. Intraction diagram

An interaction diagram shows an interaction, consisting of a set of objects and their relationships, including the messages that may be dispatched among them. A sequence diagram is an interaction diagram that emphasizes the time ordering of messages. Graphically, a sequence diagram is a table that shows objects arranged along the X-axis and messages, ordered in increasing time, along the Y-axis. A collaboration diagram is an interaction diagram that emphasizes the structural organization of the objects that send and receive messages. Graphically, a collaboration diagram is a collection of vertices and arcs. We have discussed sequence and collaboration diagrams in detail, in the chapter 4.

5.8. PACKAGE DIAGRAM

Large systems involve manipulating potentially large numbers of classes, interfaces, components, nodes, diagrams, and other elements. As you scale up to systems such as these, you will find it necessary to organize these things into larger chunks. In the UML, the package is a general purpose mechanism for organizing modeling elements into groups.

Concept

A package groups model elements and forms a namespace.

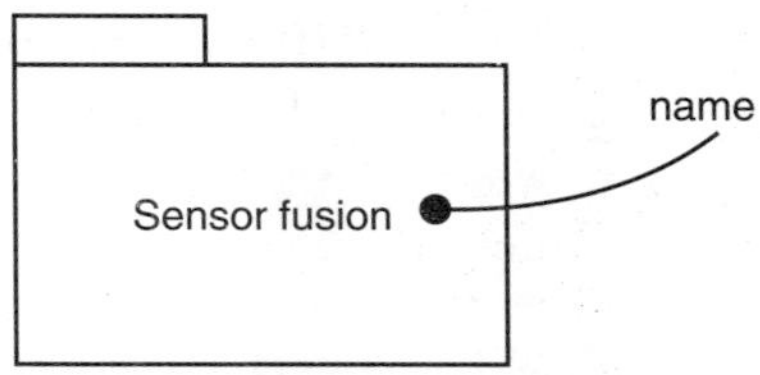

Fig. 5.22. Package

You use packages to arrange your modeling elements into larger chunks that you can manipulate as a group. You can control the visibility of these elements so that some things are visible outside the package while others are hidden. You can also use packages to present different views of your system's architecture.

Well-designed packages group elements that are semantically close and that tend to change together. Well-structured packages are therefore loosely coupled and very cohesive, with tightly controlled access to the package's contents.

In the UML, the chunks that organize a model are called packages. A package is a general purpose mechanism for organizing elements into groups. Packages help you organize the elements in your models so that you can more easily understand them. Packages also let you control access to their contents so that you can control the seams in your system's architecture.

The UML provides a graphical representation of package, as Figure 5.22 shows. This notation permits you to visualize groups of elements that can be manipulated as a whole and in a way that lets you control the visibility of and access to individual elements.

SUMMARY

- A use case describes a set of sequences, in which each sequence represents the interaction of the elements outside the system with the system itself.
- The purpose of a use case is to define a piece of coherent behavior without revealing the internal structure of the system.
- Activity diagrams are one of the five diagrams in the UML for modeling the dynamic aspects of systems. An activity diagram is essentially a flowchart, showing flow of control from activity to activity.
- A state machine diagram describes this behavior. It contains state machines with states and state transitions that are triggered by events.
- A trigger references exactly one event and establishes a relationship to a behavior. And an event specifies some occurrence that can be measured with regard to location and time.

SOLVED PROBLEMS

Q. 5.1. List out the all Behavioral Elements in UML.

Ans: There are two primary kinds of behavioral Elements:

(*a*) Interaction (*b*) State Machine

Q. 5.2. What is the use case?

Ans: A use case describes a set of sequences, in which each sequence represents the interaction of the elements outside the system (its actors) with the system itself (and its key abstractions). A use case represents a functional requirement of your system as a whole.

The purpose of a use case is to define a piece of coherent behavior without revealing the internal structure of the system.

Q. 5.3. What is state?

Ans: A state represents a set of value combinations for the underlying element. It has a name, and may have an internal behavior that is executed based on defined events.

Q. 5.4. What is State Machine?

Ans: A state machine describes behavior of a structure that is generally a class. With regard to the concrete execution, we often speak of a context object. It means a state machine describes the states and state transitions of a structure.

Q. 5.5. Draw a State Machine diagram for Rail Road Crossing.

Ans:

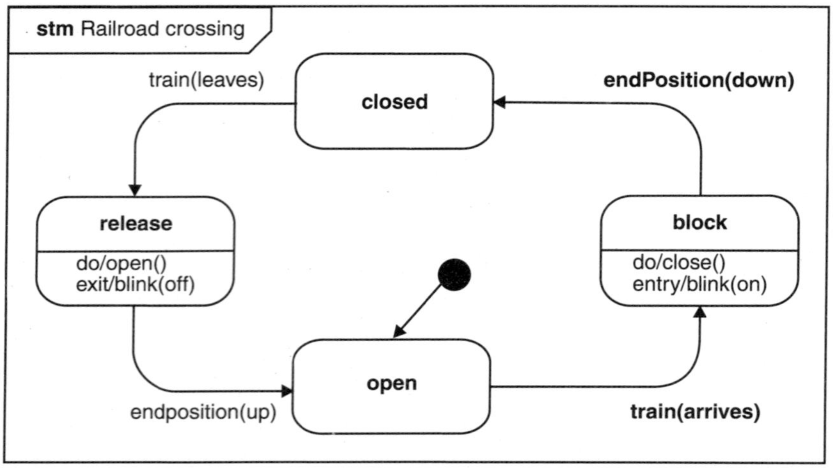

Fig. 5.23.

Q. 5.6. What are the types of events?

Ans: There are four type of events:

1. A call event occurs when an operation is invoked.
2. A change event occurs when a Boolean expression tests to true. This is the case when a value has changed accordingly in the system.
3. A signal event occurs when a signal has been received.
4. A time event is further distinguished.
   (*a*) A relative time event occurs when the state from which the corresponding transition originates has been active over a certain period of time (keyword after(<time>)).
   (*b*) An absolute time event occurs when the defined absolute time occurs (keyword at(<time>)).

Q. 5.7. What is pseudo-state? Explain in detail.

Ans: A pseudo-state is a control element that influences the sequence of events in a state machine. It is not a real state, and a pseudo-state does not represent a value combination either.

UML knows 10 different pseudo-states. Each of them describes a special property that controls state transitions. For example, the initial state is a pseudo-state. It is responsible for marking the initial state of a state machine.

All ten pseudo-states are given below: Initial State, Shallow history, Deep history, Join, Fork, Junction, Choice, Exit point, Entry point, Terminate.

Q. 5.8. What is the difference between Fork and Join?

Ans: Fork and join are similar to the elements of the same names from the activity diagram. A fork has one incoming transition and several outgoing transitions that lead to different states, positioned orthogonally to one another. This means that all target states are activated. The outgoing transitions may have neither conditions nor triggers.

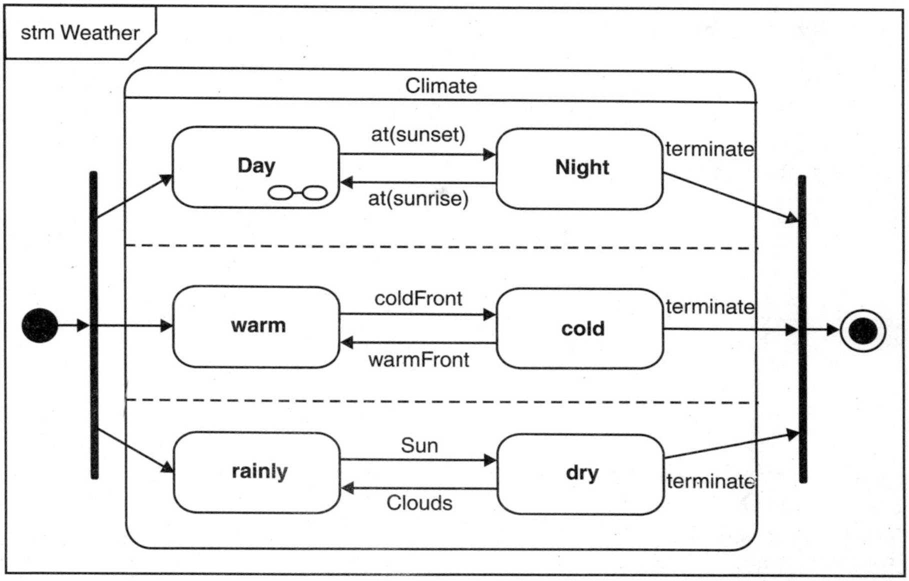

Fig. 5.24. Example for fork and join

In contrast, a join has several incoming transitions, which originate from states that are positioned orthogonally to one another, and one outgoing transition (Fig. 5.24).

Both elements are denoted as filled bars. They differ in the context of incoming and outgoing transitions.

Q. 5.9. Define Process and Thread in short.

Ans: Process: Specifies a heavyweight flow that can execute concurrently with other processes.

Thread: Specifies a lightweight flow that can execute concurrently with other threads within the same process.

Q. 5.10. What are the types of interaction diagram?

Ans: There are two type of Interaction diagram:

- Sequence diagrams and
- Collaboration diagrams.

Q. 5.11. The direction central for some of the first toy electric trains was accomplished by interrupting the power to the train. Prepare state diagrams for the head light and wheels of the train, corresponding to the following scenario.

Power is off ; train is not moving.
Power is turned on ; train moves forward and train headlight shines.
Power is turned off ; train stops and headlight goes out.
Power is turned on ; headlight shines and train does not move.
Power is turned off ; headlight goes out.
Power is turned on ; train runs backward with its headlight shining.
Power is turned off ; train stops and headlight goes out.
Power is turned on ; headlight shines and train does not move.
Power is turned off ; headlight goes out.
Power is turned on ; train runs forward with its headlight shining.

Ans. State Machine of Headlight:

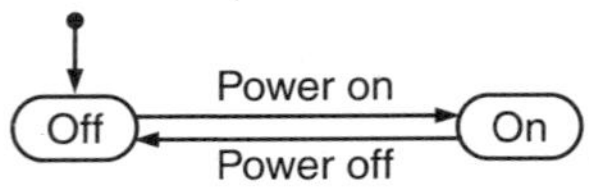

State Machine for wheels:

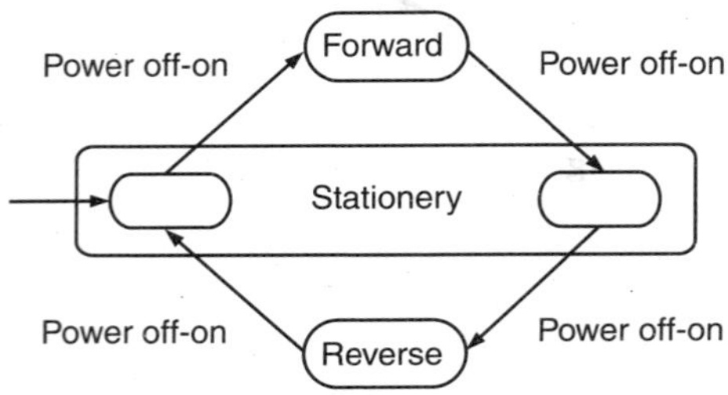

Q. 5.12. Figure Q. 5.12 shows a partially completed state diagram for the control of telephone answering machine. Calls are automatically answered as follows:

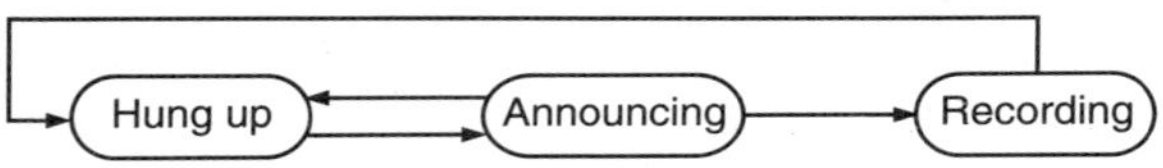

Fig. Q.5.12

- **An incoming call is detected on the first ring and the machine answers the call with a pre-recorded announcement.**
- **When the announcement is complete, the caller's message is recorded.**
- **When the caller hang up, the machine hangs up and shuts off.**

Place the following activity in the diagram: Call detected, answer call, play announcement, record, message, caller hangs up, announcement complete.

Ans. State Diagram of Telephone Recording Machine:

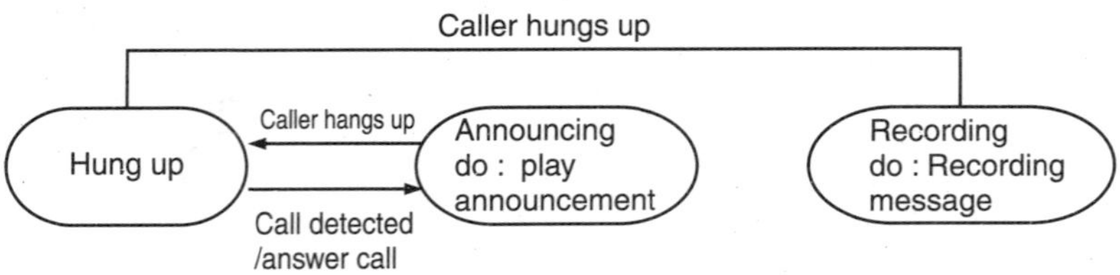

Q. 5.13. Prepare again the state diagram for the telephone answering machine. Machine should answer after five rings. If the telephone is answered before five rings, the machine should do nothing. Be careful to distinguish between five calls in which the telephone is answered on the first ring and one call that rings five times.

Ans. State Diagrams for Telephone Answering Machine:

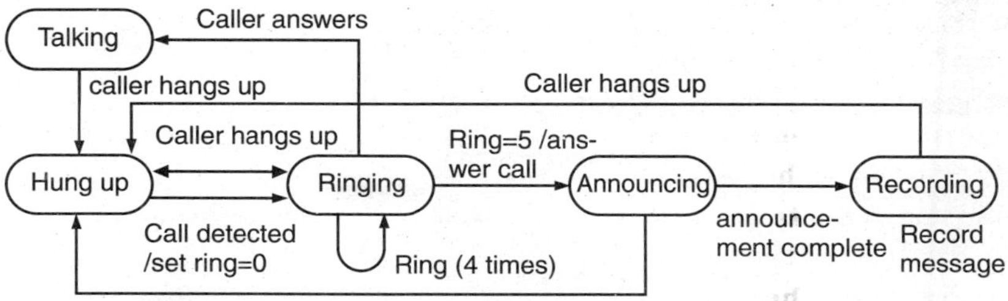

Q. 5.14. Figure Q. 5.14 shows the partially completed state diagram for data transmission control in the computer. The disk controller signals to host (such as CPU or DMA) each time, when a new byte of data is available. The data must then be read and stored before another byte is ready. When the disk controller senses the data has been read, it indicates that data is not available in preparation for the next byte. If any byte is not read before the next one comes along, the disk controller asserts a data lost error signal until the disk controller is reset. Add the following activity to the diagram: reset, indicate data not available, indicate data available, data read by host, new data read, indicate data lost.

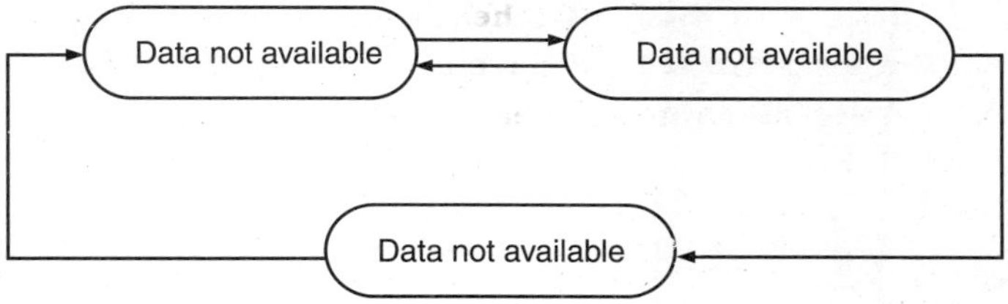

Fig. Q. 5.14.

Ans.

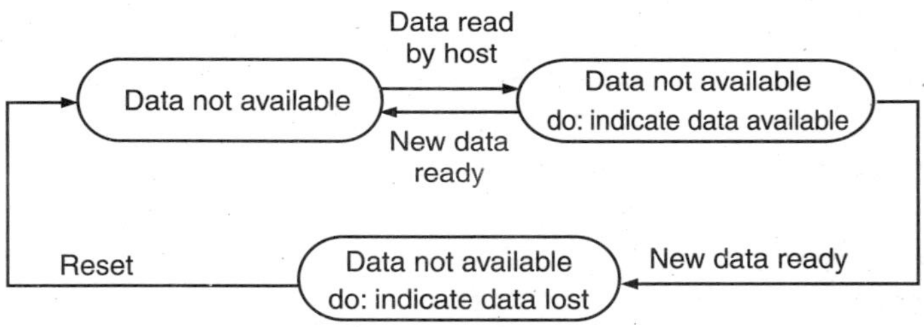

Q. 5.15. Figure Q.5.15 is a partially completed state diagram for one kind of motor control that is commonly used in household appliances. A separate appliance control determines when the motor should be on and continuously inserts "on" as an input to the motor control when the motor should be running.

When "on" is inserted, the motor control should start and run the motor. Starting is accomplished by applying power to both the "start" and the "run" windings of the motor.

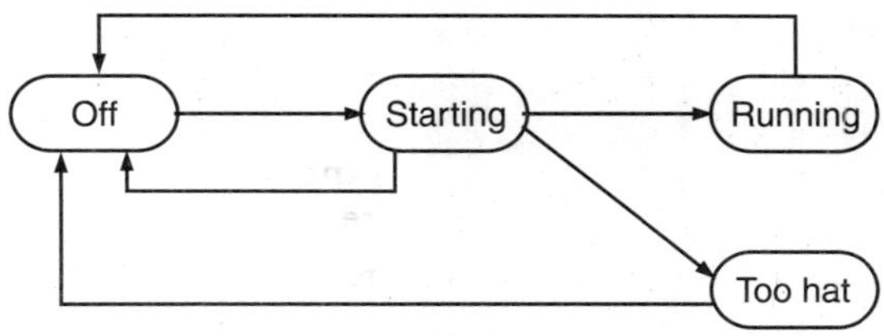

Fig. Q. 5.15.

An sensor, called a "starting relay", determines when the motor has started, at which point the "start" winding is turned off leaving only the "run" winding powered. Both windings are shut off when "on" is not inserted.

Motor could be damaged by overheating if they are overloaded or fail to start.

If the motor becomes too hot, the motor control removes power from both windings and ignores any "on" insertion for the appliance control until a reset button on the motor control is pressed and the motor has cooled off.

Add the following to the diagram:

Activities: Apply power to run winding, apply power to start winding.

Condition: Motor is overheated, on is inserted, motor is running.

Event: Reset.

Ans. State Diagram for the Motor Control:

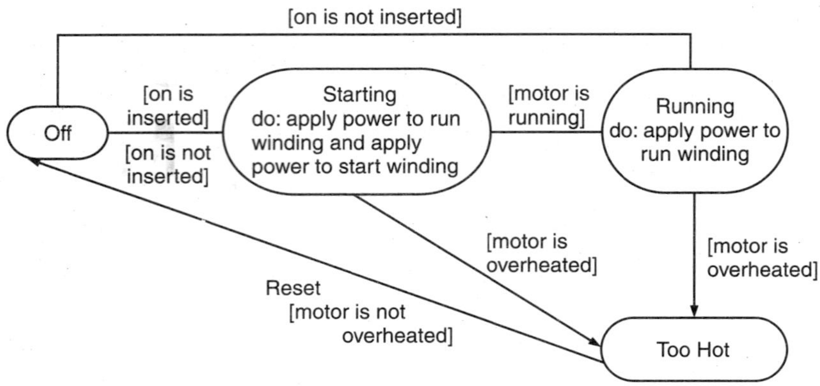

EXERCISES

Q. 5.1. What is the role of Use Cases?

Q. 5.2. What are relations used in Use Cases?

Q. 5.3. Explain activity diagrams.

Q. 5.4. What do you mean by state and state machine?

Q. 5.5. What do you mean by exit and entry behavior in the state machine?

Q. 5.6. Explain Transition in detail.

Q. 5.7. What are the role of trigger and events.

Q. 5.8. What is the mean of pseudo-state?

Q. 5.9. What is the difference between Fork and join?

Q. 5.10. What is the difference between Process and Thread?

Q. 5.11. What is the difference between normal class and active class?

Q. 5.12. Explain all about the interaction diagram.

Q. 5.13. What is package? Explain with package diagram.

Chapter 6

Architectural Modeling

CHAPTER OUTLINE

6.1. COMPONENT

A component is a physical and replaceable part of a system that conforms to and provides the realization of a set of interfaces.

Components are used to model the physical things that may reside on a node, such as executables, libraries, tables, files, and documents. A component typically represents the physical packaging of otherwise logical elements, such as classes, interfaces, and collaborations.

Good components define crisp abstractions with well-defined interfaces, making it possible to easily replace older components with newer, compatible ones.

In the UML, all these physical things are modeled as components. A component is a physical thing that conforms to and realizes a set of interfaces. Interfaces therefore bridge your logical and physical models. For example, you may specify an interface for a class in a logical model, and that same interface will carry over to some physical component that realizes it.

In software, many operating systems and programming languages directly support the concept of a component. Object libraries, executables, COM+ components, and Enterprise Java Beans are all examples of components that may be represented directly in the UML by using components. Not only can components be used to model these kinds of things, they can also be used to represent other things that participate in an executing system, such as tables, files, and documents. Graphically, a component is rendered as a rectangle with tabs.

> **Concept**
>
> *A component is a physical and replaceable part of a system that conforms to and provides the realization of a set of interfaces.*

In many ways, components are like classes: Both have names; both may realize a set of interfaces; both may participate in dependency, generalization, and association relationships; both may be nested; both may have instances; both may be participants in interactions. However, there are some significant differences between components and classes.

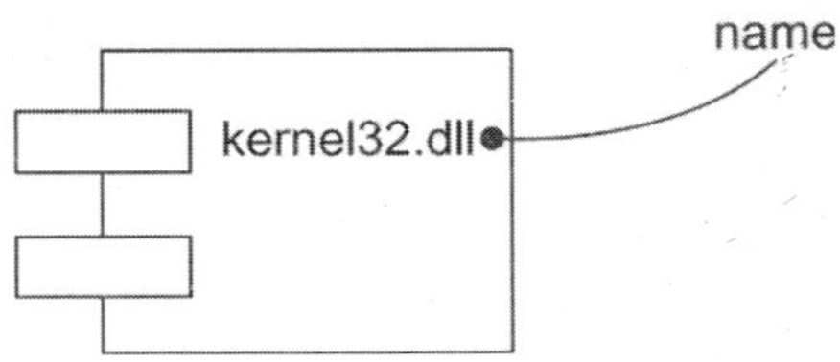

Fig. 6.1. Component

- Classes represent logical abstractions; components represent physical things that live in the world of bits. In short, components may live on nodes, classes may not.
- Components represent the physical packaging of otherwise logical components and are at a different level of abstraction.
- Classes may have attributes and operations directly. In general, components only have operations that are reachable only through their interfaces.

6.2. DEPLOYMENT

Nodes are used to model the topology of the hardware on which your system executes. A node typically represents a processor or a device on which components may be deployed.

Good nodes crisply represent the vocabulary of the hardware in your solution domain.

When you develop a software-intensive system, you have to consider both its logical and physical dimensions. On the logical side, you'll find things such as classes, interfaces, collaborations, interactions, and state machines. On the physical side, you'll find components (which represent the physical packaging of these logical things) and nodes (which represent the hardware on which these components are deployed and execute).

The components you develop or reuse as part of a software-intensive system must be deployed on some set of hardware in order to execute. This is in effect what a software-intensive system is all about—such a system encompasses both software and hardware.

The UML provides a graphical representation of node, shown in Fig. 6.2. This canonical notation permits you to visualize a node apart from any specific hardware. Using stereotypes—one of the UML's extensibility mechanisms—you can (and often will) tailor this notation to represent specific kinds of processors and devices.

A node is a physical element that exists at run time and represents a computational resource generally having at least some memory and, often, processing capability. Graphically, a node is rendered as a cube.

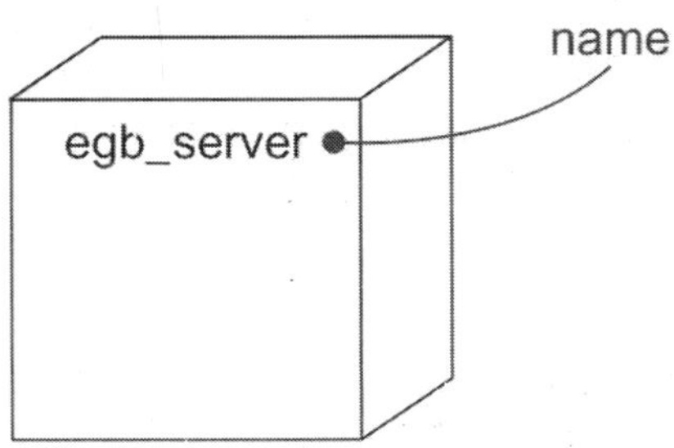

Fig. 6.2. Node

In many ways, nodes are a lot like components: Both have names; both may participate in dependency, generalization, and association relationships; both may be nested; both may have instances; both may be participants in interactions. However, there are some significant differences between nodes and components.

- Components are things that participate in the execution of a system; nodes are things that execute components.
- Components represent the physical packaging of otherwise logical elements; nodes represent the physical deployment of components.

6.3. COMPONENT DIAGRAMS

There are two kind of diagrams used in modeling the physical aspects of an object-oriented system:

1. Component Diagram 2. Deployment Diagram

Component diagrams are one of the two kinds of diagrams found in modeling the physical aspects of object-oriented systems. A component diagram shows the organization and dependencies among a set of components.

Component diagrams are used to model the static implementation view of a system. This involves modeling the physical things that reside on a node, such as executables, libraries, tables, files, and documents. Component diagrams are essentially class diagrams that focus on a system's components.

Component diagrams are not only important for visualizing, specifying, and documenting component-based systems, but also for constructing executable systems through forward and reverse engineering.

A component diagram shows a set of components and their relationships. Graphically, a component diagram is a collection of vertices and arcs.

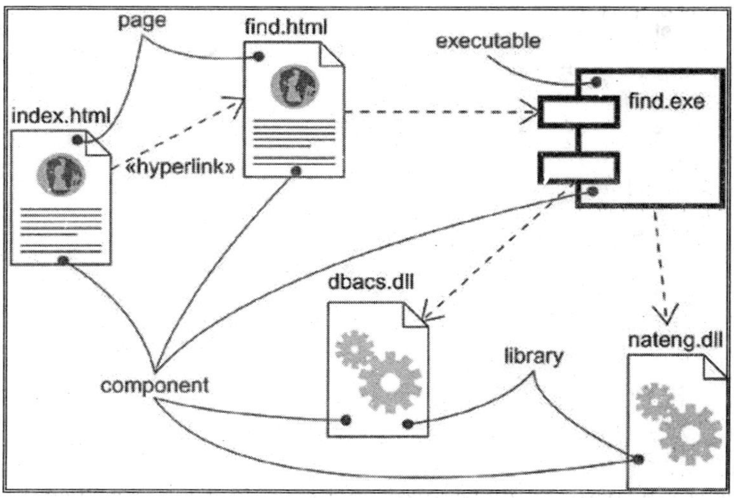

Fig. 6.3. Component Diagram

Component diagrams commonly contain:

- Components
- Interfaces

- Dependency, generalization, association, and realization relationships

Like all other diagrams, component diagrams may contain notes and constraints.

6.4. DEPLOYMENT DIAGRAMS

A deployment diagram is a diagram that shows the configuration of run time processing nodes and the components that live on them.

This is the second kind of diagrams used in modeling the physical aspects of an object-oriented system.

A deployment diagram shows the configuration of run time processing nodes and the components that live on them.

Deployment diagrams are used to model the static deployment view of a system. For the most part, this involves modeling the topology of the hardware on which your system executes.

Deployment diagrams are essentially class diagrams that focus on a system's nodes. Deployment diagrams are not only important for visualizing, specifying, and documenting embedded, client/server, and distributed systems, but also for managing executable systems through forward and reverse engineering.

At the edge of the system's software and hardware, deployment diagrams are used to reason about the topology of processors and devices on which your software executes.

In UML deployment diagrams are used to visualize the static aspect of these physical nodes and their relationships and to specify their details for construction, as in Fig. 6.4.

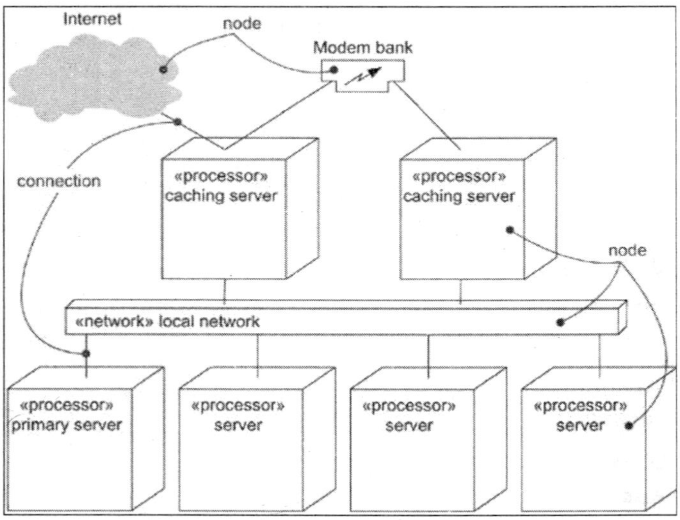

Fig. 6.4. Deployment Diagram

A deployment diagram is just a special kind of diagram and shares the same common properties as all other diagrams—a name and graphical contents that are a projection into a model. What distinguishes a deployment diagram from all other kinds of diagrams is its particular content.

Deployment diagrams commonly contain

- Nodes
- Dependency
- Association relationships

Like all other diagrams, deployment diagrams may contain nodes and constraints. Deployment diagrams may also contain components, each of which must live on some node. Deployment diagrams may also contain packages or subsystems, both of which are used to group elements of your model into larger chunks. Sometimes, you'll want to place instances in your deployment diagrams, as well, especially when you want to visualize one instance of a family of hardware topologies.

Basically deployment diagrams are used to model the static deployment view of a system. This view primarily addresses the distribution, delivery, and installation of the parts that make up the physical system.

There are some kinds of systems for which deployment diagrams are unnecessary. If you are developing a piece of software that lives on one machine and interfaces only with standard devices on that machine that are already managed by the host operating system (for example, a personal computer's keyboard, display, and modem), you can ignore deployment diagrams. On the other hand, if you are developing a piece of software that interacts with devices that the host operating system does not typically manage or that is physically distributed across multiple processors, then using deployment diagrams will help you reason about your system's software to-hardware mapping.

SUMMARY

- A component is a physical and replaceable part of a system that conforms to and provides the realization of a set of interfaces.
- In software, many operating systems and programming languages directly support the concept of a component such as Java and C#.
- In many ways, components are like classes: Both have names; both may realize a set of interfaces; both may participate in dependency, generalization, and association relationships; both may be nested; both may have instances; both may be participants in interactions.
- Nodes are used to model the topology of the hardware on which your system executes.
- A node typically represents a processor or a device on which components may be deployed.

SOLVED PROBLEMS

Q. 6.1. What is Component?

Ans: A component is a physical and replaceable part of a system that conforms to and provides the realization of a set of interfaces.

Components are use to model the physical things that may reside on a node, such as executables, libraries, tables, files, and documents. A component typically represents the physical packaging of otherwise logical elements, such as classes, interfaces, and collaborations.

Q. 6.2. What is Node?

Ans: A node is a physical element that exists at run time and represents a computational resource generally having at least some memory and, often, processing capability. Graphically, a node is rendered as a cube.

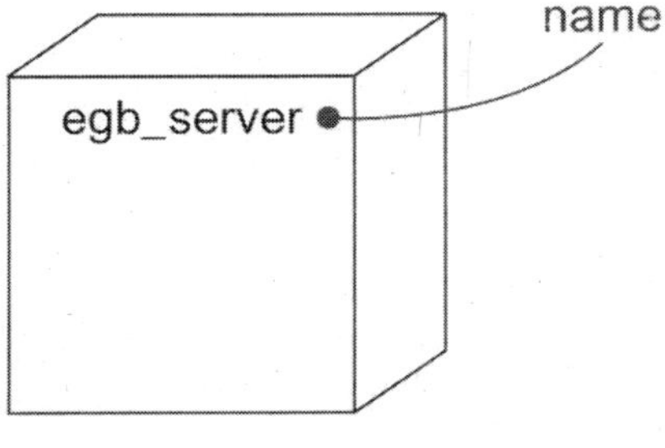

Fig. 6.5.

Q. 6.3 What is the difference between Components and Nodes?

Ans: There are some significant differences between nodes and components.

Components	*Nodes*
Components are things that participate in the execution of a system.	Nodes are things that execute components.
Components represent the physical packaging of otherwise logical elements.	Nodes represent the physical deployment of components.

EXERCISES

Q. 6.1. What is the main role of Components?

Q. 6.2. What is Node? What they represents?

Q. 6.3. How Nodes are differ from Components?

Q. 6.4. What the Component diagrams contain? Give an Example.

Q. 6.5. What do you mean by deployment?

Q. 6.6. What is the use of deployment diagram?

❑❑❑

Chapter 7

OMT Methodology

7.1. INTRODUCTION

In previous chapters, we have been presented the concepts and notations which support the OMT methodology. In this chapter we will discuss the process for devising the three OMT models.

The OMT methodology is a methodology of software engineering, which is a process for the organized production of software, using a collection of predefined techniques and notational conventions. The steps of software production are usually organized into a life cycle (called software development life cycle) consisting of several phases as follows:

- Problem formation
- Analysis
- Design
- Implementation
- Testing
- Maintenance and enhancement

The OMT methodology supports the entire software life cycle. This chapter covers life cycle phases from problem formulation through requirements analysis, design and implementation. This chapter does not cover testing and maintenance. Because testing and maintenance are simplified by an object-oriented approach, but the traditional methods used in these phases are not altered.

Thus we have explained following phases of OMT methodology:

CHAPTER OUTLINE

1. Analysis phase
2. System design phase
3. Object design phase
4. Implementation phase

7.2. ANALYSIS

The purpose of analysis phase is to state and understand the problem and modeling the application and the domain within which is operates.

> **Concept**
>
> *Analysis phase takes problem statement and produce object, dynamic and functional model for a system.*

The analysis phase accept input as a problem statement describing the problem to be solved along with a conceptual overview of the proposed system. The output from the analysis phase, is a formal model that captures the following aspects of the system:

1. Objects and their relationship (object model)
2. Dynamic flow of control (dynamic model)
3. Functional transformation of data subject to constraints (functional model)

All three sub-models are not equally important in every problem. Almost all problems have useful object models derived from real-world entities. Problems concerning interactions and timing, such as user interfaces and process control, have important dynamic models. Problems containing significant computation, such as compilers and engineering calculations, have important functional models. All three sub-models contribute operations which are summarized on the object model.

7.2.1. Problem Statement

The first step of developing anything is to understand the problem and its requirements. A problem statement defines what problem is, without any reference to the possible solutions, The problem statement should state what is to be done and not how it is to be done. It should be a statement of needs, not a proposal for solutions.

> **Concept**
>
> *Problem statement is used to understand the problem and requirements.*

The problem definition may be in user language and the problem should be described from the user's point of view. After understanding the problem definition, the developers should study the input, output and processing requirements of the problem that form the basis for program requirements analysis. The requirements should describe what needs to be done, not how it will be implemented.

The problem statement may be incomplete, ambiguous, and erroneous—it is just a starting point.

7.2.2. Object Modeling

The first step in analyzing the requirements is to construct an object model. The object model shows the static data structure of the real-world system. The object model describes real-world object classes and their relationships with each other.

The information gathered for object model from:

- Problem statements
- User interviews
- Expert knowledge of the application domain
- General knowledge of the real-world

The following steps are performed in constructing an object model:

(*i*) Identify objects and classes.
(*ii*) Prepare a data dictionary.
(*iii*) Identify association (including aggregation) between objects.
(*iv*) Identify attributes of objects and links.
(*v*) Organize and simplify object classes using inheritance.
(*vi*) Verify that access paths exist for likely queries.
(*vii*) Iterate and refine the model.
(*viii*) Group classes into modules.

(*i*) Identify Objects and Classes

The first step in constructing an object model is to identify relevant objects and classes from the application domain. In this step write down every class that comes to mind. Classes often correspond to nouns. Objects include physical entities (such as houses, employees, and machine) as well as concepts (such as seating arrangements, payment schedule and department). All classes must make sense in the application domain.

After listing the objects and classes, now discard unnecessary objects, attributes and classes according to the following criteria.

- **Redundant Classes:** If two or more classes express the same information, then the most descriptive name should be kept.
- **Irrelevant Classes:** If a class has little or nothing to do with the problem, it should be eliminated.
- **Vague Classes:** A class should be specific. Some tentative classes may have ill-defined (not understandable) boundaries or be too broad in scope.
- **Attributes:** Names that primarily describe individual objects should be restated as attributes. If the independent existence of a property is important, then make it a class and not an attribute.
- **Operation:** If a name describes an operation that is applied to object and not manipulated in its own right, then it is not a class.
- **Rules:** The name of a class should reflect its intrinsic nature and not a role that is played in an association.
- **Implementation Construct:** Constructs extraneous to the real world should be eliminated from the analysis model. They may be needed later during design, but not now.

(ii) Preparing a Data Dictionary

If you have used isolated words, they have too many interpretation so we prepare a data dictionary for all modeling entities. We write a paragraph for precisely describing each object class. The description may include:

- Scope of assumptions on its membership or use.
- Any assumptions on its membership or use.
- Any restrictions on use.
- Association description.
- Attributes.
- Operations.

(iii) Identify Associations

In the next step, identify associations between classes. Any dependency between two or more classes is an association. A reference from one class to another is an association. Attribute should not refer to class rather than an association should be used.

Association often corresponds to stative verbs or verb phrases. These verb phrases may include implicit for some associations. Moreover some association adds from real world knowledge or assumptions.

Now discard unnecessary and incorrect association, using following criteria:

- **Association between eliminated classes:** If one of the class in the association have been eliminated then the association must either be eliminated or restated in terms of other existing classes.
- **Irrelevant or implementation association:** Eliminate any association that are outside the problem domain or deal with implementation constructs.
- **Actions:** An association should describe a structural property of the application domain, not a transient event. Also delete temporary relationships.
- **Ternary association:** The association between three or more classes can be decomposed into binary association or phrased as qualified association.
- **Derived association:** An association in terms of other association should be omitted because they are redundant.

(iv) Identifying Attributes

Attributes are properties of individual objects but they should not be objects. It corresponds to noun and adjective followed by possessive phrases.

Link attributes should also be identified. A link attribute is a property of the link between two objects, rather than being a property of an individual object.

Derived attributes should be omitted or clearly specified.

(v) Refining with Inheritance

The next step is to organize classes by using inheritance to share common structure.

Inheritance can be added in two directions:

1. By generalizing common aspects of existing classes into a superclass (bottom up approach: generalization)
2. By refining existing class into specialized subclasses (top down approach: specialization).

(*vi*) Testing Access Paths

Trace access paths through the object model diagram and check whether they yield suitable sensible results or not. When multiplicity is there, then check a way to pick out for needed unique value. If you can't get any answer it indicates that diagram is missing some information.

(*vii*) Iterating Object Modeling

The entire software development process is one of the continual iteration; different parts of a model are often at different stages of completion. If a deficiency is found, go back to an earlier stage if necessary to correct it. Some refinements can only come after the dynamic and functional models are completed.

The following table shows indications for missing, incorrect and unnecessary and their solutions.

Indications	Solution
For Missing Objects	
asymmetries in associations and generalization	Add new classes by analogy.
disparate attributes and operations on a class	split a class so that each part in coherent
difficulty in generalizing cleanly	one class may be playing two roles. Split it up and one part may then fit in cleanly.
an operation with no good target class	Add the missing target class.
duplicate associations with the same name and purpose.	Generalize to create the missing super class that unites them.
A role substantially shapes the semantics of a class	May be it should be a separate class. This often means converting an association into a class.
For unnecessary classes	
lack of attributes, operations, associations on a class	Remove these type of classes and adjust attributes, operations and associations in other classes.
For Missing Associations	
missing access paths for operations	Add new associations so that queries can be answered.
For Unnecessary Associations	
reductant information in the associations	Remove association that do not add new information or mark them as derived.
lack of operations that traverse an association	It no operations use a path, may be the information not needed. The test must wait until operations are specified.
For Incorrect placement of Associations	
role names that are too broad or too narrow for their class	Move the association up or down in the class hierarchy.
For Incorrect placement of Attributes	
need to access an object by one of its attribute values	Consider a qualified association.

(*viii*) Grouping Classes into Modules

In the last step of object modeling you have to group classes into sheets and modules. Tightly-coupled classes should be grouped together, But since a sheet holds a fixed amount of information the breakdown is occasionally arbitrary.

A module is a set of classes that capture some logical subset of the entire model. For example, a model of a computer operating system might contain modules for process control, device control, file maintenance and memory management. Module may vary in size.

7.2.3. Dynamic Modeling

The dynamic modeling shows the behavior of the system, especially sequencing of interactions. The major concepts of dynamic modeling are events and states. The dynamic model is insignificant for a purely static data repository, like as database, but it is very important for interactive systems.

A dynamic model is constructed by following number of steps:

(*i*) Prepare scenarios of typical interaction sequences.

(*ii*) Identify events between objects.

(*iii*) Prepare an event trace for each scenario.

(*iv*) Build a state diagram.

(*v*) Match events between objects to verify consistency.

(*i*) Prepare a Scenario

A scenario is a sequence of events. An event occurs whenever information is exchanged between an object in the system and an outside agent, (such as user, a sensor, or another task). The information values exchanged are parameters of the event.

First prepare scenarios for 'Normal' cases, and then consider "Special cases". Then after consider user error cases, including invalid values and failures to respond. At last consider various other kinds of interactions such as help request and states queries.

(*ii*) Identify Events

An event is an action by an object that transmits information. It includes all signals, inputs, decisions, interrupts, transitions and actions to or from users or external devices. Events can be identified by carefully examining the scenario.

Events that have the same effect on flow of control, are grouped under a single name irrespective to values of their parameters. On the other hand events that have different effect on the flow of control should be distinguishable. For example, "enter password" or "dispense cash" should be an event class as the value of password or the cash dispense does not affect the flow of control, but bad account, bad password are all different events and should not be grouped under a single event name called status.

(*iii*) Prepare Event Trace Diagram

An event trace refers to an ordered list of events between different objects assigned to columns in a table. By scenario a particular column in the trace diagram, we can see the events that directly affect a particular object. Only these events can appear in the state diagram for the objects.

Events between a group of classes on an event flow diagram, which summarizes events between classes, regardless for sequence. The event flow diagram is a dynamic counter part to an object diagram, in which the paths shows possible control flows.

(iv) Building a State Diagram

A state diagram is graphical presentation of event transition from one state to another state. Prepare a state diagram for each object class with non-trivial dynamic behaviour, showing the events that objects receives and sends. Ever scenario or event trace corresponds to a path through the state diagram.

After finishing state diagram of a class when the diagram covers all scenarios, and the diagram handles all events that can affect an object of the class in each of its states. We can use the state diagram to suggest new scenarios by considering how some events not already handled should affect the state of the object. Posing "what, if" questions is a good way to test completeness and error-handling capabilities of a class.

(v) Matching Events Between Objects

In the last check for completeness and consistency at the system level when the state diagrams for each class are complete. Every event should have a sender and receiver states with any predecessors or successors are considered and it make sure they represent starting and terminal points of the interaction sequence. The consistency of the corresponding events on different state diagram should be checked.

7.2.4. Functional Modeling

The functional model shows the functional derivation of values, regardless of when they are computed. Means it shows how values are computed, regardless for sequencing, decisions, or object structure. It shows the dependency between values and function that relate them. Data flow diagram is best tool of functional modeling, is used for showing functional dependencies. The processes on a data flow diagram correspond to activities or actions in the state diagrams of the classes.

The following steps are performed in constructing a functional model:

(*i*) Identify input and output values.
(*ii*) Build data flow diagrams showing functional dependencies.
(*iii*) Describe functions.
(*iv*) Identify constraints.
(*v*) Specify optimization criteria.

(i) Identify Input and Output Values

In the first step of functional model, list out all input and output values, which are the parameters of events between the system and outside world.

For example, for a ATM machine there are some possible input and outputs.

Input for ATM System

— Bank code
— Password
— Account type
— Card code
— Transaction kind
— Amount

Output for ATM System

— Cash
— Messages
— Receipt

(ii) Building Data Flow Diagrams

After identifying input output values, construct a data flow diagram, sharing how each value is computed from input values. A data flow diagram is usually constructed in layers. Top layer may consist of a single process, or perhaps one process each together inputs, compute values, and generate outputs.

Expand each non-trivial process in the top level diagram into a lower level data flow diagram. If second level process still contain nontrivial processes, they can be expanded recursively.

(*iii*) Describing Functions

In the next step, write description of each function after data flow diagram has been refined. The description can be in natural language, mathematical equations, pseudocode, decision tables, or some other appropriate form. The description can be declarative or procedural, But focus on what the function does, not how to implement it. A declarative description specifies the relationships between input and output values and the relationship among the output values.

(*iv*) Identifying Constraints Between Objects

Constraints are functional dependencies between objects that are not related by an input-output dependency. Constraints can be on two objects at the same time, between instances of the same object at different times, or between instances of different objects at different times.

Constraints are two types:

(*i*) **Precondition constraints:** These are the conditions on function that satisfy by the input values.

(*ii*) **Post condition constraints:** These are the conditions on function that output values are guaranteed to hold.

(*v*) Specifying Optimization:

In this step, we specify values to be maximized, minimized, or otherwise optimized. If there are several optimization criteria that conflict indicates how the trade-off is to be decided.

7.2.5. Verify, Iterate and Refine the Three Models

In this step we are checking and considering the all three models which are prepared.

— Add key operations that were discovered during preparation of the functional model to the object model. Do not show all operations during analysis as this would clutter the object model; just show the most important operations.

— Verify that the classes, association, attributes and operations are consistent and complete at the particular level of abstraction. Compare the three models with the problem statement and relevant domain knowledge, and test the models using scenarios.

— Develop more detailed scenario (include all error conditions) as variations on the basic scenario. Use "What if" question on these scenario to further verify the three models.

— Iterate the above steps as needed to complete the analysis.

— Verify the final analysis with the requestd or and application domain experts.

7.3. SYSTEM DESIGN

During the design phase, decisions are made about how the problem will be solved, first at a high level, then an increasingly detailed level. The system design is the first design stage in which the basic approach for solving the problem is selected.

The system architecture is the overall organization of the system into components called subsystems.

To design a system the system designer must make the following decisions:

1. Organize the system into subsystems.
2. Identify concurrency inherent in the problem.
3. Allocate subsystem to processors and tasks.
4. Choose an approach for management of data stores.
5. Handle access to global resources.
6. Choose the implementation of control in software.
7. Handle boundary conditions.
8. Set trade-off priorities.

7.3.1. Organize the System into Subsystems

A subsystems is a package of classes, associations, operations, events and constraints that are interrelated and that have reasonably well-defined and small interface with other subsystems. A subsystems is usually identified by the service it provides. A service is a group of related functions that share some common purpose (such as I/O processing, drawing picture, or performing arithmetic).

A subsystem defines a coherent way of looking at one aspect of the problem. The lowest level subsystems are called modules.

The relationship between two subsystems can be client-supplier or peer-to-peer.

Client-supplier Relationship: The client calls on the supplier, which perform some service and replies with a result. The client must know the interface of the supplier, but the supplier does not have to know the interface.

Peer-to-peer Relationship: In this relationship each of the subsystems may call on the others. It is more complicated because the subsystems must know each other's interfaces.

The decomposition of system into subsystems can be organized using following two schemes:

(*i*) Layered Architecture.

(*ii*) Partitions.

(*i*) Layered Architecture

It was first introduced by the developers of the "THE" operating system. In this scheme subsystem knows about the layers below it, but has no knowledge of the upper layer. Thus a client supplier relationship exists between lower layers and upper layers.

Each layer is implemented in terms of the classes and operations of lower layers. The layer architecture can be open or closed.

Closed Architecture: In this architecture each layer is built in terms of the immediate lower layer. This reduces the dependencies between layers and allows changes to be made most easily because a layers interface only affects the next layer.

Open Architecture: In this architecture, a layer can use features of any lower layer up to any depth. This reduces the need to redefine operations at each level, which can result in a more efficient and compact code. However, an open architecture does not observe the principle of information hiding.

(*ii*) Partitions

Partitions vertically divide a system into several independent or weakly coupled subsystems, each providing one kind of service. The subsystem may have some knowledge of each other, but this knowledge is not deep, so major design dependencies are not created.

In practice most large systems required a mixture of layers and partitions.

7.3.2. Identifying Concurrency

Two object are said to be concurrent if they can receive events at the same time without interacting. An important goal of system design is to identify which objects must be active con-currently and which objects have activity that is mutually exclusive.

7.3.3. Allocating Subsystem to Processors and Tasks

Each concurrent subsystem must be allocated to a hardware unit, either a general purpose processor or a specialized functional unit.

The system designer must consider following aspects:

- Estimate performance needs and the resources needed to satisfy them.
- Choose hardware or software implementation for subsystems.
- Allocate software subsystems to processors to satisfy performance needs and minimize interprocessor communication.
- Determine the connectivity of the physical units that implement the subsystems.

7.3.4. Data Store Management

The internal and external data stores in a system provide clean separation points between subsystems with well-defined interfaces. In general each data store may combine data structure, data files and data bases.

Data Structure: Data structure uses main memory to store data, this type data available temporary till the application is executing, after the termination of application data will be lost (such as stack, queues and link list).

Data Files: Data files are cheap, simple, and permanent form of data store. But file operations are low level and applications must include additional code to provide a suitable level of abstraction. File implementation vary for different computer systems, so portable applications must carefully isolate file system dependencies. Implementations for sequential files are mostly standard, but commands and storage formats for random access files and indexed files vary widely.

Databases: Databases managed by database management system (DBMS). Various types of DBMS are available from vendors:

- Hierarchical DBMS
- Network DBMS
- Relational DBMS
- Object oriented DBMS
- Logical DBMS

DBMS provide fast accessing of data from secondary storage. Database are powerful and make applications easier to port to different hardware and operating system platforms, since the vendor ports the DBMS code.

7.3.5. Identifying and Handling of Global Resources

There are many global resources such as processors, tape drives, communications satellites, disk space, a work station screen, and the buttons on a mouse; logical names, such as object IDs, file names and class names; and access to shared data, such as database etc. The system designer must identify global resources and determine mechanism for controlling access to them.

— If the resource is physical object, then it can control itself by establishing a protocol for obtaining access within a concurrent system.

— If the resource is logical entity, such as an object ID or a database, then there is danger of conflicting access in a shared environment.

Each global resource must be owned by a "guardian object" that control access to it. One guardian object can control several resources.

A *lock* is a logical object associated with some defined subset of a resource that gives the lock holder the right to access the resource directly. A guardian object must still exist to allocate the locks, but after one interaction with the guardian to obtain a lock the user of the resource can access the resource directly.

7.3.6. Selection of Software Control Implementation

There are several ways to implement control in software. Although there is no logical necessity that all subsystems use the same implementation usually the designer chooses a single control style for the whole system.

There are two kinds of control flow in a software system:

(*i*) External control (*ii*) Internal control

(*i*) External Control

External control is the flow of externally-visible events among the objects in the system. There are three kinds of control for external events:

(*a*) Procedure-driven sequential system

(*b*) Event-driven sequential system

(*c*) Concurrent system

The control style adopted depends on the resources available and on the pattern of interactions in the application.

Procedure-driven Sequential Systems: In these systems control resides within the program code. Procedures issue requests for external input and then wait for it; when input arrives, control resumes within the procedure that made the call. The location of the program counter and the stack of procedure calls and local variables define the system state.

Event-driven Sequential Systems: In this systems, control resides within a dispatcher or monitor provided by the language, subsystems, or operating system. Procedures are attached to events and are called by the dispatcher, when the corresponding events occur. This process is called "call back". All procedure returns control to the dispatcher, rather than retaining control until input arrives. Events are handled directly by the dispatcher.

Concurrent Systems: In these systems, control resides concurrently in several independent objects, each a separate task. A task can wait for input, but other task continue execution, The operating system resolves conflicts among tasks.

(*ii*) Internal Control

Internal control is the flow of control within a process. It exists only in the implementation and therefore, is not inherently concurrent or sequential. Internal interaction between objects can be viewed similarly to external interaction among objects, and the same implementation mechanism can be used. The difference is that: External interaction inherently involve waiting for events because different objects are independent and cannot force other objects to respond, and internal operations are generated by objects as part of the implementation algorithm, so their response patterns are predictable.

There are three common kinds of internal control flow are common:

(*a*) Procedure calls. (*b*) Quasi-concurrent inter-task calls.

(*c*) Concurrent inter-task calls.

Quasi-concurrent inter-task calls, such as co-routines or light weight processes, are programming conveniences in which multiple address spaces or call stacks exist but in which only a single thread of control can be active at once.

7.3.7. Handling Boundary Conditions

Apart from the steady-state of a system designer must consider the boundary conditions, such as:

(*i*) Initialization (*ii*) Termination

(*iii*) Failure

(*i*) Initialization: The system must be initialize with all the required things in initial state, such as constant data, parameters, global variables, tasks, guardian objects, and possibly the class hierarchy itself. During initialization only a subset of the functionality of the system is usually available.

(*ii*) Termination: The system must contain the termination statements, and conditions. Abnormal terminations are harmful for the system. At the point of termination, system must release all the initialization things, and occupied space in main memory.

(*iii*) Failure: The abnormal termination is called failure, which was not planned by the system. Failure can arise from user errors, from the exhaustion of system resources, or from an external break-down. The good system designer plans for orderly failure.

7.3.8. Setting trade-off Priorities

The system designer must set priorities that will be used to guide trade-off during the rest of design.

For example, sometimes it is necessary to scarifice complete functionality to get a piece of software into use earlier or a system can often be made faster using extra memory.

The designer must determine the relative importance of the various criteria as a guide for making trade-off decision during design. All the trade-offs are not made during system design, but the priorities for making them are established.

7.3.9. Common Architectural Frameworks

There are several prototypical architectural frameworks that are common in existing systems. Each of these is well-suited to a certain kind of system.

These includes same kinds of system:

(*i*) Functional Transformation system

- (*a*) Batch computation system
- (*b*) Continuous transformation system

(*ii*) Time dependent system

- (*a*) Interactive interface (*b*) Dynamic simulation
- (*c*) Real-time.

(*iii*) Database system

(*a*) Transaction manager

Batch Computation: A data transformation executed once on an entire input set.

Continuous Transformation: A data transformation performed continuously as inputs changes.

Interactive Interface: A system dominated by external interactions.

Dynamic Simulation: A system that simulates evolving real-world objects.

Real-time System: A system dominated by strict-timing constraints.

Transaction Manager: A system concerned with storing and updating data, often including concurrent access from different physical locations.

7.4. OBJECT DESIGN

The object design phase determines the full definitions of the classes and associations used in the implementation, as well as the interfaces and algorithms of the methods used to implement operations. The object design phase adds internal objects for implementation and optimize data structure and algorithms. Object design is analogues to the preliminary design phase of the traditional software development life cycle.

During the object design phase, the designer must perform the following steps:

(*i*) Combine the three models to obtain operations on classes.

(*ii*) Design algorithm to implement operations.

(*iii*) Optimize access paths to data.

(*iv*) Implement control for external interactions.

(*v*) Adjust class structure to increase inheritance.

(*vi*) Design associations.

(*vii*) Determine object representation.

(*viii*) Package classes and associations into modules.

7.4.1. Grouping the Three Models

After completion of analysis phase, we have the three models: Object model, dynamic model and functional model, but the object model is the main framework around which the design is constructed.

The designer must convert the actions and activities of the dynamic model. In making this conversion, we start the process of mapping the logical structure of the analysis model into a physical organization of a program.

The designer must transfer operations from the functional and dynamic model onto the object model for implementation. A process from the functional model becomes an operation on an object. An event from the dynamic model may also become an operation on an object, depending on the implementation of control.

7.4.2. Designing Algorithms

Each operation from the analysis model must be assigned an algorithm that implements it clearly and efficiently, according to the optimization goals selected during system design. The analysis specification tells *what* the operation does from the viewpoint of its client, but the algorithm shows *how* this operation is does.

An algorithm may be subdivided into calls on simpler operations and so on recursively, until the lowest level operations are simple enough to implement directly without further refinements.

The designer of algorithm must:

- Choose algorithms that minimize the cost of implementing operations.
- Select data structures appropriate to the algorithms.
- Define new internal classes and operation as necessary.
- Assign responsibility for operations to appropriate classes.

7.4.3. Design Optimization

In the design optimization, we optimize the model which is made with the help of system design and analysis. The analysis model contains the logical information about the system, while the design model must add details to support efficient information access. For the efficiency and clarity, we must extend and restructure the model. The original information is not discarded, but new redundant information is added to optimize access paths and preserve intermediate results that would otherwise have to be recomputed. Algorithm can be rearranged to reduce the number of operations to be executed.

During the design optimization, the designer must:

- Add redudant associations to minimize access cost and maximize convenience of accessing.
- Rearrange the computation for greater efficiency and rearrange the order of execution of appropriate algorithm.
- Save derived attributes to avoid recomputation of complicated expressions.

7.4.4. Implementation of Control

The designer must refine the strategy for implementing the state-event models present in the dynamic model. In the system design, we have chosen the basic strategy for realizing the dynamic model and in the object design we must flesh out this strategy.

There are three basic approaches to implementing the dynamic model (as explained in Section 7.3.6).

(*i*) Using the location within the program to hold state (procedure-driven system)

(*ii*) Direct implementation of a state machine mechanism (event-driven system).

(*iii*) Using concurrent system.

7.4.5. Adjustment of Inheritance

During object design, the definitions of internal classes and operations can be adjusted to increase the amount of inheritance. These adjustment include modifying the argument list of a method, moving attributes and operations from a class into a superclass, defining an abstract superclass to cover the shared behavior of several classes, and splitting an operation into an inherited part and a specific part. Delegation should be used rather than inheritance when a class is similar to another class but not truly a subclass.

Thus to increasing the amount of inheritance designer should:

- Rearrange and adjust classes and operations to increase inheritance.
- Abstract common behavior out of groups of classes.
- Use delegation to share behavior when inheritance is symmetrically invalid.

Delegation is also technique to share behavior among the classes.

7.4.6. Design of Association

Association provides access paths between objects. There are conceptual entities useful for modeling and analysis. To make intelligent decisions about association, we first need analyze the way they are used and the implement each association as a distinct object or by adding object valued attributes to one or both classes in the association.

Associations subsum many implementation techniques under a single uniform notation during analysis, but they can be implemented as pointer with in objects or distinct objects depending on their access patterns. An association traversed in a single direction can be implemented as an attribute pointing to another objects or a set of objects, depending on the multiplicity of the association.

A bidirectional association can be implemented as a pair of pointers, but operations that update the association must always modify both directions of access. Associations can also be implemented as association objects.

7.4.7. Object Representation

For the exact representation, the designer must choose when to use primitive types in representing objects and when to combine groups of related objects.

Classes can be defined in terms of other classes, but everything must be implemented in terms of built-in primitive data types (such as integers, strings and enumerated type) supplied by the programming language. Some classes can be combined.

7.4.8. Physical Packaging

Programs are made of discrete physical units that can be edited, compiled, imported, or otherwise manipulated. Programs must be packaged into physical modules for editors and compilers as well as for the convenience of programming terms. Object-oriented languages have various degrees of packaging.

Packaging involves the following issues:

- Hiding internal information from outside view (using private and protected data).
- Coherence of entities (entities should be organize in a coherent way).
- Constructing physical modules (entities should be organize in a common theme).

7.4.9. Object Design Documentation

Design decisions should be documented by extending the analysis model, by adding detail to the object, dynamic, and functional models. Implementation constructs are appropriate, such as pointers (in the object model), structured pseudocode (in the dynamic model) and functional expressions (in the functional model).

The design document should be an extension of the Requirements Analysis Document. Thus the design document will include a revised and much more detailed description of the object model, in both graphical form (object model diagrams) and textual form (class descriptions).

The functional model will also be extended during the design phase, and it must be kept current.

The dynamic model is implemented using an explicit state control or concurrent tasks then the analysis model or its extension is adequate. If the dynamic model is implemented by location within program code, then structured pseudocode for algorithms is needed.

7.5. IMPLEMENTATION

In this section we will discuss how can you implement your designed model in software development. The software development of system can be implemented using:

(*i*) Programming languages

(*ii*) Database management system

Writing code is an extension of the design process. Writing code should be straightforward, almost technical, because all the difficult decisions should already have been made during design. The code should be a simple translation of the design decisions into the syntaxes of a particular language.

7.5.1. Using Programming Languages

Implementation of the system using programming language is a traditional way, most of the programming languages are capable of expressing the three aspects of software specification:

- Data structure
- Dynamic flow of control
- Functional transformation

Data structure provides a way of data declaration, is a subset of a language. The statements that are used to declare data structures are commonly non-procedural part of the languages.

Flow of control may be expressed either procedurally (conditionals, loops and calls) or non- procedurally (rules, constraints, tables and state machines). Traditional languages are purely procedural, although the programmer can implement non procedural constructs as data.

Functional transformations are expressed in terms of the primitive operations of the language, as well as calls to subprograms. Most procedural languages are similar in the kinds of functionality they supported do not differ greatly from FORTRAN.

There are two types of programming languages for implementation of an object-oriented design:

(*i*) Object-oriented languages (*ii*) Non-object oriented languages

(*i*) Object-Oriented Languages

Implementation of an object-oriented design is easiest using an object-oriented language, but even these languages vary in their degree of support for object-oriented concepts. The object-oriented language is the most natural implementation target for an object-oriented design.

Good programming style is important to maximize the benefits of object-oriented design and programming; most benefits came from greatly reduced maintenance and enhancement costs and in reuse of the new code on future projects. Object-oriented programming style guidelines include conventional programming style as well as uniquely applicable to object-oriented concepts such as inheritance. There are number of languages that support object concepts such as inheritance. There are number of languages that support object oriented programming such as C++, small talk and Java.

Advantages of Object-oriented Languages

- **Reusability**

Once a class or code has been written, created and debugged, it can be distributed to other programmers for use in their own programs. This is called reusability. It is similar to the way a library of functions in a procedural language can be incorporated into different programs. The idea of reusability can be active by inheritance.

- **Extensibility**

  Most software is eventually extended. Extensibility is enhanced by
  - Encapsulation of classes and methods.
  - Minimizing dependencies between classes and methods.
  - Using methods to access attributes of other classes.
  - Distinguishing public from private operations on a class.

  The techniques for reusability enhance extensibility.

- **Robustness**

A program is said to robust, if it does not fail even if it receives improper parameters. Robustness against internal bug (error) may be trade-off against efficiency. Robustness against user errors should never be sacrificed. Program should always protect against user and system errors.

It provides the facility of:

- Protection against errors.
- Validation of arguments.
- Service of debugging.
- Optimization after the program execution.
- Avoidation of predefined limits.
- Performance monitoring.

- **Programming-in-the-Large**

Programming-in-the-large refers to writing large, complex programs with teams of programmers. The writing of large programs with team of programmer requires more discipline, best documentation and communication than one-person or small projects. It provides the use of exactly the same names as in the object model. You should make methods more readable and understandable.

(*ii*) Non-Object-Oriented Language

Object-oriented concept can be mapped into non-object-oriented language constructs. There is not a big problem (excluding issue of expressiveness) to implement object oriented design, because programming languages are eventually converted to machine language. Use of a non-object-oriented language requires greater care and discipline to preserve the object oriented structure of the program. The most non-object-oriented languages, like C, Ada, Pascal and Fortran provides the implementation of object-oriented design.

Implementing an object-oriented design in a non-object-oriented language requires basically the same steps as implementing a design in an object oriented language.

The programmer using a non-object-oriented language must map. Object-oriented concepts into the target language, whereas the compiler for a object-oriented language performs such a mapping automatically. The steps required to implement a design are:

- Translate classes into data structure (or records).
- Pass arguments to methods (or functions).
- Allocate storage for objects.
- Implement inheritance in data structure.
- Implement method resolution.
- Implement associations.
- Deal with concurrency.
- Encapsulate internal details of classes.

7.5.2. Using Database System

A database management system (DBMS) is a computer program that is designed to provide general purpose functionality for storing, retrieving and controlling access to permanent data. A DBMS protects data against accidental loss and makes it available for sharing. It provides the management of a permanent, self-descriptive repository of data, called a *database* and is stored in one or more files.

The Advantage of DBMS:

- **Crash Recovery:** The database is protected from hardware crashes, disk media failures and some user errors.
- **Security:** Data can be protected against unauthorized users to read and write access.
- **Integrity:** You can specify the rules and conditions that data must satisfy. A DBMS can control the quality of its data over and above facilities that may be provided by application programs.
- **Sharing between Users:** The same database can be access by multiple users at the same time.
- **Sharing between Applications:** Multiple application programs can read and write data from/to the same database. A database is natural medium that facilitates communication between free-steady programs.
- **Extensibility:** Without interrupting the existing program data can be add to the database. Data can be reorganized for faster performance.
- **Data Distribution:** The database may be partitioned across various users, sites, organization and hardware platforms.

Several DBMS paradigms are available:

- Hierarchical DBMS
- Network DBMS
- Relational DBMS
- Object-oriented DBMS

Hierarchical and Network DBMS brings the conceptual DBMS close to the underlying physical data structures.

Relational DBMS present the database at a higher level of abstraction than hierarchies and networks are easier to use. Relational DBMS implementations are improving in performance as the mature and use smaller optimization techniques.

The object-oriented DBMS provide full supports for all the three models of the object design. The object classes used by the programming language are the classes used by the DBMS, because their models are consistent, there is no need to transform the program's object model to something unique for the database manager.

Object-oriented database have the ability to model all the models directly with in the database supporting a complete problem/solution modeling capability. It also provides the features of inheritance, polymorphism and dynamic binding.

7.6. COMPARISON OF METHODOLOGIES

There are several popular software engineering approaches for developing the softwares. And most of the approaches are based on the data flow diagrams. In this section we are presenting two other approaches to compare with OMT methodology, which are:

- Structured Analysis/Structured Design (SA/SD)
- Jackson Structured Design (JSD)

Basically our aim is to clearly identify the major differences and similarities between the OMT and other methods logics, because each approach has their own strengths and weaknesses.

7.6.1. Comparison with SA/SD

SA/SD methodology contains much common features as OMT. Both methodologies use similar modeling constructs and support the three orthogonal views of a system. SA/SD includes a variety of notations for formally specifying software. During the analysis phase, data flow diagrams, process specifications, a data dictionary, state transition diagrams, and entity relationship diagram are used to logically describe a system. In the design phase, details are added to the analysis models and the data flow diagrams are converted into structure chart descriptions of programming language code.

The difference between SA/SD and OMT is primarily a matter of style and emphasis.

- In SA/SD approach, the functional model dominates, the dynamics model is next important, and the object model least important.
- SA/SD organizes a system around procedures, while OMT organizes a system around real-world objects, or conceptional objects that exist in the user's view of the world.
- SA/SD is useful for problems where functions are more important and complex than data.
- An SA/SD design has a clearly-designed system boundary, across which the software procedures must communicate with the real world, so it can be difficult to extend a SA/SD design to a new boundary. While it is much easier to extend an object oriented design; one merely adds objects and relationship near the boundary to represent objects that existed previously only in the outside world.
- An object oriented design is more extensible, provides better traceability, and better integrates database and programming code.

7.6.2. Comparison with JSD

Jackson Structured Development (JSD) is another methodology of software engineering. JSD has a different development style than SA/SD or OMT. JSD does not distinguish between analysis and design and instead emphasizes on both phases together as specification. JSD divides system development into two stages: specification, then implementation. A JSD model describes the real-world in terms of entities, actions and ordering of actions.

JSD software development consists of six sequential steps:

(*i*) Entity action
(*ii*) Entity structure
(*iii*) Initial model
(*iv*) Function
(*v*) System timing
(*vi*) Implementation

— JSD approach uses graphical models, but JSD is less graphically oriented than SA/SD and OMT.

— JSD approach is complex and difficult to fully comprehend. JSD is complex because it was specifically designed to handle difficult real time problems. For these problems, JSD may produce a superior design and be worth the effort.

— JSD places more emphasis on action and less on attributes than OMT.

SUMMARY

In this chapter we have explained on OMT methodology of software engineering. OMT methodology supports entire software development life cycle. But in this chapter we skip the phase of testing and maintenance, because these phases uses traditional methods. So we have explained all the phases except these phases. Our main emphasis was on the system design and object design because analysis phase contains all three models (object model, dynamic model and functional model) and we have already explained these models in previous chapters.

In analysis phase we have also described problem statement for describing the actual problem which you need to solve using OMT methodology.

All steps of system design are explained in terms of software engineering as well as object-oriented design.

In the object design we have also described the way of designing algorithm, optimizing and adjustment of inheritance.

After the explanation of analysis, system design and object design phase we have also provided the techniques of implementation of the model on the software coding and computer implementation.

Software engineering also provides several methodologic to implement this design. Thus we have compared OMT methodology with the SA/SD and JSD methodology.

SA/SD and OMT modeling have much in common. Both methodologies support the three orthogonal views of a system. The difference is that SA/SD emphasizes the functional model while OMT emphasizes the object model. A JSD model begins with consideration of the real world, it emphasizes on entities and actions in the real world.

SOLVED PROBLEMS

Q. 7.1. What is data dictionary?

Ans. Data dictionary is collection of content and their description of the system. The data dictionary contents includes description of all modeling outlines, description of each object class, scope of the class with in the current problem, description of associations, attributes, and operations.

Q. 7.2. What do you mean by testing and maintenance of the system ?

Ans. Basically testing and maintenance are the phases of software life cycle.

Testing: Testing is a phase in which we test the software, before the implementation. There are some levels of testing in which we complete the task of testing.

The basic levels are unit testing, integration testing, and system and acceptance testing. Figure Q.7.2 shows the relation of the faults introduced indifferent phases, and the different levels of testing.

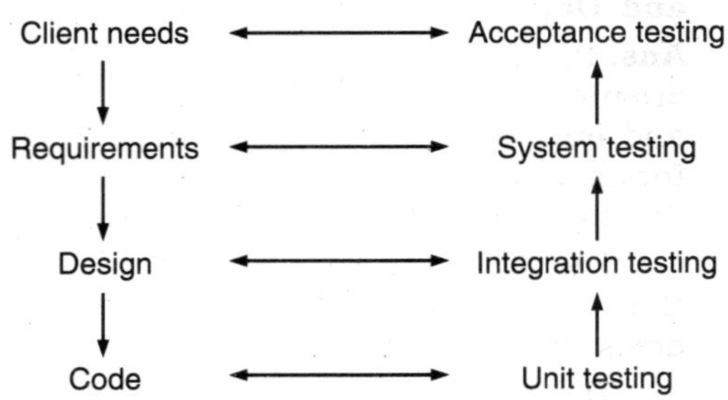

Fig. Q. 7.2.

Unit Testing: Testing the code of modules.

Integration Testing: Unit tested modules are combined into subsystem and then testing these subsystem to check interfaces between modules.

System and Acceptance Testing: Here the entire software system, is tested also tested on the client site.

Maintenance: Once the software is delivered and deployed, it enters the maintenance phase. All system need maintenance. Software needs to be maintained not because some of its components wear out and need to be replaced, but because there are often some residual errors remaining in the system that must be removed as they are discovered. There are some categories of maintenance:

(1) Corrective maintenance
(2) Adaptive maintenance
(3) Perfective maintenance
(4) Some extra maintenance

Q. 7.3. Define Delegation.

Ans. When classes are not in true generalization relationship but methods are required to share among them. In this case we should use delegation instead of inheritance.

Delegation provides a proper mechanism to achieve the desired code reuse. The operation (method) is caught in the desired class and forwarded to another class for actual execution. Since each operation must be explicitly forwarded, unexpected side effects are less likely to occur. The names of operations in the catching class may differ from these in the class supplying the operations. Each class should choose names which are most appropriate for its purpose.

Q. 7.4. Why we classify data and attributes operations as public and private ?

Ans. Public attributes and operations are visible outside a class and have published interfaces. Once a public operations and attributes are used by other classes, these are very hard and typical to change their interface, so public operations and attributes should be defined very carefully.

Private attributes and operations are internal to a class and are used to help implement the public operations. Private members can be deleted or their interfaces can be changed to modify the implementation of the class with impact limited to other methods on the class.

Q. 7.5. What is Software Engineering ?

Ans. Software engineering is a discipline of computer science, whose aim is the production of quality software, software that is delivered on time, within the budget, and that satisfies its requirements. The objective of software engineering is that solving the problems by producing good quality maintainable software, on time, with in budget.

Q. 7.6. What do you mean by "Design" ?

Ans. Design is the highly significant phase in the software development where the designer plans "How" a software system should be produced in order to make it functional, reliable and reasonably easy to understand, modify and maintain.

Q. 7.7. What is relationship and difference between OOA and OOD (Object Oriented Analysis and Object Oriented Design) ?

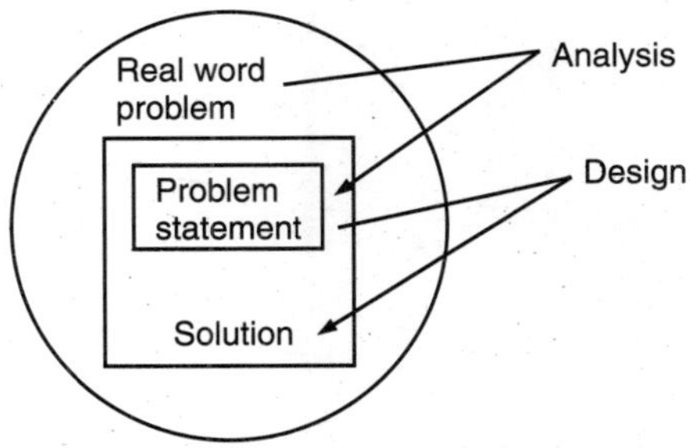

Ans. Pure object-oriented development requires that object-oriented techniques be used during the analysis, design and implementation of the system. However, much of the focus of the object-oriented approach to software development has been on OOA and OOD.

The fundamental difference between OOA and OOD is that, the OOA deals with the problem domain, while OOD deals with the solution domain.

The similarity if OOA and OOD is basic contracts (*i.e.*, objects and classes). The objective of path OOA and OOD is to model some domain, frequently the OOA and OOD methodologies and the representations look quite similar.

EXERCISES

Q. 7.1. What are the phases of OMT methodology ?

Q. 7.2. What is software development life cycle ?

Q. 7.3. Explain the Analysis phase of the OMT methodology.
Q. 7.4. Explain system design phase of the OMT methodology.
Q. 7.5. Explain the object design phase of the OMT methodology.
Q. 7.6. What is data dictionary ?
Q. 7.7. What is the benefit to organizing the system into subsystem ?
Q. 7.8. What do you mean by trade off priorities ?
Q. 7.9. What are the advantages and disadvantages of using a database in the system ?
Q. 7.10. What are the common architectural frameworks ?
Q. 7.11. What do you mean by system topology ?
Q. 7.12. Define delegation.
Q. 7.13. Explain the layered architecture schemes for decomposition of system into subsystem.
Q. 7.14. Explain the relationship of subsystems.
Q. 7.15. What are the techniques of data stores?
Q. 7.16. What is the difference between external and internal control?
Q. 7.17. What are the boundary conditions?
Q. 7.18. What are the languages available for the implementation of models?
Q. 7.19. What are the advantages of object oriented languages ?
Q. 7.20. What are the advantages of DBMS?
Q. 7.21. Compare the OMT methodology with SA/SD methodology.
Q. 7.22. Compare the OMT methodology with JSD methodology.
Q. 7.23. Which is the best methodology for the software development?

❑❑❑

Chapter 8

INTRODUCTION TO JAVA

8.1. INTRODUCTION

Java is a powerful object-oriented programming language introduced by Sun Microsystems in 1991. Java was originally called Oak, developed by James Gosling.

Java has built-in support to create programs with a graphical user interface (GUI), utilize the Internet, create client-server solutions, and much more. Programs written in Java can run, without change, on any of the common computer operating systems Windows 95/NT, Macintosh, and UNIX. A variant of Java programs called applets can be embedded inside a web page and execute on the computer that is viewing the page, automatically and in a secure environment.

As a language, Java is closely related to C++, which is also object-oriented but retains a lot of idiosyncrasies inherited from its predecessor language C. Java has removed the inconsistent elements from C++, is exclusively object-oriented, and can be considered a modern version of C++.[1] Because of its logical structure Java has quickly become a popular choice as a teaching language, and because of its extensive Internet support and the promise of writing programs once and using them on every operating system Java is becoming more and more accepted in industry.

CHAPTER OUTLINE

1 Java does have disadvantages. For example, programs written in Java are generally slower than those in C++ and it is difficult to accomplish system-level tasks in Java.

Sun's marketing people refer to all Java versions 1.2 + as JAVA 2 and 1.5 + as JAVA 5.

8.2. FEATURES OF JAVA

The basic features of java are given below by which, java have more fame in the computer world.

1. **Simple:** All the Syntax of java is like C++. There is no preprocessor, and much larger library. Confusing features such as: operator overloading, independent functions, global variables, go-to statements, structure, pointers, are *eliminated* from java.
2. **Object-Oriented:** All the functions are defined inside the classes. You cannot write any independent function. This forces you to think Object Oriented, java also refers as a pure object-oriented language.
3. **Network Savvy:** Java designed with internet in mind. So it supports maximum, network protocol and there component such as switch, socket, port etc. Java standard library provides URL classes.
4. **Platform Independent:** It is platform independent programming language. Java compiles to byte-code (more details section 9.4). This is interpreted (platform = system architecture + operating system).
5. **Robust:** Java is graceful in the presence of software or hardware errors. Java has very good *exception handlers*. Programmers are forced to think about potential sources of error.
6. **Secure:** Java has no pointers. All pointer-related security problems are gone. Java applets have limited access to local machine. It is more secure.
7. **Architecture Neutral:** Java programs can runs on PC, Macintosh, UNIX machines, etc. it means system with any architecture of mother board.
8. **Portable:** Java hides architecture-specific concepts. Byte-codes are the same regardless of the machine on which it is run, means you can port your program from one system to other system of different architecture.
9. **High Performance:** Byte-codes are interpreted – making it slow. But we have "Just-in-time" (JIT) compilers. These compile the Byte-codes into the machine code of target machine. This improves the performance.
10. **Multithreaded:** Java programs are capable for easy to set up multiple threads of execution and coordinate parallel processes.
11. **Dynamic:** Because Java programs move through the Internet they can exhibit dynamic behavior.
12. **Internet:** Applets are the most popular programs in java which basically used for Internet web browsers.
13. **Garbage Collection:** Java has an automatic garbage collector which release the objects which are not in use from long time.

8.3. IDENTIFIERS AND VARIABLES

Identifiers refer to the names of variables, functions, arrays, classes, etc. created by programmer. They are fundamental requirement of any language. Each language has its own rules for naming these identifiers. No reserve words can be used for Identifier name.

Names for Variable

Variable names can start with a letter, '$' (dollar symbol), or '_' (underscore); it cannot start with a number; cannot be a reserved word (or key word).

Names for Methods

Method name should be verbs or verb phrases with first letter in lowercase, and the first letter of subsequent words capitalized; cannot be reserved words.

Example: setColor()

Names for Classes and Interfaces

The class and interface names should be descriptive names that begin with a capital letter, by convention; cannot be a reserved word.

> **Concept**
>
> *Identifiers basically used to identify the name of variables, function, classes, arrays etc.*

Names for Constant

They are in capitals letter.

Example: Font.BOLD, Font.ITALIC

A *variable* is just a memory location (or several locations treated as a unit) that has been given a name so that it can be easily referred to and used in a program.

Variable Declaration

The syntax of variable declaration is as follows:

<datatype> <variable name>

Example: `int num1;`

Like C++ , java allows the declaration of variable anywhere in the scope. This means that a variable can be declare right at the place of its first use.

Variable Initialization

You can initialize the variable at the time of declaration, the syntax of initialization is as follows:

<datatype> <variable name> = value

Example: `double num2 = 3.1419;`

Java Reserved Words

These are explicitly reserved identifiers and cannot be used as names for the program variables or other user-defined program elements. The following table gives the complete set of java keywords.

abstract	default	if	package	this
boolean	do	implements	private	throw
break	double	import	protected	throws
byte	else	instanceof	public	transient
case	extends	int	return	null

try	Const	for	new	switch
continue	while		synchronized	super
catch	final	interface	short	void
char	finally	long	static	volatile
class	float	native		

8.4. DATA TYPES

The Java programming language has many built-in data types. These fall in two broad categories:

- Primitive Types and
- Class Types

Primitive types are simple values, are not objects. Class types are used for more complex types, including all of the types that you declare yourself. Class type are used to create objects. These also known as reference types.

8.4.1. Primitive Types

The java programming language defines eight primitive data types, which can be considered in four categories:

- Logical – boolean
- Textual – char
- Integral – byte, short, int, and long
- Floating point – double and float

The following table shows the size, default value, minimum value maximum value of all data types of java.

Data Type	Size	Default	Min Value Max Value
byte (Signed integer)	8 bits	0	–128 +127
short (Signed integer)	16 bits	0	–32,768 +32,767
int (Signed integer)	32 bits	0	–2,147,483,648 +2,147,483,647
long (Signed Integer)	64 bits	0	–9,223,372,036,0 854, 775, 808 +9,223,372,036,0 854, 775, 807
float (IEEE 754 floating-point)	32 bits	0.0	1.4E–45 3.4028235E38
double (IEEE 754 floating-point)	64 bits	0.0	4.9E–324 1.7976931348623157E308
char (Unicode character)	16 bits	\u0000	\u0000 \uFFFF
boolean	1 bit	false	

8.4.2. Reference Type

All data types discussed above, boolean, char, byte, short, int, long, float and double, are called primitive data types. All other types refer to objects rather than primitives. Variables that refer to objects are reference *variables*. For example, you can define a class Person:

```
public class Person
{
    ....
}
public class Test
{
    public static void main(String[] args)
    {
        Person emp = new Person();
    }
}
```

The variable **emp** is a reference variable holding one Person Object.

8.5. STATEMENTS IN JAVA

Like C and C++, Java has all three type of statements for creating the programs. The types of statements are:

1. Simple Statements
2. Compound Statements
3. Control Statements

8.5.1. Simple Statements

A simple statement is any expression that terminates with a semicolon. For example

```
var1 = var2 + var3;
var3 = var1++;
var3++;
```

8.5.2. Compound Statements

Related statements can be grouped together in braces to form a compound statement or block. For example:

```
{
    boolean flag = true;
    System.out.println(" i = " + i);
    i++;
}
```

Semantically, a block behave like a statement and can be used anywhere a single statement is allowed. There is no semicolon after the closing braces. Any variable declared in a block remain in scope up to the closing brace. Once the block is exited, the block variables cease to exist. Blocking improves readability of program code and can help make your program easier to control and debug.

8.5.3. Control Statements

Again control statements are divided into three statements.

1. Conditional statements: if, switch
2. Looping statements: for, while, do-while
3. Jumping statements: continue, break

The if statement

The if statement is implemented in two forms:

- Simple **if** statement
- **if.....else** statement

Simple if statement

```
if (expression is true)
{
        action 1;
}
        action 2;
        action 3;
```

if.....else statement

```
if (expression is true)
{
        action 1;
}
else
{
        action 2;
}
        action 3;
```

The switch statement

This is a multiple-branching statement where, based on condition, the control is transferred to one of the many possible points. This is implemented as follows:

```
switch ( expression)
{
        case 1:
              {
                      action 1;
              }
        case 2:
              {
                      action 2;
              }
        case 3:
              {
                      action 3;
              }
```

```
        default:
            {
                action 4;
            }
    }
    action 5;
```

The do-while statement

The **do-while** is an exit-controlled loop. Based on a condition, the control is transferred back to a particular point in the program. The syntax is as follows:

```
do
{
    action 1;
}
while(condition is true);
```

The while statement

This is also a loop structure, but is an entry-controlled one. The syntax is as follows :

```
while(condition is true)
{
    action 1;
}
    action 2;
```

The for statement

The **for** statement is an entry-controlled loop and is used when an action is to be repeated for a predetermined number of times. The syntax is as follows:

```
for(initial value; condition; expression)
{
    action 1;
}
action 2;
```

New form of 'for' loop is introduced in Java 5. The current 'for' loop construct is powerful, but somewhat ill-suited to iteration over collections (collections discussed in chapter10). The standard idiom to iterate over a collection is containing too many words, such as:

```
for (Iterator i = c.iterator(); i.hasNext(); ) // No ForUpdate
    {
        String s = (String) i.next();
        ...
    }
```

The situation is made little better with the addition of generics:

```
for (Iterator<String> i = c.iterator(); i.hasNext(); ) {
    String s = i.next();
    ...
}
```

The second form of the for loop specifically designed for iteration over collections and arrays. Some time this loop is refer as a foreach loop. Under the syntax of this, the above code could be replaced by this:

```
for (String s: c)
{
    ...
}
```

Similarly, the following code could be used to calculate the sum of an int array.

```
int sum = 0;
for (int j: a)
    sum += j;
```

This loop affects only the compiler; it demands no support from the VM. It does not require a new keyword and is fully compatible with all preexisting programs.

The break statement

The **break** statement can be used to exit any kind of loop when a specific condition is met. Control is transferred immediately to the first statement following the closing brace of the loop body. The use of break statement is useful where you need to abort a loop if some even occurs while executing the loop body. The break statement is also used in switch statements to prevent falling through to a subsequent case label.

The continue statement

The **continue** statement can be used to transfer control to the end of the loop body, prior to the 're-evaluation' of the text expression. If continue is used in for loop, the iteration expression is executed before the re-evaluation of the test.

SUMMARY

This chapter gives you the introduction of the java, with its history, and features

— The main feature of java are object-oriented, platform independent, Robust, recure, Portable, and Multithreading.

— Java has all the features of object oriented programming like inheritance, abstraction, encapsulation and polymorphism.

— Java is platform independent, means java programs can be execute under the environment of any operating system.

— Java has no pointers, that why it is called secure.

— Java can set multiple threads in a single program.

— We have also given the method of defining identifiers and variables.

— We have also given a short introduction of operators in this chapter.

— If you are a programmer or student who already known that C and C++, will be filmier with these operators.

— Java also contain that all three types of statements like C and C++ but it removes the 'goto' statements.

SOLVED PROBLEMS

Q. 8.1. Consider the following code:

```
arrayOfInts[j] > arrayOfInts[j+1]
```

What operators does the code contain?

Ans. [], >, [], +

Q. 8.2. Consider the following code:

```
int i = 10;
int n = i++%5;
```

What are the values of i and n after the code is executed?

Ans. i is 11, and n is 0.

Q. 8.3. What are the final values of i and n if instead of using the postfix increment operator (i++), you use the prefix version (++i))?

Ans. i is 11, and n is 1.

Q. 8.4. To invert the value of a boolean, which operator would you use?

Ans. The logical complement operator "!".

Q. 8.5. Which operator is used to compare two values, = or == ?

Ans. The == operator is used for comparison, and = is used for assignment.

Q. 8.6. Explain the following code sample: `result = someCondition ? value1: value2;`

Ans. This code should be read as: "If `someCondition` is `true`, assign the value of `value1` to `result`. Otherwise, assign the value of `value2` to `result`."

Q. 8.7. What is the meaning of "final" keyword?

Ans. final keyword is used to declare the variable, method and class. A final variable is a variable whose value cannot be changed once set. A final method cannot be overridden. A final class cannot be subclassed further.

Q. 8.8. Why java is called 100% pure object oriented language?

Ans. Java is 100% pure object oriented language because java obeys all the principles of OOAD(Object oriented analysis and design). We can't write java program without using at least one class or object.

Q. 8.9. Does JAVA has goto?

Ans. No, Java does not have goto statements Because:

1. goto statements form infinite loops which are drawbacks in a program.
2. goto statements decrease the reliability of a program.
3. goto statements make documentation of a program very difficult. (Documentation means preserving a copy of a program for future use.)

Q. 8.10. What are the disadvantages/Limitations of Java?

Ans. Limitations of java:

1. Development time is much longer than using scripting languages.
2. It's a bit slower than C.
3. String manipulation in Perl takes much less lines of code.

EXERCISES

Q. 8.1 Who is the developer of Java? And what is the original name of Java?

Q. 8.2. What are the main features of Java programming language?

Q. 8.3. What do you main by identifiers? And what are the rules for identifiers?

Q. 8.4. What the data types are available in Java programming language?

Q. 8.5. What are the operators available in Java programming language?

Q. 8.6. What do you mean by precedence rules of operator? Give an example.

Q. 8.7. Explain Type-Casting. And how it is done in Java programming language?

Q. 8.8. What are the statements are available is Java programming language?

Q. 8.9. What is the role of conditional operator? Give an example.

Q. 8.10. What do you mean by pure object oriented language? Is Java pure object oriented language? If yes How?

Q. 8.11. Is Java is similar to C++ ? Give the advantages of Java over C++.

Q. 8.12. What is the difference between simple and compound statements?

Q. 8.13. Give the syntax of following statements:

(*i*) while (*ii*) do-while

(*iii*) for (*iv*) switch

(*v*) if-else

Q. 8.14. Differentiate the following statements:

(*a*) while and do-while (*b*) for and while

(*c*) continue and return

Q. 8.15. What is literal?

Q. 8.16. Give the meaning of each of the following Java operators:

(*a*) ++ (*b*) &&

(*c*) !=

Q. 8.17. What does the computer do when it executes a variable declaration statements. Give an example.

Q. 8.18. Explain what is meant by an assignment statement, and give an example. What are assignment statement used for?

Q. 8.19. One of primitive type in java is boolean. What is the boolean type? Where are boolean values used? What are its possible values?

Q. 8.20. Write a for loop that will print out all the multiple of 3 from 3 to 36, that is 3, 6, 9, 12, 15, 18, 21, 24, 27, 30, 33, 36.

Q. 8.21. Show the exact output that would be produced by the following main () routine:

```
public static void main (String [] args)
{
    int x, y;
    x = 5;
    y = 1;
    while (x > 0)
    {
        x = x - 1;
        y = y*x;
        System.out.println (y);
    }
}
```

□□□

Chapter 9

Beginning with Java Programs

CHAPTER OUTLINE

9.1. GUIDE LINES FOR JAVA PROGRAMS

Java is an Object-Oriented Programming, all functions are defined in the class and every Java program must follow these guidelines:

- Java is case sensitive, i.e., the word *Program* is different from *program.*
- Curly brackets { and } are used to group statements together.
- An executable Java program must contain at least the following lines as a framework:

```
public class Name
{
      public static void main(String args[])
            {
            ... program code ...
      }
}
```

- Every statement must end with a semicolon.
- A Java program containing the above framework must be saved using the filename *Name.java,* where *Name* (including correct upper and lower cases) is the word that follows the keywords public class and the file extension is *.java.*

9.2. OBJECTS, CLASSES AND METHODS

Java is an object oriented programming, all programs are based on concept of objects and classes.

9.2.1. Object

Objects are the basic run time entities in an object oriented system. They may represent a person, a place, a bank account, a table of data or any item that the program has to handle. They may also represent user-defined data such as vectors, time and lists. Objects take up space in the memory and have an associated address like a record in Pascal, or a structure in C.

When a program is executed, the objects interact by sending messages to one another. For example, if "customer" and "account" are two objects in a program, then the customer object may send a message to account object requesting for the bank balance. Each object contains data, and code to manipulate the data. Objects can interact without having to know details of each other's data or code. It is sufficient to know the type of message accepted, and the type of response returned by the object.

Objects have unique identity; even if two or more objects have identical attribute values, the system always treats them as different objects. When you create objects in software, you are given a handle to the objects. In java this handle is the name of the objects. Irrespective of what the handle is, it always uniquely identifies the object.

9.2.2. Class

We just explained that objects contain data, and code to manipulate that data. The entire set of data and code of an object can be made a user-defined data type with the help of a *class.* Once a class has been defined, we can create any number of objects belonging to that class.

> **Concept**
>
> *A class is a user defined data type, and the object are also known as class variables.*

A class is a template or blueprint used in creating multiple objects. A class encapsulates all of the data and behaviors that make up an object. When you ask a java to create or instantiate an object, java uses the class definition as a template for building the new object. Each object is associated with the data of type class with which they are created. A class is thus a collection of objects of similar type. For example, apple, orange and mango are members of the class fruit. Classes are user defined data types and behave like the built-in types of programming language.

Class Declaration

A class must be declared using the keyword class followed by the class name.

Syntax

```
class <classname>
{
      — Body of the class
}
```

A typical class declaration is as follows:

```
< access modifier> <classname>
{
        — Member variable declarations;
        — Method declarations and definitions
}
```

9.2.3. Method

When you defines a class for an object-oriented program, you implement all of the behavior of that class in one or more methods. A java method is equivalent to a function, procedure, or subroutine in other languages, except that it must be defined inside a class definition. In other words, there is no support for global methods in java; every method must be defined within a class.

Concept

A java method is equivalent to a function, procedure, or subroutine in other language, but it must be defined within a class.

Method declarations

```
<access modifier> <static/final> <return type> <method name>
<arguments list>
{
- - Method body;
}
```

Example:

```
public static void main(String args[])
{
}
```

9.2.4. Access Modifiers

Java controls access to variables, methods and classes through the use of access modifiers. The access modifiers are **private, public** and **protected**.

A **public** is the least restrictive access modifier. Variables and methods declared as public may be seen and used by any other class.

If an access modifier is not specified (default), the variables and methods may be used by any other class within the same package (a container of classes).

Variables and methods declared as **protected** may be seen from any subclass of that class. It may also be seen from any class with the package in which it exists.

The most restrictive access modifier is **private**. A private method or variable may not be accessed by any other class. The restriction applies to all other classes regardless of their package.

9.3. A SIMPLE PROGRAM

To create a Java program you first create a text file containing the lines

```
public class Name
{
      public static void main(String args[])
      {
            ... more lines ...
      }
}
```

The file containing our code is called the source code file. This source code file can be create any text editor like notepad, MS-Word, WordPad or DOS Editor. But java source code files should be saved as *Name.java*, where *Name* is the name that appears in the first line of the program: *public class Name*. That *Name* is referred to as the name of the class, or program. By convention its first letter is capitalized.

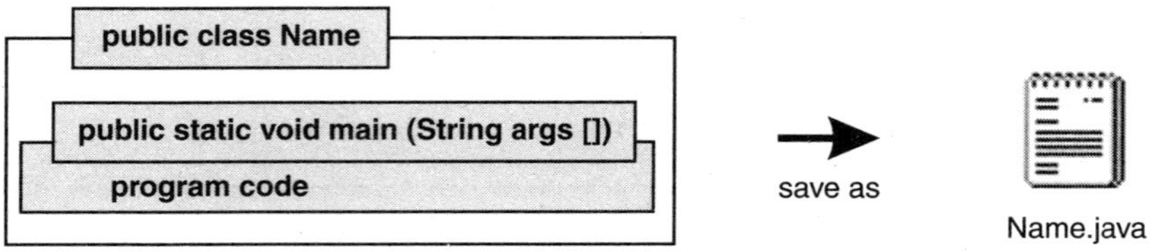

Fig. 9.1. Saving a Java source code file

Example 9.1: The first Program

Create a program containing the necessary Java code to get the computer to write on the screen.

Our first Java program looks as follows:

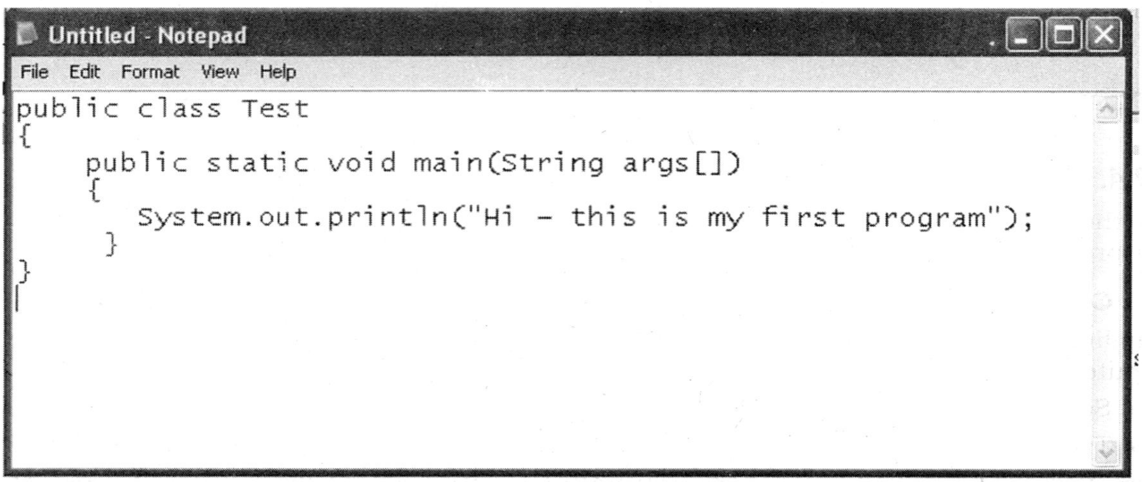

This program, or class, is called and must be saved under the file name Test.java you should create a folder for java programs like as:

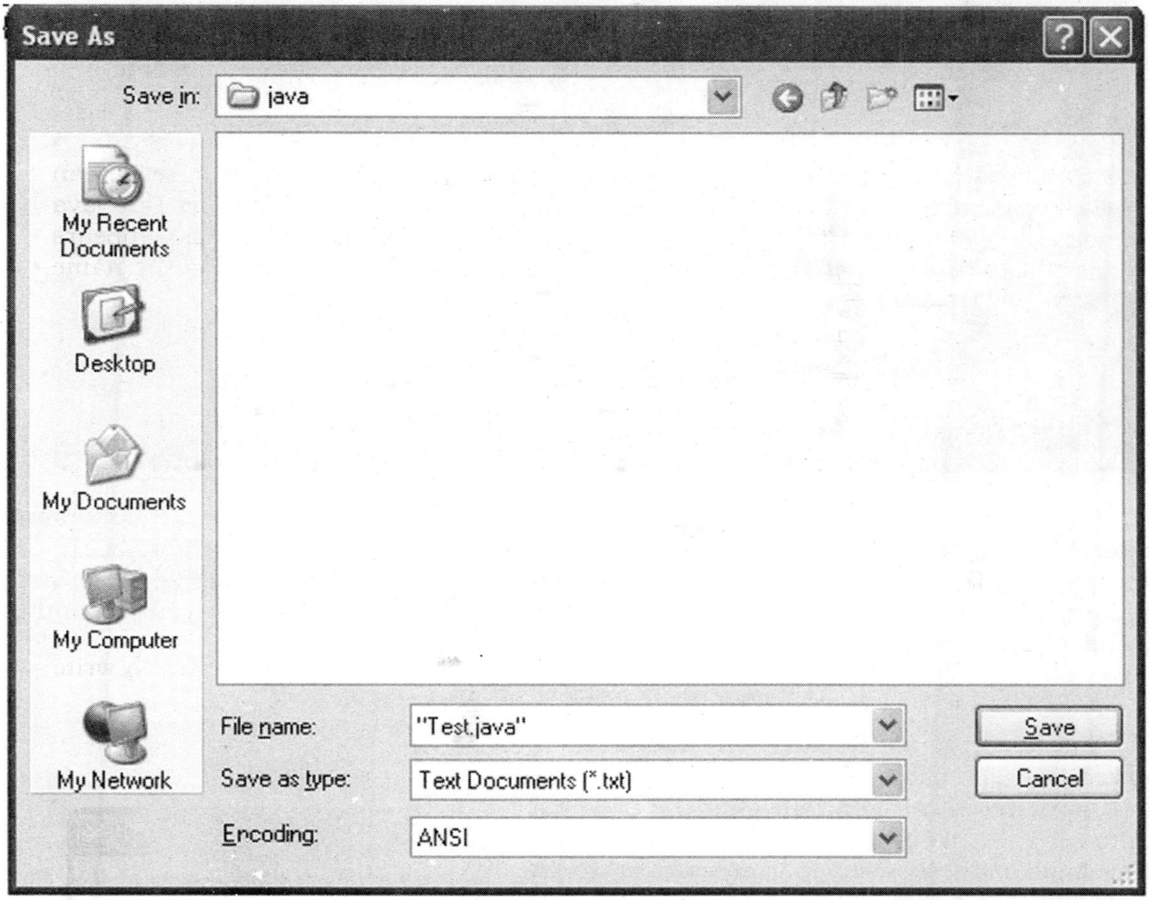

9.4. COMPILING AND EXECUTION

After the program creation the next step is compilation and execution of the program. Java provides the Java Developer's Kit for this purpose.

9.4.1. Compiling a Java Program

Compiling is the process of transforming the source code file into a format that the computer can understand and process. To compile the program,

- Select **Start → Programs → MS-DOS Prompt**
- Type the command " **cd c:\java"**
- Type **"javac Test.java"** and Press Enter

For example,

If you find an error **"javac is not recognized as an internal or external command, operable program or batch file"**. It means path for the java compiler is not set. To Set the path of the compiler, type following command (or wherever javac.exe is located, according to my computer it is in c:\ jdk1.4\bin it may be different in your computer) and press Enter. The path of the compiler will set.

```
c:\java> set path = c:\jdk1.4\bin
```

The resulting file is called the ***byte-code***, or ***class file***. The name of the class file is the same as the name of the program plus the extension *.class*. The program *javac* from the Java Developer's Kit is used to transform a source code file into a class file.

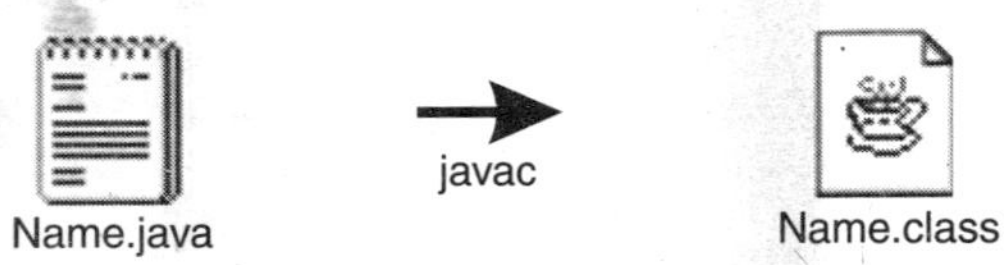

Fig. 9.2. Compiling and creating class file

If a source code contains any errors, they are flagged by the compiler. You need to fix them and re-compile until there are no further errors. In case of an error, the javac compiler shows the line number and position of where *it thinks* the error occurred in your source code.

- If the compiler points out an error, then there is an error *at or before* the indicated position.
- If the compiler reports a certain number of errors, than this is *the least* amount of errors.
- If one error is fixed, other errors may automatically disappear or new ones may appear.

Fix your source code a few errors at a time. Recompile often to see if the number of errors and the error messages change until no errors are reported. If you can not find an error at the position indicated by the compiler, look at the code *before* that position.

9.4.2. Executing a Java Program or Class

The Java compiler does not produce an executable file, so Java programs can not execute under the operating system of your machine. Instead they execute inside a *Java Virtual Machine*, which is invoked using the java program of the JDK.

To execute a Java program the Java Developer's Kit provides a program called 'java'. When executing that program with your class file as parameter the following happens:

- the Java Virtual Machine (JVM) must be present on your computer
- the JVM locates and reads your class files
- the JVM inspects your class file for any security violations
- the JVM executes, or interprets, your class file according to its instructions if possible

Under Windows and UNIX, execute a program by typing at the command prompt, where is the name of the program (no extension). For example,

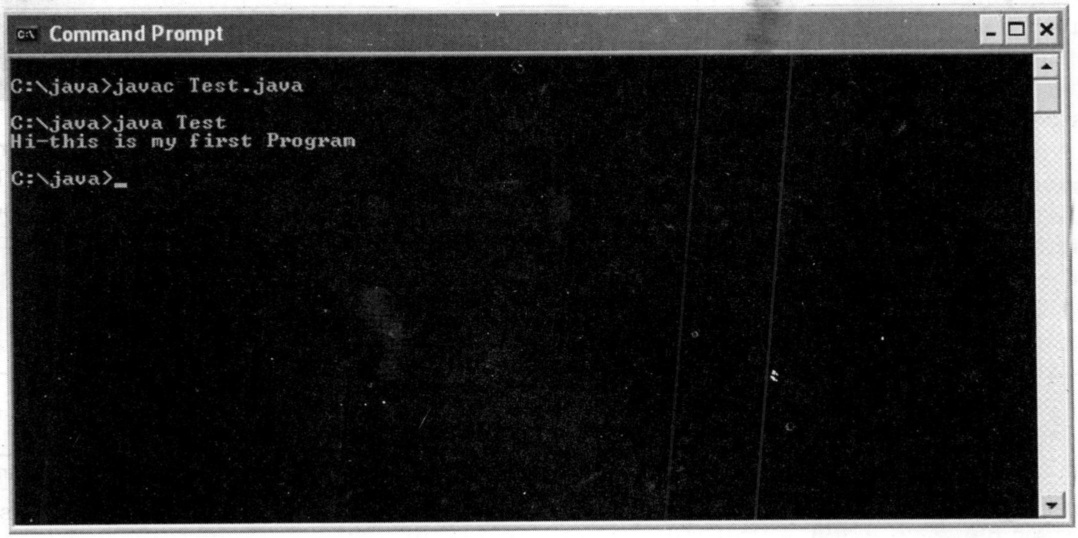

On a Macintosh, double-click the icon and select the appropriate class file.

A good question at this point is which line in a Java program executes *first*. The default program entry point is that part of a class (or program) where execution begins. For every Java class (or program), the standard program entry point consists of the line:

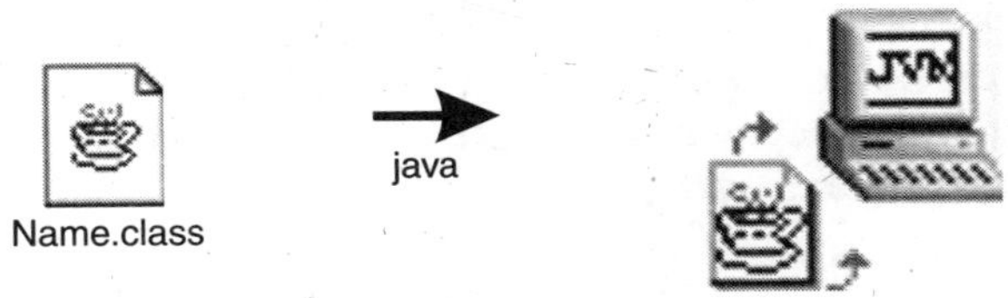

Fig. 9.3. Executing a class file

```
public static void main (String args [])
```

If that line is not present in your source code, the JVM can not execute your program and displays an error message.

9.5. JAVA VIRTUAL MACHINE (JVM)

The Java Virtual Machine (JVM) is a platform-independent engine used to run Java applets and applications. The JVM knows nothing of the Java programming language, but it does understand the particular file format of the platform and implementation independent class file produced by a Java compiler. Therefore, class files produced by a Java compiler on one system can execute without change on any system that can invoke a Java Virtual Machine.[1]

When invoked with a particular class file, the JVM loads the file, goes through a verification process to ensure system security, and executes the instructions in that class file.

> **Concept**
>
> *The Java Virtual Machine (JVM) is a platform-independent engine used to run Java applets and applications.*

The JVM, in other words, forms a layer between the operating system and the Java program that is trying to execute. That explains how *one* Java program can run without change on a variety of systems: it can not! A Java program runs on only *one* system, namely the **Java Virtual Machine**. That virtual system, in turn, runs on a variety of operating systems and is programmed quite differently for various systems. To the Java programmer, it provides a unified interface to the actual system calls of the operating system.[2]

You can include graphics, graphical user interface elements, multimedia, and networking operations in a Java program and the JVM will negotiate the necessary details between the class file(s) and the underlying operating system. The JVM produces exactly the same results – in theory – regardless of the underlying operating system. In the Basic (or C, or C++) programming language, for example, you can create code that specifies to multiply two integers 1000 and 2000 and store the result as another integer. That code works fine on some systems, but can produce negative numbers on others.[3] In Java, this can not happen: either the code fails on all platforms, or it works on all platforms.

Because the JVM is in effect its own computer, it can shield the actual computer it is running on from potentially harmful effects of a Java program. This is especially important because Java programs known as *applets* can *automatically* start executing on your machine when you are surfing the web if the appropriate feature of your web browser is enabled. If these programs were allowed to interfere with your system, you could accidentally execute a program that would proceed to erase your entire disk. That, of course, would prompt people to disable Java on their web browser, which in turn would be bad news for anyone who supports the Java concept.

---

1 In general, the Java Virtual Machine is an abstractly specified class file interpreter that can be realized by different software makers. The JVM that comes with the JDK was created by SUN Microsystems, but any other JVM is also able to run the same class files. Different JVM's can vary in efficiency, but they all must run the same class files.

2 This is somewhat similar to old Basic programs: a simple Basic program can run on virtually any system that has a Basic interpreter installed since the interpreter mediates between the program trying to run and the operating system.

3 Programming languages have a *largest possible* integer whose value can differ on different systems. A C++ program executing on a machine with a largest integer bigger than 2,000,000 produces the correct result, but on a system where the largest integer is, say, 32,767 it fails. The JVM has the *same* largest integer on every platform.

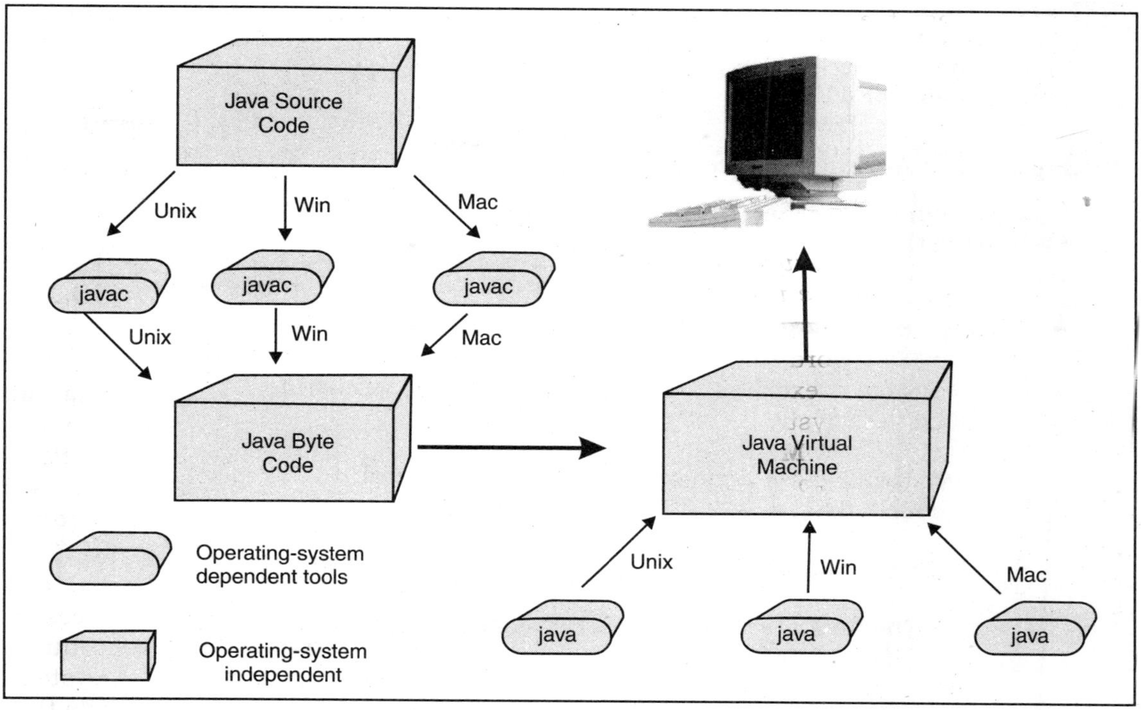

Fig. 9.4. Illustrating the machine dependent/independent parts of Java programs

9.6. SOME EXAMPLE OF PROGRAMS

Example 9.2 : A Simple Average Program

Create a simple program to compute the average of 3 positive integers.

```
public class ComputeAverage
{
    public static void main(String[] args)
    {
        int a, b, c ;
        int sum, avg ;
        a = 6;
        b = 4;
        c = 8;
        sum = a + b + c ;
        avg = sum / 3;
        System.out.println("The values of the positive integers.");
        System.out.println("First integer: " + a + ", Second integer
        : " + b + ", Third integer: " + c);
        System.out.println ("Their average is:" + avg + ".");
        }
}
```

OUTPUT:

```
The values of the positive integers.
First integer: 6, Second integer: 4 , Third integer: 8
Their average is: 6.
```

Example 9.3. A Temperature Conversion Program

Create a complete program that converts a temperature from degrees Fahrenheit in degrees Celsius.

```
public class Converter
{
        public static void main(String args[])
        {
                System.out.println("\nProgram to convert Fahrenheit to
                Celcius.\n");

                // Defining the temperature value in Fahrenheit
                double temp = 212.0;

                double convertedTemp = 5.0 / 9.0 * (temp - 32.0);

                System.out.println(temp + " Fahrenheit = " + convertedTemp
                + " Celcius.");
        }
}
```

OUTPUT:

```
Program to convert Fahrenheit to celcius.
212.0 Fahrenheit = 100 Celcius
```

Example 9.4. A Simple interest Program

Create a simple program to compute the amount of interest that is earned on Rs. 17,000 invested at an interest rate of 0.07 for one year.

```
public class Interest
{
        public static void main(String args[])
                {

                double principal; // The value of the investment.
                double rate;             // The annual interest rate.
                double interest; // Interest earned in one year.

                principal = 17000;
                rate = 0.07;
                interest = principal * rate;     // Compute the interest.
                principal = principal + interest;

                System.out.println("The interest earned is Rs. " + interest);
                System.out.println("The value of the investment after one
                year is Rs. " + principal);

                }
```

```
}
```

OUTPUT:

```
The interest earned is Rs. 1190
The value of the investment after one year is Rs. 18190s
```

Example 9.5. A Simple interest Program for 5 years

Create a simple program to compute the amount of interest that is earned on an investment over a period of 5 years. The initial amount of the investment is Rs. 17,000 and the interest rate is 0.07. Display the value of the investment at the end of each year.

```
public class Interest3
{
     public static void main(String args[])
          {

               double principal=17000;
               double rate=0.07;

               /* Simulate the investment for 5 years. */
               int years;
               years = 0;
               while (years < 5)
               {
               double interest; // Interest for this year.
               interest = principal * rate;
               principal = principal + interest;
               years = years + 1;
               System.out.println("The value of the investment after "
               + years);
               System.out.print(" years is Rs.+ principal);
                    }
          }
}
```

OUTPUT:

```
The value of the investment after 1 year is Rs. 18190
The value of the investment after 2 year is Rs. 19463.3
The value of the investment after 3 year is Rs. 20825.73
The value of the investment after 4 year is Rs. 22283.53
The value of the investment after 5 year is Rs. 23843.4
```

SUMMARY

This chapter is very important chapter for every Beginner of Java language. In this we have explained the way of programming. We have also given the pictures of required steps which will helpful for every programmer.

— You must save your program with class name, and extension must be "Java".

— Java has a new concept of Byte code, which is intermediate level code and interpreted by JVM.

— JVM (Java Virtual Machine) is a software used to run java applications and applets.

— We have also explained the whole procedure of compilation and execution of the java programs.

SOLVED PROBLEMS

Q. 9.1. What is BYTE Code?OUTPUT:

Ans. A BYTE Code is intermediate level code, which is interpreted by the JVM. It is not directly executable on the machine. A BYTE code is a mediator code between java program and machine code of Operating System (O.S.) and is available in *'.class'* file. A *'.class'* file is generated when the java program or application gets compile. Byte code is same to any O.S. but machine code varies to every O.S.

Q. 9.2. What is the Java Virtual Machine?

Ans. The Java Virtual Machine is a software that can be ported onto various hardware-based platforms.

Q. 9.3. What gives java it's "write once and run anywhere" nature?

Ans. When java program is written and compiled then it will create a (*.class*) file which consists of byte code instructions, understandable to JVM. This class file is system independent. Every system has it's own JVM. So JVM will convert this byte code into machine language understandable to that system. So it has run write once and run anywhere nature. Java is 'platform independent' programing language. The Byte code is same for any Operating System. That's why we can execute an application in one OS which was already compiled in other OS we can directly copy the application from one OS to other OS without writing again.

Q. 9.4. What is native code?

Ans. The native code is code that after you compile it, the compiled code runs on a specific hardware platform.

Q. 9.5. What is a Java compiler?

Ans. A Java compiler is a program which converts **Java source code** into Java bytecode.

A basic Java compiler is included as part of the JDK (Java Development Kit). This Java compiler is called "javac".

Q. 9.6. What is the JRE?

Ans. The JRE is the Java Runtime Environment.

The JRE consists of the Java Virtual Machine, the Java libraries, and all other components necessary to run Java applications and applets.

Sun distributes the JRE separately, or as part of J2EE, J2SE, and J2ME.

Q. 9.7. What is special about static variables?

Ans. A static variable is shared by all instances of a class. Every instance has a reference to the same variable and can modify it directly. Static instances are declared with the static modifier.

Q. 9.8. Write a program that produces the following output:

```
Hello World!
It's been nice knowing you.
Goodbye world!
```

Ans. This is three lines of text so it needs three System.out.println() statements. In other words:

```
class HelloWorld {
public static void main (String args[]) {
```

```
public class TestProg
{
public static void main(String args[])
{
    try {
        String s;
        StringBuffer sb = new StringBuffer("A StringBuffer");
        // create a char array
        char ca[] = "A character array!".toCharArray();
        // create a byte array containing the default encoding
        // ISO-Latin-1
        byte ba[] = "A byte array!".getBytes();
        // create a byte array containing Unicode
        byte bau[] = "A Unicode byte array!".getBytes("Unicode");

        // create an empty String
        s = new String();
        System.out.println(s); // prints out nothing

        // create a String from a byte array
        s = new String(ba);
        System.out.println(s); // prints out "A byte array!"

        // create a String from a byte array from index 2 (inclusive),
        // take 1 byte
        s = new String(ba, 2, 1);
        System.out.println(s); // prints out "byte"

        // create a String from a byte subarray containing Unicode
        s = new String(bau, "Unicode");
        System.out.println(s); // prints out "A Unicode byte array!"

        // create a String from a byte subarray containing Unicode
        s = new String(bau, 0, 20, "Unicode");
        System.out.println(s); // prints out "A Unicode"

        // create a String from a StringBuffer
        s = new String(sb);
        System.out.println(s); // prints out "A StringBuffer"

        // create a String from a character subarray
        s = new String(ca, 2, 4);
        System.out.println(s); // prints out "char"

        // create a String from a character array
        s = new String(ca);
```

```
                System.out.println(s); // prints out "A character array!"
            }
            catch(java.io.UnsupportedEncodingException e) {
                System.out.println(e);
            }
        }
}
```

- Notice that you need to catch java.io.UnsupportedEncodingException for the methods in which you have to specify an encoding.

10.7.2. Important String Operations

Java has several string operations because strings are a common and important part of programming. These operation includes:

String Length

Use the length() method from the String class. The following example is self-explanatory:

```
String s = "hello, world!";
System.out.println(s.length()); // prints out 13
```

Character Array to String conversion

The String class has a useful constructor to convert an array to String, as you can see in the following example:

```
public class TestProg
{
        public static void main(String args[])
        {
        char [] array = {'H','e','l','l','o',',',' ','w','o','r',
        'l','d','!' };
        String s = new String(array);
        System.out.println(s); // prints out: Hello, world!
        }

}
```

String to Character Array conversion

And you can convert a string to character array also, the String class has a useful method doing just that:

```
public class TestProg
{
        public static void main(String args[])
        {
            String s = "Hello, world!";
            char array[] = s.toCharArray();

```

```
            for (int i=0; i<array.length; i++)
            System.out.print(array[i]);
            System.out.println();

        }

    }
```

OUTPUT

```
Hello, world!
```

String to lowercase or uppercase

The following example is self-explanatory:

```
public class TestProg
{
    public static void main(String args[])
    {
        String s = "Hello, World!";
        String supper = s.toUpperCase();
        String slower = s.toLowerCase();

        System.out.println(supper); // prints out: HELLO, WORLD!
        System.out.println(slower); // prints out: hello, world!
    }
}
```

Note: for internationalization purposes, the methods *toUpperCase()* and *toLowerCase()* perform some checking for languages as German and Turkish to do the correct conversion. If you need these functions in time-critical applications, you might want to consider rewriting them without these checking.

Concatenate two Strings

There are two ways to concatenate one string to another. The most common is to use the (rather ambiguous) + operator:

```
String s1 = "Hello, ";
String s2 = "world!";
String s3 = s1 + s2;
System.out.println(s3); // prints out: Hello, world!
```

The String class also provides a method *concat* that does just the same:

```
String s1 = "Hello, ";
String s2 = "world!";
String s3 = s1.concat(s2);
System.out.println(s3); // prints out: Hello, world!
```

In the previous example, the method concat is invoked on s1. s1 remains unchanged and a completely new String will be created. A String is immutable, Once you assigned it a value, it cannot change anymore during its lifetime. You will always have to create a new String object.

Comparison of two Strings

Use the method java.lang.String.equals(Object). Will return true if it matches and false if it doesn't.

```
String a = "some text";
String b = "some text";

if (a.equals(b)) {
      System.out.println("String a is equal to String b");
} else {
      System.out.println("String a is NOT equal to String b");
}
```

public int compareTo(String str)

where, str is the String being compared to the invoking String. The compareTo() method returns an int value as the result of String comparison. If the value is less than zero, then the invoking string is less than the argument string. If value is zero means the invoking string and the argument string are same. If the value is greater than zero it means the invoking string is greater than the argument string. The String class also has the compareToIgnoreCase() method that compares two strings without taking into consideration their case difference.

10.7.3. All String Classes

String Class:

The String class represents character strings. All string literals in Java programs, such as "abc", are implemented as instances of this class.

Strings are constant; their values cannot be changed after they are created. String buffers support mutable strings. Because String objects are immutable they can be shared. For example:

```
String str = "abc";
```

is equivalent to:

```
char data[] = {'a', 'b', 'c'};
String str = new String(data);
```

Here are some more examples of how strings can be used:

```
System.out.println("abc");
String cde = "cde";
System.out.println("abc" + cde);
String c = "abc".substring(2,3);
String d = cde.substring(1, 2);
```

The class String includes methods for examining individual characters of the sequence, for comparing strings, for searching strings, for extracting substrings, and for creating a copy of a string with all characters translated to uppercase or to lowercase.

The Java language provides special support for the string concatenation operator (+), and for conversion of other objects to strings. String concatenation is implemented through the StringBuffer class and its append method. String conversions are implemented through the method toString, defined by Object and inherited by all classes in Java.

StringBuffer Class:

The StringBuffer class is a thread-safe, mutable sequence of characters. A string buffer is like a String, but can be modified. At any point in time it contains some particular sequence of characters, but the length and content of the sequence can be changed through certain method calls.

String buffers are safe for use by multiple threads. The methods are synchronized where necessary so that all the operations on any particular instance behave as if they occur in some serial order that is consistent with the order of the method calls made by each of the individual threads involved.

The principal operations on a StringBuffer are the append and insert methods, which are overloaded so as to accept data of any type. Each effectively converts a given datum to a string and then appends or inserts the characters of that string to the String-Buffer. The append method always adds these characters at the end of the buffer; the insert method adds the characters at a specified point.

For example, if z refers to a string buffer object whose current contents are “start”, then the method call z.append(“le”) would cause the string buffer to contain “startle”, whereas z.insert(4, “le”) would alter the string buffer to contain “starlet”.

In general, if sb refers to an instance of a StringBuffer, then sb.append(x) has the same effect as sb.insert(sb.length(), x).

Whenever an operation occurs involving a source sequence (such as appending or inserting from a source sequence) this class synchronizes only on the string buffer performing the operation, not on the source.

Every string buffer has a capacity. As long as the length of the character sequence contained in the string buffer does not exceed the capacity, it is not necessary to allocate a new internal buffer array. If the internal buffer overflows, it is automatically made larger.

StringBuilder Class:

The StringBuilder class is mutable sequence of characters. This class provides an API compatible with StringBuffer, but with no guarantee of synchronization. This class is designed for use as a drop-in replacement for StringBuffer in places where the String-Buffer was being used by a single thread (as is generally the case). Where possible, it is recommended that this class be used in preference to StringBuffer as it will be faster under most implementations.

The principal operations on a StringBuilder are the append and insert methods, which are overloaded so as to accept data of any type. Each effectively converts a given datum to a string and then appends or inserts the characters of that string to the string builder. The append method always adds these characters at the end of the builder; the insert method adds the characters at a specified point.

For example, if z refers to a string builder object whose current contents are “start”, then the method call z.append(“le”) would cause the string builder to contain “startle”, whereas z.insert(4, “le”) would alter the string builder to contain “starlet”.

In general, if sb refers to an instance of a StringBuilder, then sb.append(x) has the same effect as sb.insert(sb.length(), x). Every string builder has a capacity. As long as the length of the character sequence contained in the string builder does not exceed the capacity, it is not necessary to allocate a new internal buffer. If the internal buffer overflows, it is automatically made larger.

Differences:

String is immutable whereas StringBuffer and StringBuilder can change their values.

The only difference between StringBuffer and StringBuilder is that StringBuilder is unsynchronized whereas StringBuffer is synchronized. So when the application needs to be run only in a single thread then it is better to use StringBuilder. StringBuilder is more efficient than StringBuffer.

Criteria to choose among String, StringBuffer and StringBuilder

- If your text is not going to change use a String Class because a String object is immutable.
- If your text can change and will only be accessed from a single thread, use a StringBuilder because StringBuilder is unsynchronized.
- If your text can changes, and will be accessed from multiple threads, use a StringBuffer because StringBuffer is synchronous.

10.8. COLLECTIONS

A collection is a single object representing a group of objects. The objects in the collection are called *elements*. Typically, collections deal with many types of objects, all of which are of a particular kind (that is they all descend from a common parent type).

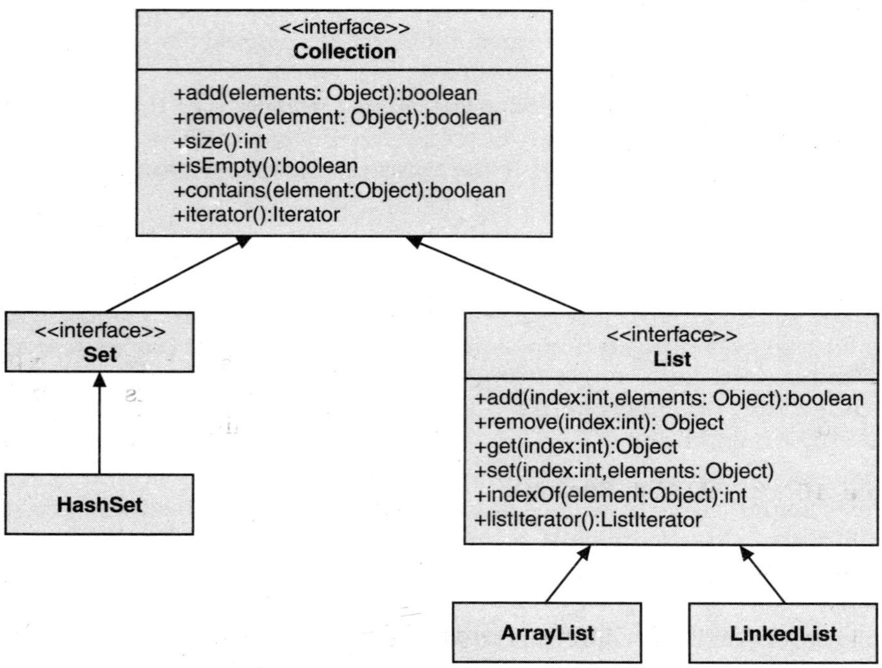

- Collection – A group of objects known as elements; implementations determine whether there is specific ordering and whether duplicates are permitted.

- Set – An unordered collection; no duplicates are permitted
- List – An orderd collection; duplicates are permitted

Collections are maintain references to objects of type Object. This enables any objects to be sorted in the collection. It also necessitates the use of correct casting before you can use the object, after retrieving it from the collection.

Following figure shows hierarchy of the Collection API, where HashSet class supplies an implementation of the Set interface and the ArrayList and LinkedList classes supply an implementation of the List interface.

In the following example, the program declares a variable (set) of type Set and is initialized to new HashSet object. It then adds few elements and prints the set to standard output. The attempt to add duplicate items in code fails, the add method returns false.

Example 10.12: A Set Example

```
import java.util.*;

public class SetExample
{
    public static void main(String [] args)
    {
        Set set = new HashSet();
        set.add("Ram");
        set.add("Geeta");
        set.add("3rd");
        set.add(new Integer(420));
set.add(new Float(5.0F));
set.ad d("Geeta");              //duplicate, not added
set.add(new Integer(420));      // duplicate, not added
        System.out.println(set);
    }
}
```

OUTPUT

```
[Ram, Geeta, 3rd, 420, 5.0]
```

In the next example, the program declares a variable (list) of type List and is initialized to new ArrayList object. It then adds a few elements and prints the set to standard output. Because lists allow duplicates, so addition of duplicate will succeed.

Example 10.13: A List Example

```
import java.util.*;
public class ListExample
{
    public static void main(String [] args)
    {
        List list = new ArrayList();
```

```
        list.add("Ram");
        list.add("Geeta");
        list.add("3rd");
        list.add(new Integer(420));
        list.add(new Float(5.0F));
        list.add("Geeta");                //duplicate, is added
        list.add(new Integer(420));       // duplicate, is added
        System.out.println(list);
    }
}
```

Output

```
[Ram, Geeta, 3rd, 420, 5.0, Geeta, 420]
```

SUMMARY

— In this chapter we have explained how to deal with methods and objects.

— In java there is no support for global methods, every method must be defined within a class.

— You already knows that methods are simple operations which are associated with objects or class. The methods are provides some useful functionality.

— Methods may be return type of void, method may take arguments or without arguments.

— Method overloading is a concept of overloading a method with same name and different arguments.

— Java has given a new concept of garbage collection, which is the process of releasing the memory occupied by the objects that are no longer needed. This is an automatic process.

— We have also explained the concept of Arrays, how to create, how to initilize and how to access the elements of array in the program.

SOLVED PROBLEMS

Q. 10.1. What is an object in Java?

Ans. In Java the Object class is the ultimate superclass of every other object type. All objects are extended from the Object class, either directly or by inheritance through any number of parent classes. If a class does not explicitly extend any named class, it implicitly extends the Object class. An object with a small o is the word used to describe an instance of a Java class.

Q. 10.2. What's the difference between an object and a class?

Ans. An class merely defines what an object will be if it is created. An object is an actual, created instance of a class. For example, A class is like a blueprint for a house, while the object is the actual house. A single blueprint can be used to make many different houses.

Q. 10.3. What's the difference between a local variable and a member variable?

Ans. A local variable is only accessible within the method where it's declared. A member variable, a field, is accessible in any method in the class.

Q. 10.4. Can objects be used in place of arrays?

Ans. Yes, sometimes it is sensible to use an object to carry other object references instead of an array. For instance, you could issue an object as the return value of a method that must return multiple object references.

Q. 10.5. Can we use the same variable name in two methods?

Ans. It is possible to use the same variable names for method local variables in two separate methods. However, all variables in a single method must have different names.

```
public class MethodLocalVariables
{
      void testOne()
      {
            int test;
      }
      void testTwo()
      {
      int test;
      }
}
```

Q. 10.6. Explain Garbage collection mechanism in Java.

Ans. Garbage collection is a complex mechanism, but to put it in a simple way, it is the process of identifying objects which are no longer referenced by the program and discarded. The finalize() methods of these objects are called and resources occupied by the objects are released to the system. Garbage collection is carried on a separate thread and runs continuously.

Q. 10.7. What do you know about the garbage collector?

Ans. Garbage collector is used to recollect memory from the unused objects. They are the objects that are no longer needed because of function or class scope is going to finish.

Q. 10.8. Can the main method be overloaded?

Ans. Yes, any Java method can be overloaded, provided there is no final method with the same signature already. The Java interpreter will only invoke the standard entry point signature for the main method, with a string array argument, but your application can call its own main method as required.

Q. 10.9. What's the index of the first component of a one hundred component array of a two hundred component array? of a one thousand and twenty-four component array?

Ans. The index of the first component of an array is always 0.

Q. 10.10. What's the index of the last component of a one hundred component array? of a two hundred component array? of a one thousand and twenty-four component array?

Ans. 99, 199, and 1023 respectively. The index of the last component of an array is always one less than the length of the array.

Q. 10.11. What's the mininum number of components an array can have?

Ans. An array can have as few as zero components; that you can declare an array with no components at all. However you cannot declare an array with a negative number of components. If you try, a java.lang.NegativeArraySizeException will be thrown at runtime.

Q. 10.12. What's the difference between a brace and a bracket?

Ans. { or } is a brace. [or] is a bracket. Braces are used to initialize arrays. Brackets are used to select a component of an array by specifying its index.

Q. 10.13. Write a program that randomly fills a 3 by 4 by 6 array, then prints the largest and smallest values in the array.

Ans.

```
class array346 {

public static void main(String[] args) {

double[][][] array = new double[3][4][6];

 for (int i = 0; i < 3; i++) {
     for (int j = 0; j < 4; j++) {
     for (int k = 0; k < 6; k++) {
     array[i][j][k] = Math.random();
     }
 }
}

double max = array[0][0][0];
double min = array[0][0][0];

for (int i = 0; i < 3; i++) {
     for (int j = 0; j < 4; j++) {
     for (int k = 0; k < 6; k++) {
          if (max < array[i][j][k]) max = array[i][j][k];
          if (min > array[i][j][k]) min = array[i][j][k];
          }
     }
}
System.out.println("Maximum: " + max);
System.out.println("Minimum: " + min);

}

}
```

Q. 10.14. Write a Complex class that represents a single complex number and includes methods for all the usual operations, i.e., addition, subtraction, multiplication, division, etc.

Ans:

```
public class Complex {

     private double u;
     private double v;

     Complex (double x, double y) {

     u=x;
     v=y;
```

```
    }

    public double Real () {

        return u;

    }

    public double Imaginary () {

    return v;

  }

public double Magnitude () {

    return Math.sqrt(u*u + v*v);

  }

    public double Arg () {

    // Need to watch out for zero
    return Math.atan(v/u);

  }

    // Add z to w; i.e. w += z
    public Complex Plus (Complex z) {
    return new Complex(u + z.u, v + z.v);

  }

    // Subtract z from w
    public Complex Minus (Complex z) {

    return new Complex(u - z.u, v - z.v);

  }

    public Complex Times (Complex z) {

    return new Complex(u*z.u - v*z.v, u*z.v + v*z.u);

  }
```

```
    // divide w by z
    public Complex DivideBy (Complex z) {

  double rz = z.Magnitude();

    return new Complex((u * z.u + v * z.v)/(rz*rz),(v * z.u - u * z.v)
    /(rz*rz));

    }

}
```

Notice especially that x and y are now private. They cannot be accessed by external code even if we want them to be.

EXERCISES

Q. 10.1. What do you mean by invocation of methods?

Q. 10.2. Differentiate between actual and formal parameters?

Q. 10.3. What are the advantages of methods?

Q. 10.4. What is the scope of any variable?

Q. 10.5. What is concept of constructor? Why it is needed?

Q. 10.6. Can we use constructors with parameters?

Q. 10.7. What do you mean by method overloading? Is it possible with constructors?

Q. 10.8. Explain the concept of Garbage collection. What is the alternative to garbage collection?

Q. 10.9. What is the difference between garbage collector and finalization?

Q. 10.10. What is meant by the terms subclass and superclass?

Q. 10.11. What does the computer do when it executes the following statement? Try to give as complete an answer as possible.

color[ ] palette = new color [12];

Q. 10.12. What does it mean to sort an array?

Q. 10.13. What is meant by basetype of an array?

Q. 10.14. What is meant by a dynamic array? What is the advantage of a dynamic array over a regular array?

Q. 10.15. Show the exact output produced by the following code segment:

```
char [ ] [ ] pic = new char [6] [6];
for (int i = 0; i < 6; i++)
for (int j = 0; j < 6; j ++)
{
     if (i = = j || i = = 0 || i = = 5)
          pic [i] [j] = '*';
     else
          pic [i] [j] = 'a'
}
     for (int i = 0; i < 6; i++)
     {
     for (int j = 0; j < 6; j ++)
```

```
        System.out.print (pic [i][j]);
        System.out.println();
    }
```

Q. 10.16. Write a program that finds the largest value in an array of integers.

Q. 10.17. Suppose that temperature measurements were made on each day of 1999 in each of 100 cities. The measurements have been stored in an array

int [ ] [] temps = new int [100] [365];

where temps [c][d] holds the measurement for city number c on the dth day of the year. Write a code segment that will print out the average temperature, over the course of the whole year, for each city. The average temperature for a city can be obtained by adding up all 365 measurements for that city and dividing the answer by 365.0.

Q. 10.18. Suppose that a class, Employee is defined as follows:

```
class Employee
{
    String lastName;
    String FirstName:
    double hourlywage:
    int yearswithCompany;
}
```

Suppose that data about 100 employees is already stored in an array.

Employee [ ] employeeData = new Employee [100];

Write a code segment that will output the firstname, lastname, and hourly wage of each employee who has been with the company for 20 years or more.

Q. 10.19. Suppose that A has been declared and initialized with the statement:

double [ ] A = new double [20];

and suppose that A has already been filled with 20 values. Write a program segment that will find the average of all the non-zero numbers in the array. Declare any variables that you needed.

Chapter 11

INHERITANCE, PACKAGES AND INTERFACES

CHAPTER OUTLINE

11.1. INHERITANCE

Inheritance means taking an existing class and adding functionality by deriving a new class from it.

Object-oriented programming allows classes to *inherit* commonly used state and behavior from other classes.

A class that is derived from another class is called a *subclass* (also a *derived class, extended class*, or *child class*). The class from which the subclass is derived is called a *superclass* (also a *base class* or a *parent class*).

Excepting `Object`, which has no superclass, every class has one and only one direct superclass (single inheritance). In the absence of any other explicit superclass, every class is implicitly a subclass of Object.

Classes can be derived from classes that are derived from classes, and so on, and ultimately derived from the topmost class, Object. Such a class is said to be *descended* from all the classes in the inheritance chain stretching back to Object.

The idea of inheritance is simple but powerful: When you want to create a new class and there is already a class that includes some of the code that you want, you can derive your new class from the existing class. In doing this, you can reuse the fields and methods of the existing class without having to write (and debug!) them yourself.

Concept

Inheritance is a mechanism of sharing the members (member data and member function) among the classes.

A subclass inherits all the *members* (fields, methods, and nested classes) from its superclass. Constructors are not members, so they are not inherited by subclasses, but the constructor of the superclass can be invoked from the subclass.

11.1.1. Defining Subclass

In this example, Bicycle now becomes the *superclass* of MountainBike, RoadBike, and TandemBike. In the Java programming language, each class is allowed to have one direct superclass, and each superclass has the potential for an unlimited number of *subclasses*:

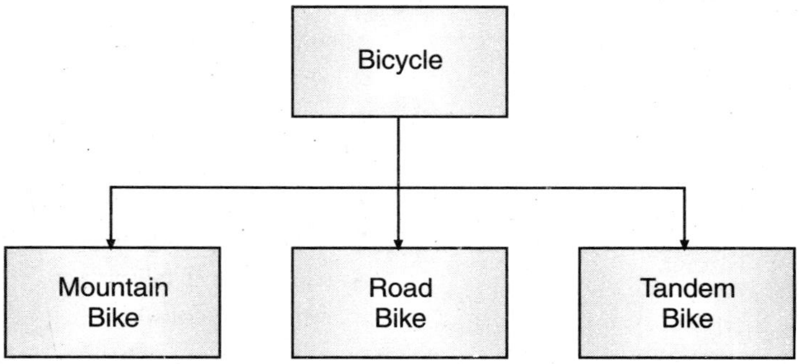

Fig. 11.1. A hierarchy of bicycle classes

The syntax for creating a subclass is simple. At the beginning of your class declaration, use the *extends* keyword, followed by the name of the class to inherit from:

```
class MountainBike extends Bicycle
{
// new fields and methods defining a mountain bike would go here
}
```

This gives MountainBike all the same fields and methods as Bicycle, yet allows its code to focus exclusively on the features that make it unique. This makes code for your subclasses easy to read. However, you must take care to properly document the state and behavior that each superclass defines, since that code will not appear in the source file of each subclass.

11.1.2. How to Use Inheritance

We will first create a parent class called Person. We will then create a child class called Student. Then we will create a program to use the Student class. The following is a Person class which has a variable for the person's name. There are also set and get methods for the name.

```
class Person
{
    private String name;
```

```
        public void setName(String n)
        {
            name = n;
        }
        public String getName()
        {
        return name;
        }
}
```

Now we will create the Student class that has a variable for the student number and the get and set methods for student number. The extends keyword is used to inherit from the Person class in the following example.

```
class Student extends Person
        {
        private String stuNum;

        public void setStuNum(String sn)
        {
            stuNum = sn;
        }

        public String getStuNum()
        {
            return stuNum;
        }
}
```

Finally we have to create a program that will instantiate a Student object, set the name and student number and then print the values using the get methods.

```
public class TestInheritance
{
        public static void main(String[] args)
        {
            Student stu = new Student();
            stu.setName("John Smith");
            stu.setStuNum("12345");
            System.out.println("Student Name: " + stu.getName());
            System.out.println("Student Number: " + stu.getStuNum());
        }
}
```

Here is another example of inheritance.

Java supports only Single Inheritance while C++ also supports Multiple Inheritance.

The Forms of Single Inheritance:

- Single Level Form
- Multiple Level Form
- Hierarchical Form

11.2. SINGLE LEVEL INHERITANCE

In the single level inheritance single class can derived from only one base class. If a class B is derived form class A, like a figure 11.2, then it is called single level inheritance.

Here we have given some example to understand the concept of single inheritance.

Example 11.1 Single inheritance

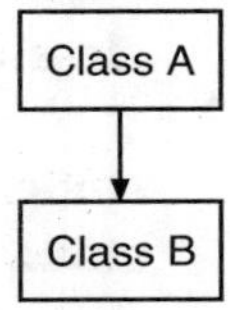

Fig 11.2. Single Level Inheritance

```
class single
{
static class A
{
      int a1, a2;
      void getdata()
      {
            a1 = 10;
            a2 = 20;
      }
      void putdata()
      {
            System.out.println("\nThe value of a1 is:"+ a1 + " and a2:" + a2);
      }
}
static class B extends A          // class B is derived by class A

{
      int b1, b2;
      void indata()
      {
            b1 = 30;
            b2 = 40;
      }
      void outdata()
      {
            System.out.println("\nThe value of b1 is:"+ b1 + " and b2:" + b2);
      }
}
public static void main(String [] args)
{
B Obj=new B();
Obj.getdata();          // calling base class member function
Obj.indata();           // derived class member function
```

```
Obj.putdata();
Obj.outdata();
}
}
```

OUTPUT

```
The value of a1 is: 10  and a2 is: 20

The value of b1 is: 30  and b2 is: 40
```

Example 11.2 Single inheritance with Input Method

```
import java.util.*;
class single
{
static Scanner sc=new Scanner(System.in);
static class A
{
        int a1, a2;
        void getdata()
        {
                System.out.println("Enter the value of a1:");
                a1 = sc.nextInt();
                System.out.println("Enter the value of a2:");
                a2 = sc.nextInt();
        }
        void putdata()
        {
                System.out.println("\nThe value of a1 is:"+ a1 + " and a2:" + a2);
        }
}

static class B extends A            // class B is derived by class A

{
        int b1, b2;
        void indata()
        {
                System.out.println("Enter the value of b1:");
                b1 = sc.nextInt();
                System.out.println("Enter the value of b2:");
                b2 = sc.nextInt()
        }
        void outdata()
        {
                System.out.println("\nThe value of b1 is:"+ b1 + " and b2:" + b2);
        }
```

```
}
public static void main(String [] args)
{
B Obj=new B();
Obj.getdata(); // calling base class member function
Obj.indata();  // derived class member function
Obj.putdata();
Obj.outdata();
}
}
```

OUTPUT

```
Enter the value of a1:
10
Enter the value of a2:
20
Enter the value of b1:
30
Enter the value of b2:
40

The value of a1 is: 10  and a2 is: 20

The value of b1 is: 30  and b2 is: 40
```

11.3. MULTI LEVEL INHERITANCE

In the Multilevel inheritance a class derived from another derived class. If a class B is derived from class A and class C derived from class B, like a Fig. 11.3, then it is called multilevel inheritance, because there are more than one levels. You can also derive class C, and then derive its derived and so on.

Here we have given some example to understand the concept of Multilevel inheritance.

Example 11.3 Program to Demonstrate Multilevel inheritance

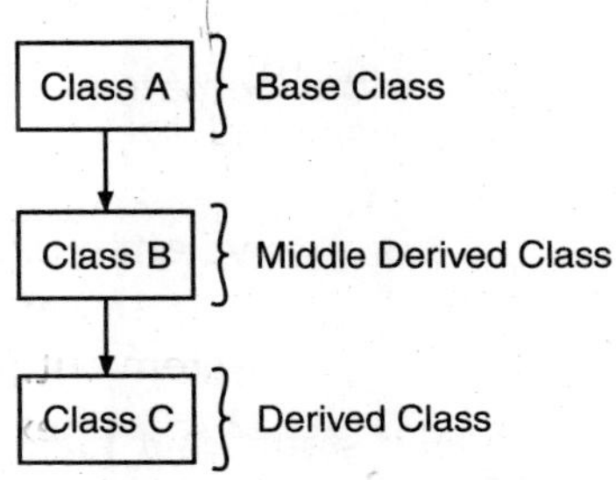

Fig. 11.3. Multilevel Inheritance

```
import java.util.*;
class multilevel
{
static Scanner sc=new Scanner(System.in);
static class A
{
    int a1, a2;
    void getdata()
    {
        System.out.println("Enter the value of a1 is:");
    a1 = sc.nextInt();
        System.out.println("Enter the value of a2 is:");
```

```
        a2 = sc.nextInt();
        }
        void putdata()
        {
            System.out.println("\nThe value of a1 is:"+ a1 + " and a2:" + a2);
        }
    }

    static class B extends A          // class B is derived by class A
    {
        int b1, b2;
        void indata()
        {
            System.out.println("Enter the value of b1 is:");
            b1 = sc.nextInt();
            System.out.println("Enter the value of b2 is:");
            b2 = sc.nextInt();
        }
        void outdata()
        {
            System.out.println("\nThe value of b1 is:"+ b1 + " and b2:" + b2);
        }
    }
    static class C extends B          // class C is derived by class B
                                      (Multi-level inheritance)
    {
        int c1, c2;
        void input()
        {
            System.out.println("Enter the value of c1 is:");
            c1 = sc.nextInt();
            System.out.println("Enter the value of c2 is:");
            c2 = sc.nextInt();
        }
        void output()
        {
            System.out.println("\nThe value of c1 is:"+ c1 + " and c2:" + c2);
        }
    }

    public static void main(String [] args)
    {
    C Obj=new C();
    Obj.getdata();  // calling base class member function
    Obj.indata();   // Middile derived class member function
```

```
Obj.input();          // Derived class member function
Obj.putdata();
Obj.outdata();
Obj.output();
}
}
```

OUTPUT

```
Enter the value of a1:
2
Enter the value of a2:
3
Enter the value of b1:
4
Enter the value of b2:
5
Enter the value of c1:
6
Enter the value of c2:
7

The value of a1 is: 2  and a2: 3

The value of b1 is: 4  and b2: 5

The value of c1 is: 6  and c2: 7
```

11.4. HIERARCHICAL INHERITANCE

In the Hierarchical inheritance more than one class derived from only single base class. If a class B and class C are derived form class A, like a Fig. 11.4, then it is called hierarchical inheritance. Because there are more than one derived classes of a single base class thus both classes inherits the features of class A.

Here we have given some example to understand the concept of Hierarchical inheritance.

Example 11.4 Program to impliment Hierarchical inheritance

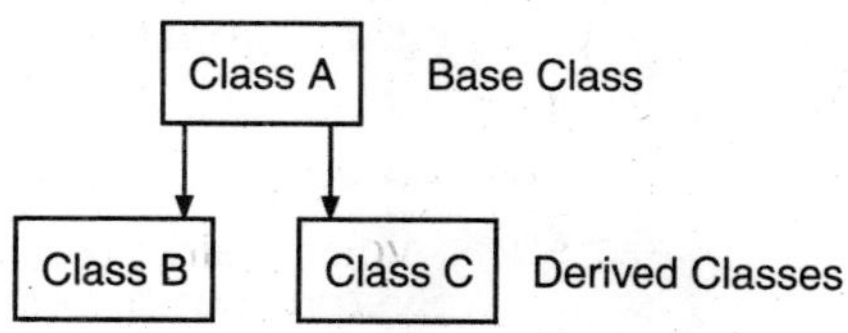

Fig. 11.4. Hierarchical Inheritance

```
import java.util.*;
class hierarchical
{
static Scanner sc=new Scanner(System.in);
static class A
{
    int a1, a2;
    void getdata()
```

```
        {
            System.out.println("Enter the value of a1 is:");
            a1 = sc.nextInt();
            System.out.println("Enter the value of a2 is:");
            a2 = sc.nextInt();
        }
        void putdata()
        {
            System.out.println("\nThe value of a1 is:"+ a1 + " and a2:" + a2);
        }
    }

    static class B extends A          // class B is derived by class A
    {
        int b1, b2;
        void indata()
        {
            System.out.println("Enter the value of b1 is:");
            b1 = sc.nextInt();
            System.out.println("Enter the value of b2 is:");
            b2 = sc.nextInt();
        }
        void outdata()
        {
            System.out.println("\nThe value of b1 is:"+ b1 + " and b2:" + b2);
        }
    }

    static class C extends A          // class C is also derived by class
                                  A (Hierarchical inheritance)
    {
        int c1, c2;
        void input()
        {
            System.out.println("Enter the value of c1 is:");
            c1 = sc.nextInt();
            System.out.println("Enter the value of c2 is:");
            c2 = sc.nextInt();
        }
        void output()
        {
            System.out.println("\nThe value of c1 is:"+ c1 + " and c2:" + c2);
        }
    }
```

```
public static void main(String [] args)
{
B Ob1 =new B();
C Ob2 =new C();

Ob1.getdata();        // calling by Object of class B
Ob1.indata();

Ob2.getdata();        // calling by Object of class C
Ob2.input();

Ob1.putdata();
Ob1.outdata();

Ob2.putdata();
Ob2.output();
}
}
```

OUTPUT

```
Enter the value of a1:
1
Enter the value of  a2:
1
Enter the value of b1:
2
Enter the value of b2:
2
Enter the value of a1:
5
Enter the value of a2:
5
Enter the value of c1:
10
Enter the value of c2:
10

The value of a1 is: 1  and a2 is: 1

The value of b1 is: 2  and b2 is: 2

The value of a1 is: 5  and a2 is: 5

The value of c1 is: 10  and c2 is: 10
```

11.5. METHOD OVERRIDING

When you call a method on an object, Java looks for that method definition in the class of that object, and if it doesn't find one, it passes the method call up the class hierarchy until a method definition is found. Method inheritance enables you to define and use methods repeatedly in subclasses without having to duplicate the code itself.

> **Concept**
>
> *Method overriding means same name methods in different classes of Inheritance hierarchy.*

However, there may be times when you want an object to respond to the same methods but have different behavior when that method is called. In this case, you can override that method. Overriding a method involves defining a method in a subclass that has the same signature as a method in a superclass. Then, when that method is called, the method in the subclass is found and executed instead of the one in the superclass.

11.5.1. Creating Methods that Override Existing Methods

To override a method, all you have to do is create a method in your subclass that has the same signature (name, return type, and parameter list) as a method defined by one of your class's superclasses. Because Java executes the first method definition it finds that matches the signature, this effectively "hides" the original method definition. Here's a simple Example 11.5 shows a simple class with a method called display(), which prints out the name of the class from where method is involved.

Example 11.5 Program to demonstrate method overriding.

```
//Declaration of super class
    class One
    {
    public void display()
    {
        System.out.println("I am from class one");
    }
    }
//Declaration of Main class
    class override extends One
    {
    public void display()
    {
    System.out.println("I am from class  override");
    }
    public static void main(String [] args)
    {
        override o = new override();
        o.display();
```

```
    }
    }
```

OUTPUT

```
I am from class override
```

In the main() method of override class we invoked the display() method, which hides the display() method of class **One** and gives the output from display method of **override** class. Note that to implement the concept the overriding you must use the same name and signature of the method in the inheritance hierarchy. Now see one more example.

Example 11.6 Program to demonstrate method overriding.

```
//Declaration of super class
    class One
    {
    public void display()
    {
        System.out.println("I am from class one");
    }
    }
//Declaration of sub class
    class Two extends One
    {
    public void display()
    {
        System.out.println("I am from class two");
    }
    }
//Declaration of Main class
    class override
    {
    public void display()
    {
        System.out.println("I am from class override");
    }
    public static void main(String [] args)
    {
        override o = new override();
        o.display();
        Two t=new Two();
        t.display();
    }
    }
```

OUTPUT

```
I am from class override
I am from class two
```

In the above example we have shows that we can also create the object of any class and invoke the method which hidden by its child class. Now question is that can we call the method of base class and super class by the object of leaf class (means last most child class of inheritance hierarchy)? Lets move to next section with more example.

11.5.2. Calling the Original Method

Usually, there are two reasons why you want to override a method that a superclass has already implemented:

- To replace the definition of that original method completely
- To augment the original method with additional behavior

You've already learned about the first one; by overriding a method and giving that method a new definition, you've hidden the original method definition. But sometimes you may just want to add behavior to the original definition rather than erase it altogether. This is particularly useful where you end up duplicating behavior in both the original method and the method that overrides it; by being able to call the original method in the body of the overridden method, you can add only what you need.

To call the original method from inside a method definition, use the super keyword to pass the method call up the hierarchy:

```
void myMethod (String a, String b)
{
        // do stuff here
        super.myMethod(a, b);
        // maybe do more stuff here
}
```

The **super** keyword, somewhat like the this keyword, is a placeholder for this class's superclass. You can use it anywhere you can use this, but to refer to the superclass rather than to the current class.

For example, see those display() methods used in the previous example.

Example 11.7 **Program to demonstrate to call original method in overriding concept.**

```
//Declaration of base class
    class One
    {
    public void display()
    {
          System.out.println("I am from class one");
    }
    }
//Declaration of sub class
    class Two extends One
    {
    public void display()
    {
          super.display();
```

```
        System.out.println("I am from class two");
    }
    }
//Declaration of Main class
    class override extends Two
    {
    public void display()
    {
        super.display();
        System.out.println("I am from class override");
    }
    public static void main(String [] args)
    {
        override o = new override();
        o.display();
    }
    }
```

OUTPUT

```
I am from class one
I am from class two
I am from class override
```

Rather than duplicating most of the behavior of the superclass's method in the sub class, you can rearrange the superclass's method so that additional behavior can easily be added to sub class, as you saw in above example.

11.5.3. Overriding Constructors

Constructors cannot technically be overridden. Because they always have the same name as the current class, you're always creating new constructors instead of inheriting the ones you've got. Much of the time, this is fine, because when your class's constructor is called, the constructor with the same signature for all your superclass is also called, so initialization of all the parts of a class you inherit can happen.

However, when you're defining constructors for your own class, you may want to change how your object is initialized, not only by initializing new variables your class adds, but also to change the contents of variables that are already there. You do this by explicitly calling your superclass's constructors, and then changing whatever you like.

To call a regular method in a superclass, you use super.methodname(arguments). Because with constructors you don't have a method name to call, however, you have to use a different form:

```
super(arg1, arg2, ...);
```

Similar to using this(...) in a constructor, super(...) calls the constructor method for the immediate superclass (which may, in turn, call the constructor of its superclass, and so on).

Example 11.6, shows a class called NamedPoint, which extends the class Point from Java's awt package. The Point class has only one constructor, which takes an x and a y argument and returns a Point object. NamedPoint has an additional instance variable (a string for the name) and defines a constructor to initialize x, y, and the name.

Example 11.8: The NamedPoint class.

```
import java.awt.Point;
class NamedPoint extends Point
{
    String name;
    NamedPoint(int x, int y, String name)
    {
    super(x,y);
    this.name = name;
    }
}
```

The constructor defined here for NamedPoint (lines 6 through 8) calls Point's constructor method to initialize Point's instance variables (x and y). Although you can just as easily initialize x and y yourself, you may not know what other things Point is doing to initialize itself, so it's always a good idea to pass constructors up the hierarchy to make sure everything is set up correctly.

11.6. ABSTRACT CLASSES AND METHODS

11.6.1. Abstract Classes

Sometimes, a class that you define represents an abstract concept and as such, should not be instantiated. Take, for example, food in the real world. Have you ever seen an instance of food? No. What you see instead are instances of carrot, apple, and (my favorite), chocolate. Food represents the abstract concept of things that we can eat. It doesn't make sense for an instance of food to exist.

Similarly in object-oriented programming, you may want to model abstract concepts but you don't want to be able to create an instance of it. For example, the Number class in the java.lang package represents the abstract concept of numbers. It makes sense to model numbers in a program, but it doesn't make sense to create a generic number object. Instead, the Number class makes sense only as a superclass to classes like Integer and Float which implement specific kinds of numbers. Classes such as Number, which implement abstract concepts and should not be instantiated, are called *abstract classes*. An abstract class is a class that can only be subclassed—it cannot be instantiated.

> **Concept**
>
> *The abstract class is a class, which can not directly instantiated, its sub-class can be instantiated.*

To declare that your class is an abstract class, use the keyword abstract before the class keyword in your class declaration:

```
abstract class Number
{
    . . .
}
```

If you attempt to instantiate an abstract class, the compiler will display an error similar to the following and refuse to compile your program:

```
    AbstractTest.java:6: class AbstractTest is an abstract class.
    It can't be instantiated.
    new AbstractTest();
    ^
1 error
```

11.6.2. Abstract Methods

An abstract class may contain *abstract methods*, that is, methods with no implementation. In this way, an abstract class can define a complete programming interface thereby providing its subclasses with the method declarations for all of the methods necessary to implement that programming interface. However, the abstract class can leave some or all of the implementation details of those methods up to its subclasses.

> **Concept**
>
> *Abstract method is a method without definition. It is implemented by subclasses.*

Let's look at an example of when you might want to create an abstract class with an abstract method in it. In an object-oriented drawing application, you can draw circles, rectangles, lines, Beziers, and so on. Each of these graphic objects share certain states (position, bounding box) and behavior (move, resize, draw). You can take advantage of these similarities and declare them all to inherit from the same parent object—GraphicObject.

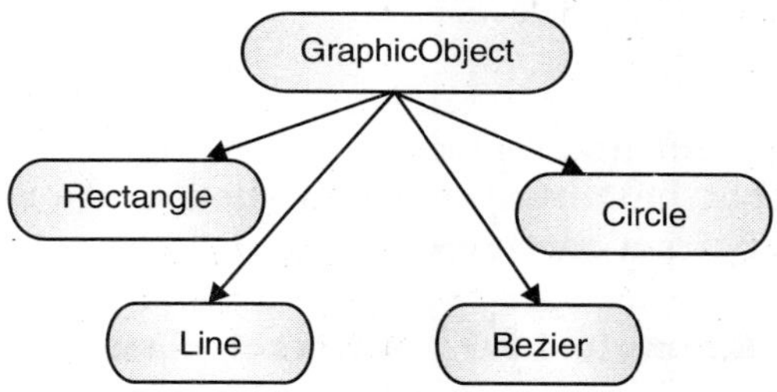

Fig. 11.5. Graphic object

However, the graphic objects are also substantially different in many ways: drawing a circle is quite different from drawing a rectangle. The graphics objects cannot share these types of states or behavior. On the other hand, all GraphicObject's must know how to draw themselves; they just differ in how they are drawn. This is a perfect situation for an abstract superclass.

First you would declare an abstract class, GraphicObject, to provide member variables and methods that were wholly shared by all subclasses such as the current position and the moveTo() method. GraphicObject would also declare abstract methods for methods, such as draw(), that needed to be implemented by all subclasses, but implemented in entirely different ways (no suitable default implementation in the superclass makes sense). The GraphicObject class would look something like this:

```
abstract class GraphicObject
{
     int x, y;
     . . .
void moveTo(int newX, int newY)
```

```
{
      . . .
}
abstract void draw();
}
```

Each non-abstract subclass of GraphicObject, such as Circle and Rectangle, would have to provide an implementation for the draw() method.

```
class Circle extends GraphicObject
{
      void draw()
      {
            . . .
      }
}
class Rectangle extends GraphicObject
{
      void draw()
      {
            . . .
      }
}
```

An abstract class is not required to have an abstract method in it. But any class that has an abstract method in it or that does not provide an implementation for any abstract methods declared in its superclasses *must* be declared as an abstract class.

Example 11.9: Abstract class

```
abstract class Cat
{
    String abstractClassName = "Cat";
    //abstract classes can have variables
    String getAbstractClassName()
    //abstract classes can have methods
 {
          return abstractClassName;
 }

    abstract String getClassName();
    //this must be implemented in any class extending Cat
}
    class Himalayan extends Cat
    {
    //you can not extend from two classes at once
```

```
    String className = "Himalayan";

    public Himalayan() {}

    String getClassName()
    //must have this method,
    // because Cat declared it as an abstract
        {
        return className;
        }

  public static void main(String[] args)
  {

        //Cat percy = new Cat();
        //can not instantiate an abstract class

        Himalayan cappuccino = new Himalayan();

        System.out.println(cappuccino.getAbstractClassName());
        //output is: Cat
        System.out.println(cappuccino.getClassName());
        //output is: Himalayan
   }
  }
```

OUTPUT

```
Cat
Himalayan
```

Example 11.10 Program to display the use of abstract class.

```
// simple abstract class
    abstract class Calc
    {
        void sum(int a,int b)
        {
            System.out.println("Sum is = "+(a+b));
        }
        abstract void sub(int a,int b);
    }
// class inherit with abstract class
    class MyCalc extends Calc
    {
        void sub(int a,int b) // method of Calc class
```

```
        {
            System.out.println("Sub is = "+(a-b));
        }
        public static void main(String aa[])
        {
            MyCalc o = new MyCalc();
            o.sum(100,20);
            o.sub(100,20);
        }
    }
```

In the previous sections we have used the classes called concrete classes which always required to use the abstract classes, because abstract classes can't create there instances. There we distinguish both of by table:

Concrete class	Abstract class
only contain concrete methods	can have abstract and concrete methods.
can create instance	can't create instance, inherit only.
can be inherit.	inherit only

The main advantage of abstract classes are this is used for memory management, memory for methods provided by the abstract class.

Note: All abstract methods of an abstract class must be override in sub classes else sub class also a kind of abstract class.

In the future, abstract classes will not in use, because using classes java support only single inheritance, than one class can extends with one concrete class or one abstract class. In the inheritance with concrete class & in the place of abstract class future will use the concept of interfaces. (interface - it is a pure abstract class.)

11.7. PACKAGES

Many times when we get a chance to work on a small project, one thing we intend to do is to put all java files into one single directory. It is quick, easy and harmless. However if our small project gets bigger, and the number of files is increasing, putting all these files into the same directory would be a nightmare for us. In java we can avoid this sort of problem by using Packages.

Packages are nothing more than the way we organize files into different directories according to their functionality, usability as well as category they should belong to. An obvious example of packaging is the JDK package from SUN (java.xxx.yyy) as shown below in figure 11.6.

Concept

The package is a container, that can contain the class files, interfaces and subpackages.

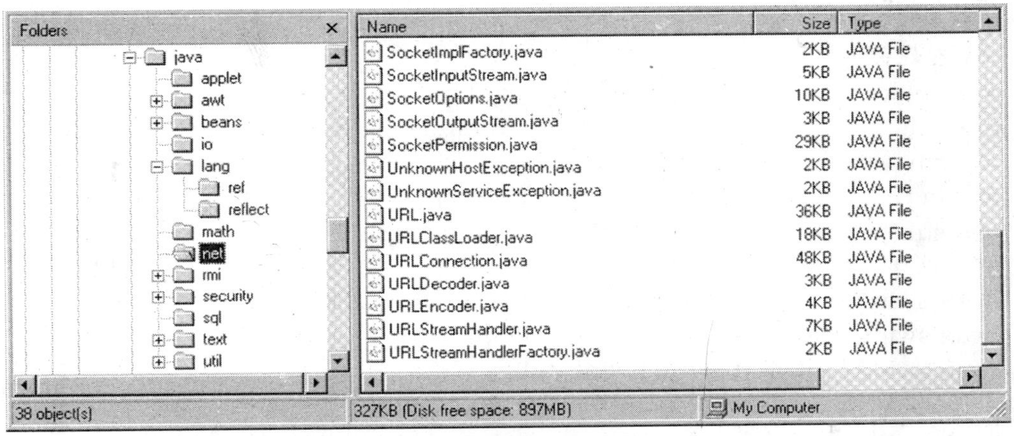

Fig. 11.6. Basic structure of JDK package

There are some comment and definition by some professionals:

- Package is a container, that can contain the class files, interfaces & sub packages.
  – *By Harsh Vardhan, M.Tech, IIT Delhi.*
- Package is a kind of folder that can contain the class files, interfaces & sub folders.
  – *By Ankit Gupta, Project Leader, Rensole Pvt Ltd, Delhi.*
- Packages are like header files in C & C++, which provides relevance functions to use.
  – *By Deepak Agarwal, Tech Lead, Portech India Info Service Pvt Ltd, Noida.*
- Packages are the bunch of related classes, interface and their functions to provide
  - Simplicity,
  - Reusability,
  - Improved performance,
  - and best for security point of view.

  – *By Dr. R.K. Bhatia, Professor, CSE, Gurukul Kangri University, Haridwar.*

Access modifiers in package

Access property	private	default	protected	public
access in same class	Y	Y	Y	Y
sub class of same package	N	Y	Y	Y
non sub class of same package	N	Y	Y	Y
sub class of another package	N	N	Y	Y
non sub class of another package	N	N	N	Y

Note: protected data members can access in sub class of another package but not access in non sub class of another package.

Packages may be pre-defined and user-defined. The most important pre-defined packages, required in most of the programs, are given below:

1. java.lang.*; – for all basic classes, this is default import.
2. java.util.*; – for Utilities classes - Date, Calendar & Collection classes

3. java.awt.*; – for Color,Font,Graphics & all UI components.
4. java.awt.event.*; – for all type of events.
5. java.applet.*; – for Applet programming.
6. java.rmi.*; – for RMI programming.
7. java.rmi.server.*; – for RMI Server programming.
8. java.net.*; – for Network programming.
9. java.sql.*; – for JDBC programming.
10. java.io.*; – for I/O classes
11. javax.swing.*; – for Swing programming. ,(x - extended)
12. javax.swing.event.*; – for Swing Events.
13. javax.swing.table.*; – for Swing table.
14. javax.servlet.*; – for Servlet programming.
15. javax.servlet.http.*; – for http Servlet programming.

11.7.1. How to Create a Package

Suppose we have a file called HelloWorld.java, and we want to put this file in a package **world**. First thing we have to do is to specify the keyword **package** with the name of the package we want to use (**world** in our case) on top of our source file, before the code that defines the real classes in the package, as shown in our HelloWorld class below:

```
package world;

public class HelloWorld {
     public static void main(String[] args) {
     System.out.println("Hello World");
     }
}
```

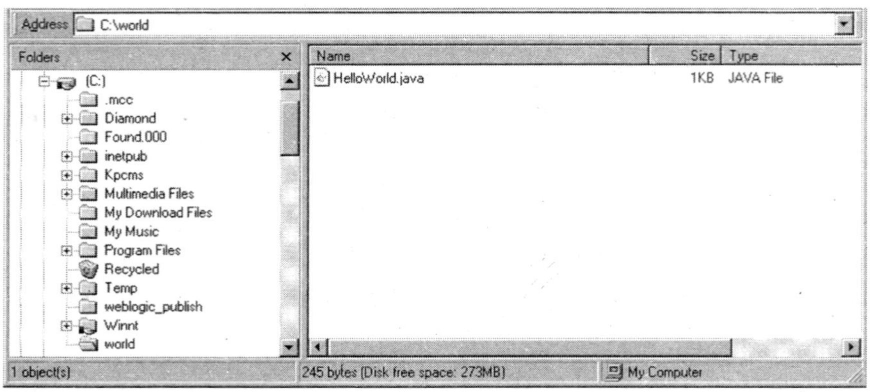

Fig. 11.7. HelloWorld in world package (C:\world\HelloWorld.java)

One thing you must do after creating a package for the class is to create nested subdirectories to represent package hierarchy of the class. In our case, we have the **world** package, which requires only one directory. So, we create a directory **world** and put our HelloWorld.java into it.

Right now we have HelloWorld class inside world package. Next, we have to introduce the world package into our **CLASSPATH**.

11.7.2. Setting up the CLASSPATH

From figure 11.4 we put the package world under C:. So we just set our CLASSPATH as:

```
set CLASSPATH=.;C:\;
```

We set the CLASSPATH to point to 2 places, . (dot) and C:\ directory.

> **Tip:** If you used to play around with DOS or UNIX, you may be familiar with . (dot) and .. (dot dot). We use . as an alias for the current directory and .. for the parent directory. In our CLASSPATH we include this . for convenient reason. Java will find our class file not only from C: directory but from the current directory as well. Also, we use ; (semicolon) to separate the directory location in case we keep class files in many places.

When compiling HelloWorld class, we just go to the world directory and type the command:

```
C:\world\javac HelloWorld.java
```

If you try to run this HelloWorld using **java HelloWorld**, you will get the following error:

```
C:\world>java HelloWorld
Exception in thread "main" java.lang.NoClassDefFoundError: HelloWorld (wrong name:
world/HelloWorld)
at java.lang.ClassLoader.defineClass0(Native Method)
at java.lang.ClassLoader.defineClass(ClassLoader.java:442)
at java.security.SecureClassLoader.defineClass(SecureClassLoader.java:101)
at java.net.URLClassLoader.defineClass(URLClassLoader.java:248)
at java.net.URLClassLoader.access$1(URLClassLoader.java:216)
at java.net.URLClassLoader$1.run(URLClassLoader.java:197)
at java.security.AccessController.doPrivileged(Native Method)
at java.net.URLClassLoader.findClass(URLClassLoader.java:191)
at java.lang.ClassLoader.loadClass(ClassLoader.java:290)
at sun.misc.Launcher$AppClassLoader.loadClass(Launcher.java:286)
at java.lang.ClassLoader.loadClass(ClassLoader.java:247)
```

The reason is right now the HelloWorld class belongs to the package world. If we want to run it, we have to tell JVM about its **fully-qualified class name** (world.HelloWorld) instead of its plain class name (HelloWorld).

```
C:\world>java world.HelloWorld
C:\world>Hello World
```

Note: fully-qualified class name is the name of the java class that includes its package name

To make this example more understandable, let's put the HelloWorld class along with its package (world) be under **C:\myclasses** directory instead. The new location of our HelloWorld should be as shown in figure 11.8.

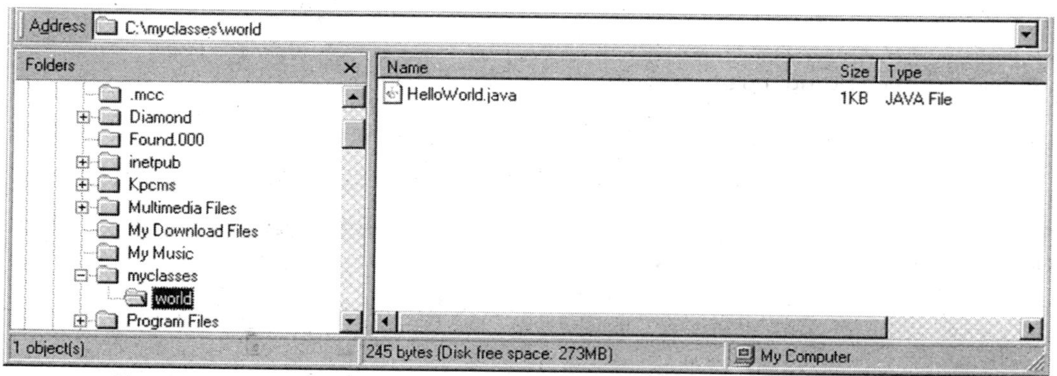

Fig. 11.8. HelloWorld class (in world package) under **myclasses** directory

We just changed the location of the package from C:\world\HelloWorld.java to C:\myclasses\world\HelloWorld.java. Our CLASSPATH then needs to be changed to point to the new location of the package world accordingly.

```
set CLASSPATH=.;C:\myclasses;
```

Thus, Java will look for java classes from the current directory and **C:\myclasses** directory instead.

Someone may ask "Do we have to run the HelloWorld at the directory that we store its class file everytime?". The answer is NO. We can run the HelloWorld from anywhere as long as we still include the package world in the CLASSPATH. For example,

```
C:\>set CLASSPATH=.;C:\;
C:\>set CLASSPATH // see what we have in CLSSPATH
CLASSPATH=.;C:\;
C:\>cd world
C:\world>java world.HelloWorld
Hello World
C:\world>cd ..
C:\>java world.HelloWorld
Hello World
```

11.7.3. Subpackage (package inside another package)

Assume we have another file called **HelloMoon.java**. We want to store it in a subpackage **"moon"**, which stays inside package **world**. The HelloMoon class should look something like this:

```
    package world.moon;

public class HelloMoon
{
    private String holeName = "rabbit hole";
```

```
    public getHoleName() {
    return hole;
    }

    public setHole(String holeName) {
    this.holeName = holeName;
    }
}
```

If we store the package world under C: as before, the **HelloMoon.java** would be **c:\world\moon\HelloMoon.java** as shown in Figure 11.9 below:

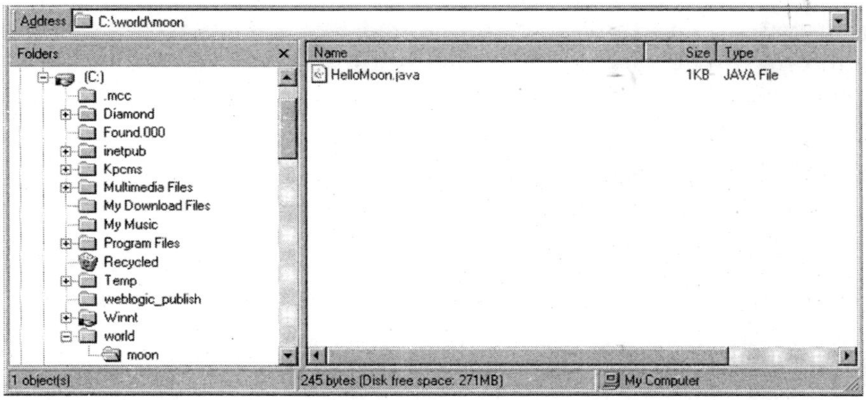

Fig. 11.9. HelloMoon in world.moon package

Although we add a subpackage under package **world**, we still don't have to change anything in our CLASSPATH. However, when we want to reference to the HelloMoon class, we have to use world.moon.HelloMoon as its fully-qualified class name.

11.7.4. How to use Package

There are 2 ways in order to use the public classes stored in package.

1. Declare the fully-qualified class name. For example,

```
...
world.HelloWorld helloWorld = new world.HelloWorld();
world.moon.HelloMoon helloMoon = new world.moon.HelloMoon();
String holeName = helloMoon.getHoleName();
...
```

2. Use an "import" keyword:

```
import world.*;          // we can call any public classes inside the
                         world package
import world.moon.*;     // we can call any public classes inside the
                         world.moon package
import java.util.*;      // import all public classes from java.util
                         package
```

```
import java.util.Hashtable;
                              // import only Hashtable class (not all classes
                              in java.util package)
```

Thus, the code that we use to call the HelloWorld and HelloMoon class should be

```
...
HelloWorld helloWorld = new HelloWorld();
    // don't have to explicitly specify world.HelloWorld anymore
HelloMoon helloMoon = new HelloMoon();
    // don't have to explicitly specify world.moon.HelloMoon anymore
...
```

Note that we can call public classes stored in the package level we do the import only. We can't use any classes that belong to the subpackage of the package we import. For example, if we import package world, we can use only the HelloWorld class, but not the HelloMoon class.

11.7.5. Using Classes Stored in jar File

Jar files are the place where we put a lot of files together. We compress these files and make them as a single bundle. Jar files may also include directories, subdirectories to represent class and package hierarchy. Normally, we can see what is inside a jar file by using the command **jar -tvf fileName.jar** as shown in figure 11.10.

From figure 11.10, we see that there is a class called javax.servlet.http.Cookie. We can call this class by

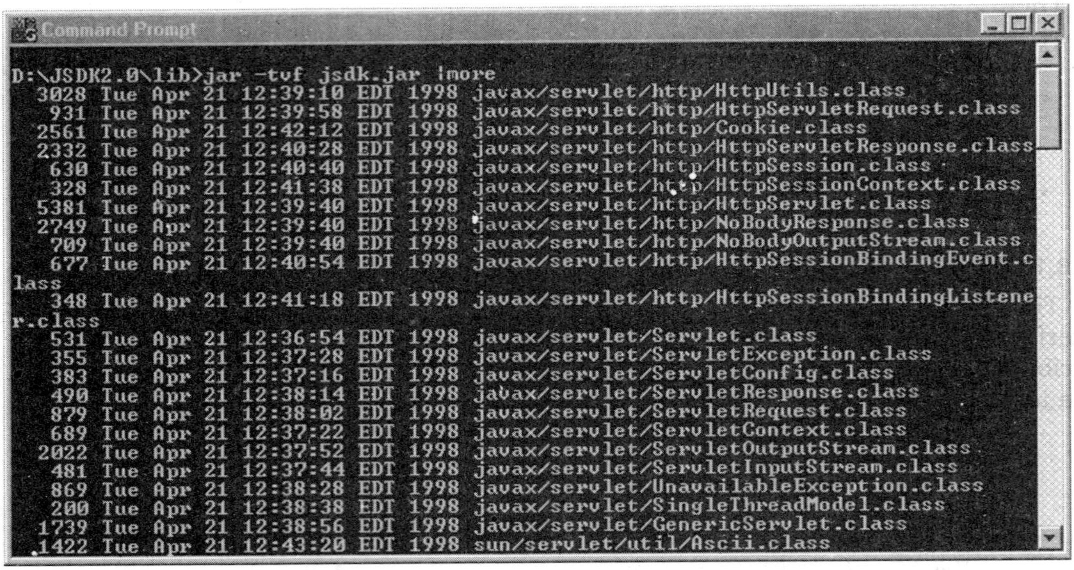

Fig. 11.10. JSDK package viewed by **jar -tvf jsdk.jar** command

```
import javax.servlet.http.Cookie; // import only Cookie class or
import javax.servlet.http.*;       // import the whole javax.servlet.http
                                      package
```

But we have to include this package in the CLASSPATH as well.

```
set CLASSPATH=.;D:\JSDK2.0\lib\jsdk.jar;
```

Note that if the package is stored inside a jar file, we have to include the jar file with its extension (.jar) in the CLASSPATH. However, if the package is a plain directory, we just put the name of directory into the CLASSPATH.

11.8. INTERFACE

In English, an *interface* is a device or system that unrelated entities use to interact. According to this definition, a remote control is an interface between you and a television set, the English language is an interface between two people, and the protocol of behavior enforced in the military is the interface between people of different ranks. Similarly, a Java interface is a device that unrelated objects use to interact with one another. Java interfaces are probably most analogous to protocols (an agreed-upon behavior). In fact, other object-oriented languages have the functionality of Java's interfaces, but they call their interfaces *protocols*.

A Java interface defines a set of methods but does not implement them. A class that implements the interface agrees to implement all of the methods defined in the interface, thereby agreeing to certain behavior.

> **Concept**
>
> *An interface is a named collection of method declarations (without definitions). An interface can also include constant declarations.*

There are some comment and definition by some professionals:

- Interface is the collection of abstract methods & final variables.
  – *By Ankit Gupta, Project Leader, Rensole Pvt Ltd, Delhi.*
- Interface is a kind of abstract class that can contain only abstract methods & final variable.
  – *By Harsh Vardhan, M.Tech, IIT Delhi.*
- Interface is a pure abstract class.
  – *By Deepak Agarwal, Tech Lead, Portech India Info Service Pvt Ltd, Noida*

11.8.1. Defining an Interface

The following figure shows that an interface definition has two components: the interface declaration and the interface body. The interface declaration declares various attributes about the interface such as its name and whether it extends another interface. The interface body contains the constant and method declarations within that interface.

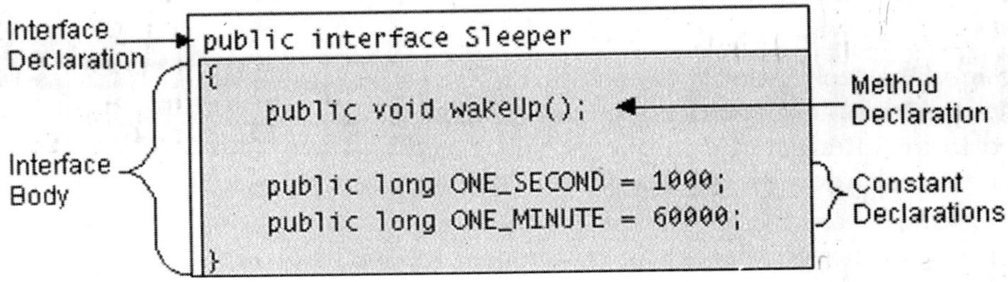

The Interface Declaration

An interface is primarily created like a class, the declaration for the Sleeper interface uses the two required elements of an interface declaration—the interface keyword and the name of the interface—plus the public access modifier:

```
public interface Sleeper
{
. . .
}
```

An interface declaration can have one other component: the public access specifier and a list of superinterfaces. The full interface declaration looks like this figure:

`public`	Makes this interface public.
`interface InterfaceName`	Class cannot be instantiated.
`Extends SuperInterfaces`	This interface's superinterfaces.

`{`
 `InterfaceBody`
 `}`	

The public access specifier indicates that the interface can be used by any class in any package. If you do not specify that your interface is public, then your interface will be accessible only to classes that are defined in the same package as the interface.

An interface can extend other interfaces just as a class can extend or subclass another class. However, while a class can extend only one other class, an interface can extend any number of interfaces. The list of superinterfaces is a comma-delimited list of all of the interfaces extended by the new interface. An interface inherits all constants and methods from its superinterfaces, unless the interface hides a constant with another of the same name or redeclares a method with a new method declaration.

The interface body

The interface body contains method declarations for all of the methods included in the interface. Here's the body of the Sleeper interface:

```
public interface Sleeper {
     public void wakeUp();

     public long ONE_SECOND = 1000;
     public long ONE_MINUTE = 60000;
}
```

The method declaration for wakeUp is followed by a semicolon (;) because an interface does not provide implementations for the methods declared within it. All methods declared in an interface are implicitly public and abstract. The use of these modifiers on a method declaration in an interface is discouraged as a matter of style.

An interface can contain constant declarations in addition to method declarations. The Sleeper interface declares two constants. All constant values defined in an interface are implicitly public, static, and final. The use of these modifiers on a constant declaration in an interface is discouraged as a matter of style.

Any class can use an interface's constants from the name of the interface, like this:

```
Sleeper.ONE_SECOND
```

An interface is mostly used to lay a foundation for other classes. For this reason, it is the prime candidate for class derivation. To derive from an interface, you must use the keyword ***implements*** as shown in example. Here is an example of a class named SportBall that derives from an interface named ISportType:

```
public class SportBall implemets ISportType
{
        int players;
        string sport;
}
```

Just as you can derive a class from an interface, you can create an interface that itself is based on another interface. Here is an example:

```
public interface ISportType extends IBall
{

}
```

Java doesn't allow multiple inheritance which is the ability to create a class based on more than one class. Multiple inheritance is allowed only if the bases are interfaces. To create multiple inheritance, separate the names of interface with a comma. Here is an example:

```
public interface ISportType extends IBall, ICourtDimensions
{

}
```

Here is some example program to implement the interface, as well as multiple inheritance:

Example 11.11 Program to use simple interface Calc in the main class.

```
// simple interface
interface Calc
{
        void sum(int a,int b);
        void sub(int a,int b);
}
// class implements with Interface
class MyCalc implements Calc
{
        public void sum(int a,int b) // method of Calc
        {
            System.out.println("Sum is = "+(a+b));
        }
```

```
        public void sub(int a,int b) // method of Calc
        {
            System.out.println("Sub is = "+(a-b));
        }
        public static void main(String aa[])
        {
            MyCalc o = new MyCalc();
            o.sum(100,20);
            o.sub(100,20);
        }
    }
```

OUTPUT

```
    Sum is = 120
    Sub is = 80
```

Now you may be little bit confuse about interface and abstract class, so don't worry here are some differences between them:

Abstract class	Interface
It contain abstract and concrete methods.	It contain only abstract methods.
Using abstract classes java support only single inheritance.	Using interfaces java support multiple inheritance also.
All attributes of class by default is normal.	All attributes of interface by default is final.
Default access modifier of abstract class is default.	Default access modifier of interface is public.
It can have constructors.	Interface doesn't have constructors.
It uses extends keyword for inheritance.	It uses implements keyword for inheritance.

Note: when you will use the concept of inheritance you must remember the which keywords to use where, see the following table:

Inheritance	Keyword
class to class	extends keyword.
interface to interface	extends keyword
interface to class	implements keyword.

There are some restrictions while we are using interface, which are given below:

- All methods of an interface must be override in child class else child class is a kind of abstract class.
- All methods of an interface must be override with public access modifier else we have error - weaker access privileges - was public.

Example 11.12: Simple Interface

```
interface Printem
{
    void printText();
}

class class1 implements Printem
{
    public void printText()
 {
    System.out.println("Hello from Java!");
    }
}

public class interfaces
{
    public static void main(String[] args)
    {
    class1 object1 = new class1();
    object1.printText();
    }
}
```

Example 11.13: Implementation of multiple interfaces

```
interface Add
{
    public int sum();
}

interface Multiply
{
    public int mul();
}

class calculate implements Add, Multiply
{
    int a,b;
    public calculate( int x, int y)
    {
        a = x;
        b = y;
    }

    public int sum()
    {
```

```
        return ( a + b );
    }
        public int mul()
    {
        return ( a * b );
    }
}

class MyInterface
{
    public static void main(String[] arg)
    {

        calculate cal = new calculate(5,10);

        Add A = (Add) cal;          // casting
        System.out.println("Sum: " + A.sum());

        Multiply M = (Multiply) cal;
        System.out.println("Multiplication:" + M.mul());
    }
}
```

OUTPUT

```
Sum: 15
Multiplication: 50
```

Example 11.14: Multiple implementations of an interface

```
interface calculate
{
    public int compute( int x, int y);
}

class Add implements calculate
{
    public int compute(int a,int b)
    {
        return ( a + b );
    }
    }

class Multiply implements calculate
{
    public int compute(int a,int b)
    {
```

```
        return ( a * b );
    }
}

class NewInterface
{
    public static void main(String[] arg)
    {

        Add A = new Add();
        Multiply M =new Multiply();

        calculate cal;

        cal = (calculate) A; //casting
        System.out.println("Sum: " + cal.compute(2,3));

        cal = (calculate) M; //casting
        System.out.println("Multiplication:" + cal.compute(2,3));
    }
}
```

OUTPUT

```
Sum: 5
Multiplication: 6
```

SUMMARY

- Java is a pure object oriented language and all object oriented programming based on the classes and objects and in we can't create any single program without class.
- Java supports Inheritance but not directly support multiple inheritance.
- Java has also the concept of abstract class and abstract methods.
- Abstract classes cannot directly be instantiated, its subclasses can be instantiated.
- Abstract method is a method with no implementation, its implementation done by the subclasses.
- Java has given a concept of package to remove the collision of class name.
- Java has no multiple inheritance but we can implement the concept of multiple inheritance using interface.

SOLVED PROBLEMS

Q. 11.1. What's the difference between abstraction and encapsulation?

Ans. Abstraction and encapsulation are two quite separate concepts in Java. Abstraction is a technique that is used to represent common functionality amongst a set of classes. An analogy is to look at a set of vehicles: Car, Truck and Bus have some common features that all road vehicles share. From a Java perspective, the mechanics of turning four wheels, an engine, could be handled in abstract form through a common superclass. Abstraction allows subclasses to share common code without duplication. Encapsulation refers to binding the attributes and behaviours in a single unit.

Q. 11.2. What do you mean by Abstract classes, abstract methods?

Ans. Abstract classes are Java classes which are declared as abstract and cannot be instantiated. They can contain non-abstract as well as abstract methods. Abstract methods are methods which are declared with the keyword "abstract" and their definition is deferred to the subclasses. A class having at least one abstract method must be declared abstract.

Q. 11.3. What are interfaces? or How they support multiple inheritance in Java?

Ans. Interfaces are classes which are declared with the keyword interface and contain only abstract methods and static variables. The abstract methods of an interface are by default public and so are the variables. Since Java does not support multiple inheritance as C++, interfaces are a way by which we can achieve it in Java. A class can implement multiple interfaces. However, it must provide the implementation of all the abstract methods of the interfaces it has inherited.

Q. 11.4. Why Java not directly supports Multiple Inheritance?

Ans. Java does not support Multiple Inheritence because, consider an example: A class named 'Super' has a method display(), we are deriving two classes 'SuperA & SuperB' from the 'Super' class. Now both SuperA & SuperB has the method display().

Now you are deriving a class 'SubOfAB' from 'SuperA' & 'SuperB', then how the class 'SubOfAB' uses the method display()-

(*a*).through the class SuperA (or)

(*b*).through the class SuperB

Due to this drawback Java does not have multiple inheritance.

Q. 11.5. If a method is declared as protected, where may the method be accessed?

Ans. A protected method may only be accessed by classes or interfaces of the same package or by subclasses of the class in which it is declared.

Q. 11.6. Is println overloading or overriding?

Ans. The println() method is an example of overloading because several methods in the class have the same name and return the same type. In this case, println(boolean), println(int) and println(String) all have the same basic method name, "println", and all return void. The only part of the method signature that varies is the type of the argument (including none), which is enough for the Java interpreter to identify the appropriate method to call at runtime.

Q. 11.7. Why can't a class be declared private?

Ans. If a class were declared private, no one could access it, including the runtime. Therefore the class could never be instantiated.

Q. 11.8. If you can add methods and fields to a subclass, why can't you remove them?

Ans. If you were to remove methods and fields from a subclass, then the subclass could no longer be used as an instance of its superclass. In fact the subclass might not have any relation at all to the superclass if all the superclass's methods and fields were removed.

Q. 11.9. What's the basic difference between an abstract class and an interface?

Ans. An interface can be implemented by a class which is already a subclass of another class. A class which is already a subclass of another class cannot extend an abstract class as well. Furthermore, an abstract class can contain non-abstract methods. An interface cannot.

Q. 11.10. What is the package?

Ans. The package is a Java namespace or part of Java libraries. The Java API is grouped into libraries of related classes and interfaces; these libraries are known as packages.

Q. 11.11. What is an enumeration?

Ans. An enumeration is an interface containing methods for accessing the underlying data structure from which the enumeration is obtained. It is a construct which collection classes return when you request a collection of all the objects stored in the collection. It allows sequential access to all the elements stored in the collection.

Q. 11.12. What is the Vector class?

Ans. The Vector class provides the capability to implement a growable array of objects.

Q. 11.13. What is the Set interface?

Ans. The Set interface provides methods for accessing the elements of a finite mathematical set. Sets do not allow duplicate elements.

Q. 11.14. What is Dictionary class?

Ans. The Dictionary class is the abstarct super class of Hashtable and Properties class.

Dictionary provides the abstarct functions used to store and retrieve objects by key-value. This class allows any object to be used as a key or value.

Q. 11.15. What is the Hashtable class?

Ans. The Hashtable class implements a hash table data structure. A hash table indexes and stores objects in a dictionary using hash codes as the objects' keys. Hash codes are integer values that identify objects.

Q. 11.16. What is the Properties class?

Ans. The properties class is a subclass of Hashtable that can be read from or written to a stream.

It also provides the capability to specify a set of default values to be used if a specified key is not found in the table. We have two methods load() and save().

Q. 11.17. Can null value be added to a List?

Ans. Yes.

A Null value may be added to any List.

Q. 11.18. What is the ResourceBundle?

Ans. The ResourceBundle class also supports internationalization.

ResourceBundle subclasses are used to store locale-specific resources that can be loaded by a program to tailor the program's appearence to the paticular locale in which it is being run. Resource Bundles provide the capability to isolate a program's locale-specific resources in a standard and modular manner.

Q. 11.19. How are Observer Interface and Observable class, in java.util package, used?

Ans. Objects that subclass the Observable class maintain a list of Observers. When an Observable object is updated it invokes the update() method of each of its observers to notify the observers that it has changed state. The Observer interface is implemented by objects that observe Observable objects.

Q. 11.20. Which java.util classes and interfaces support event handling?

Ans. The EventObject class and the EventListener interface support event processing.

Q. 11.21. Does java provide standard iterator functions for inspecting a collection of objects?

Ans. The Enumeration interface in the java.util package provides a framework for stepping once through a collection of objects. We have two methods in that interface.

```
public interface Enumeration {
boolean hasMoreElements();
```

```
Object nextElement();
}
```

Q. 11.22. The Math.random method is too limited for my needs. How can I generate random numbers more flexibly?

Ans. The random method in Math class provide quick, convienient access to random numbers, but more power and flexibility use the Random class in the java.util package.

```
double doubleval = Math.random();
```

The Random class provide methods returning float, int, double, and long values.

```
nextFloat() // type float; 0.0 <= value < 1.0
nextDouble() // type double; 0.0 <= value < 1.0
nextInt() // type int; Integer.MIN_VALUE <= value <= Integer.MAX_VALUE
nextLong() // type long; Long.MIN_VALUE <= value <= Long.MAX_VALUE
nextGaussian() // type double; has Gaussian("normal") distribution with mean 0.0 and standard
deviation 1.0)
Eg. Random r = new Random();
float floatval = r.nextFloat();
```

Q. 11.23. How can we get all public methods of an object dynamically?

Ans. By using getMethods(). It return an array of method objects corresponding to the public methods of this class.

getFields() returns an array of Filed objects corresponding to the public Fields(variables) of this class.

getConstructors() returns an array of constructor objects corresponding to the public constructors of this class.

Q. 11.24. What is CLASSPATH?

Ans. CLASSPATH is an environment variable which tells the Java compiler and the JVM where to look for Java class files.

In Java 1.2 and later, the CLASSPATH variable is less critical because Java knows how to find the default classes and because Java loads classes from the current directory by default.

If you do not want to use an environment variable, you can use the -classpath argument to the Java compiler or the JVM as such:

```
javac -classpath . JavaProg
```

To add a jar file to your CLASSPATH, include the entire pathname of the jar file — including the filename.

EXERCISES

Q. 11.1. Consider the following two classes:

```
public class classA
{
    public void methodOne (int i)
        {       }
    public void methodTwo (int i)
        {       }
    public static void methodThree (int i)
        {       }
    public static void methodFour (int i)
```

```
        {           }
}
public class classB
{
        public static void methodOne (int i)
            {           }
        public void methodTwo (int i)
            {           }
        public void methodThree (int i)
            {           }
}
```

(*a*) which method overrides a method in the superclass?

(*b*) which method hides a method in the superclass?

(*c*) what do the other methods do?

Q. 11.2. Explain Inheritance with types.

Q. 11.3. What is the mean of derived class?

Q. 11.4. What are the structures of different type of inheritance?

Q. 11.5. What is the meant by method overriding?

Q. 11.6. Can we override the constructors?

Q. 11.7. What do you mean by abstract class and Abstract method and why we need it?

Q. 11.8. What is the concept of package? What are the advantages of package?

Q. 11.9. What do you mean by CLASSPATH?

Q. 11.10. What do you mean by jar file?

Q. 11.11. What is Interface? Explain.

Q. 11.12. What are the difference between abstract class and interface?

Q. 11.13. How can we define an Interface? Give an example.

❑❑❑

Chapter 12

Advance Programming Concept

CHAPTER OUTLINE

12.1. EXCEPTION HANDLING

12.1.1. Exception Handling Fundamentals

The term *exception* is shorthand for the phrase "exceptional event."

When an error occurs within a method, the method creates an object and hands it off to the runtime system. The object, called an *exception object*, contains information about the error, including its type and the state of the program when the error occurred. Creating an exception object and handing it to the runtime system is called *throwing an exception*.

After a method throws an exception, the runtime system attempts to find something to handle it. The set of possible "somethings" to handle the exception is the ordered list of methods that had been called to get to the method where the error occurred. The list of methods is known as the *call stack* (see figure 12.1).

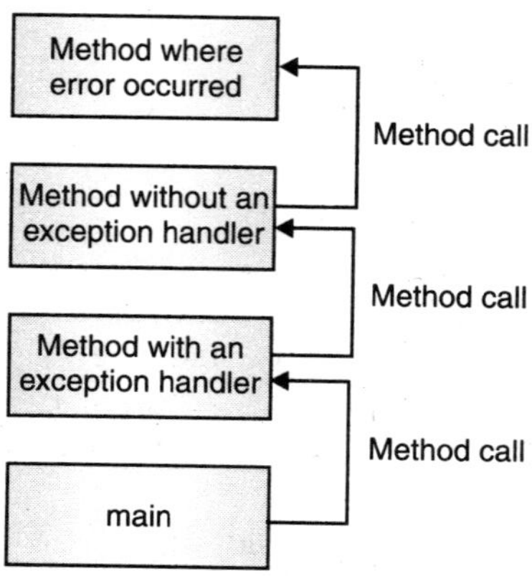

Fig. 12.1. The call stack

> **Concept**
> *An exception is an event, which occurs during the execution of a program, that disrupts the normal flow of the program's instructions.*

The runtime system searches the call stack for a method that contains a block of code that can handle the exception. This block of code is called an *exception handler*. The search begins with the method in which the error occurred and proceeds through the call stack in the reverse order in which the methods were called. When an appropriate handler is found, the runtime system passes the exception to the handler. An exception handler is considered appropriate if the type of the exception object thrown matches the type that can be handled by the handler.

The exception handler chosen is said to *catch the exception*. If the runtime system exhaustively searches all the methods on the call stack without finding an appropriate exception handler, as shown in the figure 12.2 the runtime system (and, consequently, the program) terminates.

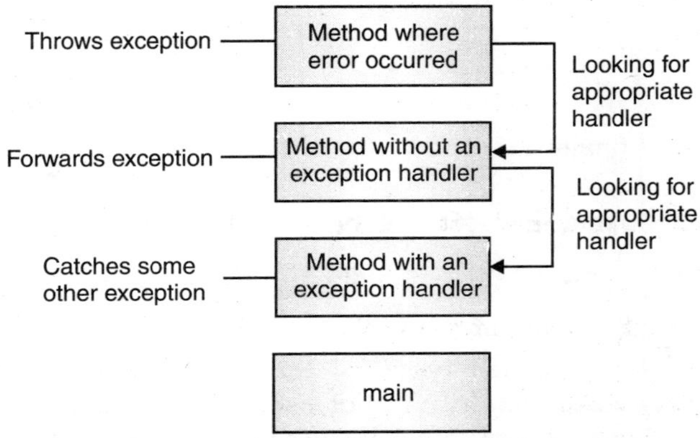

Fig. 12.2. Searching the call stack for the exception handler

> **Concept**
> *An exception is a condition that is caused by a runtime error in the program.*

A Java exception is an object that describes an exceptional (that is, an error) condition that has occurred in a piece of code. When an exceptional condition arises, an object representing that exception is created and thrown in the method that caused the error. That method may choose to handle that exception or pass it on. Either way, at some point, the exception is caught and processed.

Briefly, here is how they work. Program statements that should be monitored for exceptions are contained with in a **try block**. If an exception occurs within the try block, it is **thrown**. Our code can catch this exception (using **catch**) and handle it in some rational manner. System generated exceptions are automatically thrown by the Java run-time system. To throw an exception manually, use the keyword **throw**. An exception that is thrown out of a method must be specified as such by a throws clause. Any code that absolutely must be executed before a method returns is put in a **finally block**.

This is the general form of an exception-handling block:

```
try
{
      //block of code to be monitored for errors
```

```
    }

    catch (ExceptionType1 exOb )
    {
          //exception handler for ExceptionType1
    }
    catch (ExceptionType2 exOb )
    {
    //exception handler for ExceptionType2
    }

    //...
    finally
    {
    //block of code to be executed before try block ends
    }
```

12.1.2. Types of Exceptions

In Java, all types of exceptions are subclasses of the built-in class **Throwable**. **Throwable** has two subclasses that partition exceptions. One is Exception and the other is Error. There is an important subclass of Exception called **RuntimeException**.

Throwable serves as the base class for an entire family of classes, declared in java.lang, that your program can instantiate and throw. A small part of this family is shown in figure 12.3.

As you can see in figure 12.3, Throwable has two direct subclasses, Exception and Error. Exceptions (members of the Exception family) are thrown to signal abnormal conditions that can often be handled by some catcher, though it's possible they may not be caught and therefore could result in a dead thread. Errors (members of the Error family) are usually thrown for more serious problems, such as OutOfMemoryError, that may not be so easy to handle. In general, code you write should throw only exceptions, not errors. Errors are usually thrown by the methods of the Java API, or by the Java virtual machine itself.

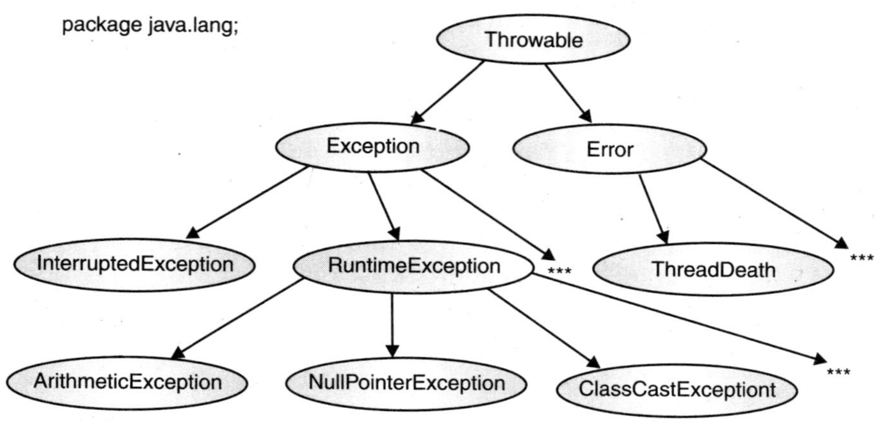

Fig. 12.3. A Partical view of the Throwable family

In addition to throwing objects whose classes are declared in java.lang, you can throw objects of your own design. To create your own class of throwable objects, you need only declare it as a subclass of some member of the Throwable family. In general, however, the throwable classes you define should extend class Exception. They should be "exceptions."

12.1.3. Divide-By-Zero Exception

This small program has an expression that causes a divide-by-zero exception.

```
class DivideByZero
{
    public static void main (String args[])
    {
        int d = 0;
        int a = 42 / d;
    }
}
```

The Java run-time system constructs a new exception object when it detects an attempt to divide-by-zero. It then throws this exception. In the above example, there is no exceptional handler to catch the exception. The default handler provided by the Java run-time system will process any exception that is not caught by a program. The output of the above program when executed by the standard Java JDK runtime interpreter:

```
java.lang.ArithemeticException:/ by zero
at DivideByZero.main(DivideByZero.java:4)
```

Although the default exception handler is provided by Java run-time system, you will usually like to handle exception yourself. The following program includes a try block and a catch clause, which processes the ArithmeticException generated by the division-by-zero:

Example 12.1: DivideByZero Exception

```
class DivideByZero
{
    public static void main (String args[])
    {
    int d, a;
    try //monitor a block of code
    {
    d = 0;
    a = 42 / d;
    System.out.println("This will not be printed. ");
    }
    catch(ArithmeticException e)
    {
    System.out.println("Division by zero.");
    }
    System.out.println("After catch statement.");
```

```
}
```

OUTPUT

```
Division by zero.
After catch statement
```

The call to println() inside the try block is never executed. Once an exception is thrown, program control transfers out of the try block into the catch block. Once the catch statement has executed, the program continues with the next line in the program following the entire try/catch mechanism.

12.1.4. Multiple Catch Clause

In some situations, we need to maintain more than one catch clause for a block of code. In such situations when an exception is thrown, each catch statement is inspected in order and the first one whose type matches that of the exception is executed. After one catch statement executes, the others are bypassed, and execution continues after the try/catch block.

Example 12.2: Multiple catch statements.

```
class Multicatch
{
    public static void main (String args[])
    {
    try
    {
    int a = args.length;
    System.out.println("a = " + a);
    int b = 42 / a;
    int c[ ] = { 1 };
    c[42] = 99;
    }
    catch (ArithmeticException e)
    {
        System.out.println("Divide by 0 " );
    }
    catch (ArrayIndexOutOfBoundException e)
    {
        System.out.println("Array index is out of bounds");
    }
        System.out.println("After try/catch blocks");
    }
}
```

SAMPLE OUTPUT

```
1. C:\> java Multicatch
a = 0
Divide by 0
```

```
After try/catch blocks
2. C:\> java Multicatch TestArg
a = 1
Array index is out of bounds
After try/catch blocks
```

When we use multiple catch statements, it is important to remember that exception subclasses must come before any of their superclasses. This is because a catch statement that uses a superclass will catch exceptions of that type plus any of its subclasses. Thus, a subclass would never be reached if it came after its superclass. Further, in Java unreachable code is an error.

12.1.5. Nested try statements

In Java, a try statement can be inside the block of another try. Each time a try statement is entered, the context of that exception is pushed on the stack. If an inner try statement does not have a matching catch statement for a particular exception, the stack is unwound and the next try statement's catch handlers are inspected for a match. This continues until one of the catch statements succeeds, or until all of the nested try statements are exhausted. If no catch statements match, then the Java run-time system will handle the exception.

Example 12.3: Nested try statements.

```
class NestTry
{
    public static void main (String args[])
    {
    try
    {
        int a = args.length;
        /* if no command-line args are present
        the following statement will generate
        a divide-by-zero exception*/
        int b = 42 / a;
        System.out.println("a = " + a);
        try //nested try block
        {
        /* if one command line arg is used,then a divide-by-zero exception
        will be generated by the following code*/
        if (a = = 1) a = a / (a - a); // division by zero

        /* if two command line arg is used ,
        then generate an out of bounds exception .*/
        if( a = 2 )
            {
            int c [] = { 1 };
            c[ 42 ] = 99; //generate an out of bound exception
            }
```

```
        }
    catch (ArrayIndexOutOfBoundsException e)
    {
        System.out.println("Array index out of bounds ");
    }
    }
    catch (ArithmeticException e)
    {
    System.out.println("Divide by zero");
    }
}
}
```

SAMPLE OUTPUT

```
1. C:\> java NestTry
Divide by zero.
2. C:\> java NestTry One
a = 1
Divide by zero
3. C:\> java NestTry One Two
a = 2
Array index out of bounds.
```

12.1.6. Throw and Throws

So far, we have been catching exceptions that are thrown by the Java run-time system.

It is possible for our program to throw an exception explicitly using **throw** statement. The usual form of throw is shown here.

```
throw throwableinstance;
```

Here *throwableinstance* must be an object of type **Throwable** or subclass of **Throwable.** The flow of execution stops immediately after the throw statement any subsequent statements are not executed. The nearest enclosing try block is inspected to see if it has a catch statement that matches the type of exception. If it does find a match, control is transferred to that statement. If not then the next enclosed try statement is inspected, and so on......If no matching catch is found, then the default exception handler halts the program.

Here is the sample program.

Example 12.4: Demonstrate throw.

```
class ThrowDemo {
    static void demoproc() {
    try {
        throw new NullPointerException("demo");
    }
    catch(NullPointerException e) {
```

```
    System.out.println("Caught inside demoproc.");
    throw e; // re-throw the exception
    }
  }

  public static void main(String args[]) {
    try {
        demoproc();
    }
    catch(NullPointerException e) {
    System.out.println("Recaught " );
    }
  }
}
```

OUTPUT

```
Caught inside demoproc.
Recaught
```

Here, **new** is used to construct an instance of **NullPointerException**. All of Java's built-in run-time exceptions have two constructors: one with no parameter, and one that takes a string parameter that describes the exception.

If a method is capable of causing an exception that it does not handle, it must specify this behavior so that callers of that method can guard themselves against that exception. We can do this by including a **throws** class in the methods declaration as follows.

```
type method-name(parameter-list) throws exception-list
{
//Body of the method.
}
```

exception-list is a list of types of exceptions that a method might throw. All exceptions except those of type **Error** or **RuntimeException**, or any of their subclasses need not be listed.

Example 12.5: Throws

```
class ThrowsDemo
{
   static void throwOne() throws IllegalAccessException
        {
        System.out.println("Inside throwOne.");
        throw new IllegalAccessException("demo");
        }
   public static void main(String args[])
   {
   try
```

```
    {
        throwOne();
    }
    catch (IllegalAccessException e)
    {
        System.out.println("Caught ");
        }
    }
}
```

OUTPUT

```
Inside throwOne.
Caught
```

12.1.7. Finally

Finally creates a block of code that will be executed after the try/catch block has completed and before the code following the try/catch block. The **finally** block will execute whether or not an exception is thrown, even if no catch statement matches the exception. Here is an example:

Example 12.6: Demonstrate finally.

```
class FinallyDemo
{
    // Through an exception out of the method.
    static void procA()
    {
        try
        {
        System.out.println("inside procA");
        throw new RuntimeException("demo");
        }
        finally
        {
        System.out.println("procA's finally");
        }
    }

    // Return from within a try block.
    static void procB()
    {
        try
            {
            System.out.println("inside procB");
            return;
            }
    finally
```

```
    {
        System.out.println("procB's finally");
    }
    }

    // Execute a try block normally.
    static void procC()
    {
        try
        {
        System.out.println("inside procC");
            }
    finally
        {
        System.out.println("procC's finally");
        }
    }

    public static void main(String args[])
    {
        try
        {
            procA();
        }
        catch (Exception e)
        {
            System.out.println("Exception caught");
        }
        procB();
        procC();
    }
}
```

OUTPUT

```
inside procA
procA's finally
Exception caught
inside procB
procB's finally
inside procC
procC's finally
```

12.1.8. Built-in Exceptions

Inside the standard package **java.lang**, Java defines several exception classes. In Java there are some exceptions (exceptions derived from **RuntimeException**) which compiler does not check to see if a method throws or handles these exceptions. Such exceptions are called unchecked exceptions. All the exceptions defined by java.lang that must be included in method's throws list are called checked exceptions.

Unchecked RuntimeException Subclasses

Exception	Meaning
ArithmeticException	Arithmetic error, such as divide by zero
ArrayIndexOutOfBoundsException	Array index is out of bounds
ArrayStoreException	Assignment to an array element of an in compatible type
ClassCastException	Invalid cast
IllegalArgumentException	Illegal argument used to invoke a method
IllegalMonitorStateException	Illegal monitor operation, such as waiting on an unlocked thread
IllegalStateException	Environment or application is in incorrect state
IllegalThreadStateException	Requested operation not compatible with current thread state
IndexOutOfBoundsException	Some type of Index is out of bounds
NegativeArraySizeExeption	Array created with the negative size
NullpointerException	Invalid use of null exception
NumberFormatException	Invalid conversion of a string to a numeric format
SecurityException	Attempt to violate security
StringIndexOutOfBounds	Attempt to index outside the bounds of a string
UnsupportedOperationException	An unsupported operation was encountered

Checked Exceptions Defined in java.lang

Exception	Meaning
ClassnotFoundException	Class not found
CloneNotSupportedException	Attempt to clone an object that does not implement the cloneable interface
IllegalAccessException	Access to a class is denied
InstantiationException	Attempt to create an Object of an abstract class or interface
InterruptedException	One thread has been interrupted by another thread
NoSuchFieldException	A requested field does not exist
NoSuchMethodException	A requested method does not exist

12.2. MULTITHREADED PROGRAMMING

12.2.1. Multitasking

The earliest computers did one thing at a time. A lab computer might calculate detonation waves for one of the physicists. Next it would calculate mixing ratios for a chemist. All programs were run sequentially, one at a time; and each had full control of the computer. However two programs couldn't run at the same time. This is called batch processing, and it's a very efficient way to get maximum usage out of a very expensive computer. Almost all the computer's time is spent actually calculating jobs and not switching between one program and another. However batch processing can be very annoying when your differential equation integration program that would take two seconds of CPU time gets stuck in line behind the physics department's nuclear structure modeling project that's going run for the next three days.

Time sharing operating systems were invented to allow multiple people to use very expensive computer at the same time. On a time sharing system many people can run programs at the same time. The operating system is responsible for splitting the time among the different programs that are running. That way you can finish integrating your differential equation while the physics department's nuclear modeling program is still churning away. The physics department nuclear modeling program might take two weeks to run instead of three days, but everyone with the shorter programs is happy (at least until the physicists figured out how to hack the computer so that it only ran their program).

> **Concept**
>
> *The multitasking is the ability of single processor to perform more than operation at the same time.*

Once systems allowed different users to run programs at the same time it was a short step to letting the same user run multiple programs simultaneously. Each running program (generally called a *process*) had its own memory space, its own set of variables, its own stack and heap, and so on. One process could produce or call another process, but afterwards the two processes behaved more or less independently. Mechanisms like inter procedure calls (IPC) were developed to allow processes to interact with each other, but such interaction was expensive and difficult. And this is where matters rested for about twenty years.

However it's not just users that want to do different things at the same time. Many programs also need to do several things at once. A web browser, for instance, can print a file in the background while it downloads a page in one window and formats the page as it downloads. The ability of an individual program to do more than one thing at the same time is most efficiently implemented through threads.

12.2.2. Thread Vs Processes

Both threads and processes are methods of parallelizing an application. However, processes are independent execution units that contain their own state information, use their own address spaces, and only interact with each other via interprocess communication mechanisms (generally managed by the operating system). Applications are typically divided into processes during the design phase, and a master process explicitly call sub-processes when it makes sense to logically separate significant application functionality. Processes, in other words, are an architectural construct.

> **Concept**
>
> *A single process may contain more than one threads.*

By contrast, a thread is a coding construct that doesn't affect the architecture of an application. A single process might contains multiple threads; all threads within a process share the same state and same memory space, and can communicate with each other directly, because they share the same variables.

Threads typically are called/invoked for a short-term benefit that is usually visualized as a serial task, but which doesn't have to be performed in a linear manner (such as performing a complex mathematical computation using parallelism, or initializing a large matrix), and then are absorbed when no longer required. The scope of a thread is within a specific code module—which is why we can bolt-on threading without affecting the broader application.

12.2.3. What is thread?

A *thread* can be loosely defined as a separate stream of execution that takes place simultaneously with and independently of everything else that might be happening. A thread is like a classic program that starts at point `A` and executes until it reaches point `B`. It does not have an event loop. A thread runs independently of anything else happening in the computer. Without threads an entire program can be held up by one CPU intensive task or one infinite loop, intentional or otherwise. With threads the other tasks that don't get stuck in the loop can continue processing without waiting for the stuck task to finish.

> **Concept**
>
> *A threads is a light-weighted process which runs concurrently with other threads. All threads of program defines a separate path of execution.*

It turns out that implementing threading is harder than implementing multitasking in an operating system. The reason, it's relatively easy to implement multitasking as individual programs are isolated from each other. Individual threads, however, are not. To return to the printing example, suppose that while the printing is happening in one thread, the user deletes a large chunk of text in another thread. What's printed? The document as it was before the deletion? The document as it was after the deletion? The document with some but not all of the deleted text? Or does the whole system go down in flames? Most often in a non-threaded or poorly threaded system it's the latter.

Threaded environments like Java allow a thread to put locks on shared resources so that while one thread is using data no other thread can touch that data. This is done with *synchronization*. Synchronization should be used sparingly since the purpose of threading is defeated if the entire system gets stopped waiting for a lock to be released. The proper choice of objects and methods to synchronize is one of the more difficult things to learn about threaded programming.

12.2.4. The Thread Classes

Java has two ways a program can implement threading. One is to create a subclass of `java.lang.Thread`. However sometimes you'll want to thread an object that's already a subclass of another class. Then you use the `java.lang.Runnable` interface.

The Thread class has three primary methods that are used to control a thread:

```
public void start()
public void run()
public final void stop()
```

The start() method prepares a thread to be run; the run() method actually performs the work of the thread; and the stop() method halts the thread. The thread *dies* when the run() method terminates or when the thread's stop() method is invoked.

You never call run() explicitly. It is called automatically by the runtime as necessary once you've called start(). There are also methods to **suspend** and **resume** threads, to put threads to sleep and wake them up, to yield control to other threads, and many more we'll discuss these later.

The Runnable interface allows you to add threading to a class which, for one reason or another, cannot conveniently extend Thread. It declares a single method, run():

```
public abstract void run()
```

By passing an object which implements Runnable to a Thread() constructor, you can substitute the Runnable's run() method for the Thread's own run() method. (More properly the Thread object's run() method simply calls the Runnable's run() method.)

12.2.5. A Simple Thread

When writing a threaded program you can pretend that you're writing many different programs, each with its own run() method. Each thread is a subclass of java.lang.Thread. The following program is a thread that prints the numbers between -128 and 127.

```
public class BytePrinter extends Thread {
public void run() {
    for (int b = -128; b < 128; b++) {
    System.out.println(b);
    }
 }
}
```

You launch this thread from another method, probably in another class, by instantiating an object of this class using new and calling its start() method. To create a thread just call the default constructor for your subclass of Thread. For instance

```
BytePrinter bp = new BytePrinter();
```

This class only has the default, no arguments constructor; but there's absolutely no reason you can't add other constructors to your Thread subclass.

Constructing a thread object puts it at the starting line. The bell goes off and the thread starts running when you call the thread's start() method like this:

```
bp.start();
```

Once the start() method is called program execution splits in two. Some CPU time goes into whatever statements follow bp.start() and some goes into the bp thread. It is unpredictable which statements will run first. Most likely they will be intermixed. The bp thread will now continue running until one of seven things happens

- bp's `run()` method completes.
- bp's `stop()` method is called.
- bp's `suspend()` method is called.
- bp's `sleep()` method is called.
- bp's `yield()` method is called.
- bp blocks waiting for an unavailable resource
- Another thread preempts this thread.

Once program control reaches the end of bp's `run()` method the thread dies. It cannot be restarted, though you can create a new instance of the same `Thread` subclass and start that.

For example, the following program launches a single `BytePrinter` thread:

Example 12.7: Simple Thread

```
public class ThreadTest
{
 public static void main(String[] args)
 {

 System.out.println("Constructing the thread...");
 BytePrinter bp = new BytePrinter();
 System.out.println("Starting the thread...");
 bp.start();
 System.out.println("The thread has been started.");
 System.out.println("The main() method is finishing.");
 return;

 }

}
```

OUTPUT

```
Constructing the thread...
Starting the thread...
The thread has been started.
The main() method is finishing.
-128
-127
-126
-125
-124
-123
-122
-121
```

```
-120
...
```

Of course this continues for a couple hundred more lines.

The following program launches three `BytePrinter` threads:

Example 12.8: Multiple threads

```
public class ThreadsTest
{

 public static void main(String[] args)
 {

 BytePrinter bp1 = new BytePrinter();
 BytePrinter bp2 = new BytePrinter();
 BytePrinter bp3 = new BytePrinter();
 bp1.start();
 bp2.start();
 bp3.start();

 }

}
```

The order of the output you see from this program is implementation dependent, and mostly unpredictable. It may look something like this:

```
-128
-127
-126
...
125
126
127
-128
-127
-126
...
125
126
127
-128
-127
-126
...
125
126
127
```

In this case the three threads run sequentially, one after the other. However on a few systems you may see something different.

In this case output from the three different threads is intermixed. Some systems use cooperative threads that require a thread to explicitly yield control before other, equal priority threads get a chance to run. Other systems use preemptive threading in which the virtual machine guarantees that all threads of the same priority get time in which to run. However, even on preemptive VMs this thread takes so little time to run it's unlikely any intermixing will occur.

Example 12.9 Program to display the alphabets one by one.

```
class MyThread extends Thread
{
    public void run()
    {
        for(int i=65;i<=90;i++)
        {
            System.out.print((char)i);
            try { sleep(1000); } catch(Exception ex){}
        }
    }
}
class MainThread
{
    public static void main(String aa[])
    {
        MyThread t = new MyThread();
        System.out.println("Thread Starting Alphabets
..............");
        t.start();
    }
}
```

OUTPUT

```
Thread Starting Alphabets ..............
A B C D E F ......... Z (The characters will display one by one using sleep method)
```

Example 12.10 Program to input your name and display in triangle form with sleep method.

```
import java.util.*;
class MyThread extends Thread
{
    char[] arr=new char[10];
    int l;
    MyThread(String s)
    {
```

```
            l = s.length();
            for(int i=0;i<l ;i++)
            {
            char s1=s.charAt(i);
            arr[i]=s1;
            }
        }

        public void run()
        {
            for(int i=0;i<l;i++)
            {
                for(int j=0;j<=i;j++)
                System.out.print(arr[j]);
                try{ sleep(1000); } catch(Exception ex){}
                System.out.println();
            }
        }
    }

    class MainThread
    {
    public static void main(String[] args)
    {
        Scanner sc = new Scanner(System.in);
        System.out.println("Enter Your Name:");
        String name = sc.next();

        MyThread t = new MyThread(name);
        System.out.println("Thread Starting to display ...............");
        t.start();

    }
    }
```

OUTPUT

```
Enter Your Name:
MIHIR
Thread Starting to display ...............
M
MI
MIH
MIHI
MIHIR
```

12.2.6. The Life Cycle of a Thread

Now that you've seen how to give a thread something to do, we'll review some details that were glossed over in the previous section. In particular, we look at the life cycle of a thread: how to create and start a thread, some of the special things it can do while it's running, and how to stop it.

The following figure 12.4 illustrates the various states that a Java thread can be in at any point during its life. It also illustrates which method calls cause a transition to another state. This diagram is not a complete finite state diagram, but rather an overview of the more interesting and common facets of a thread's life. The remainder of this page discusses a Thread's life cycle in terms of its state.

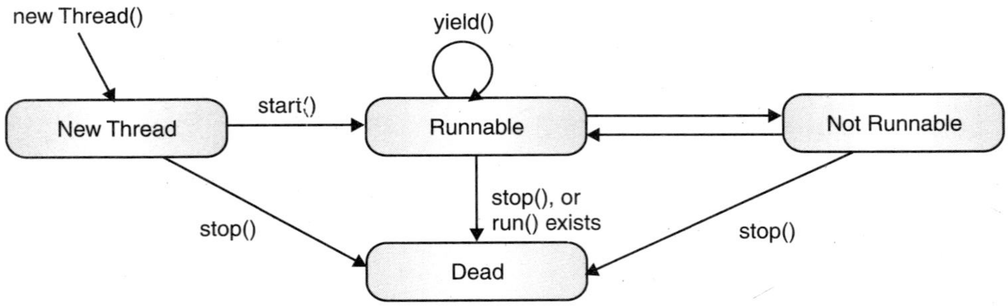

Fig. 12.4. Life cycle of thread

New Thread

The following statement creates a new thread but does not start it, thereby leaving the thread in the "New Thread" state.

```
Thread myThread = new MyThreadClass();
```

When a thread is in the "New Thread" state, it is merely an empty Thread object. No system resources have been allocated for it yet. Thus when a thread is in this state, you can only start the thread or stop it. Calling any method besides `start()` or `stop()` when a thread is in this state makes no sense and causes an IllegalThreadStateException.

Runnable

Now consider these two lines of code:

```
Thread myThread = new MyThreadClass();
myThread.start();
```

The `start()` method creates the system resources necessary to run the thread, schedules the thread to run, and calls the thread's `run()` method. At this point the thread is in the "Runnable" state. This state is called "Runnable" rather than "Running" because the thread might not actually be running when it is in this state. Many computers have a single processor, making it impossible to run all "Runnable" threads at the same time. So, the Java runtime system must implement a scheduling scheme that shares the processor between all "Runnable" threads. (See Thread Priority for more information) However, for most purposes you can think of the "Runnable" state as simply "Running". When a thread is running—it's "Runnable" and is the current thread—the instructions in its `run()` method are executing sequentially.

Not Runnable

A thread enters the "Not Runnable" state when one of these four events occurs:

- Someone invokes its sleep() method.
- Someone invokes its suspend() method.
- The thread uses its wait() method to wait on a condition variable.
- The thread is blocking on I/O.

For example, the bold line in the following code puts the current thread to sleep for 10 seconds (10,000 milliseconds):

```
try
{
      Thread.sleep(10000);
}
catch (InterruptedException e)
{
}
```

During the 10 seconds that `myThread` is asleep, even if the processor becomes available `myThread` does not run. After the 10 seconds are up, `myThread` becomes "Runnable" again and, if the processor becomes available, runs.

For each of the entrances into the "Not Runnable" state shown in the figure 12.4, there is a specific and distinct escape route that returns the thread to the "Runnable" state. An escape route only works for its corresponding entrance. For example, if a thread has been put to sleep, then the specified number of milliseconds must elapse before the thread becomes "Runnable" again. Calling `resume()` on a sleeping thread has no effect.

The following indicates the escape route for every entrance into the "Not Runnable" state.

- If a thread has been put to sleep, then the specified number of milliseconds must elapse.
- If a thread has been suspended, then someone must call its resume() method.
- If a thread is waiting on a condition variable, whatever object owns the variable must relinquish it by calling either notify() or notifyAll().
- If a thread is blocked on I/O, then the I/O must complete.

Dead

A thread can die in two ways: either from natural causes, or by being killed (stopped). A thread dies naturally when its `run()` method exits normally. For example, the `while` loop in this method is a finite loop—it will iterate 100 times and then exit.

```
public void run()
{
      int i = 0;
      while (i < 100)
      {
            i++;
            System.out.println("i = " + i);
```

```
        }
}
```

A thread with this `run()` method will die naturally after the loop and the `run()` method completes.

You can also kill a thread at any time by calling its `stop()` method. The following code snippet creates and starts `myThread` then puts the current thread to sleep for 10 seconds. When the current thread wakes up, the bold line in the code segment kills `myThread`.

```
Thread myThread = new MyThreadClass();
myThread.start();
try
{
      Thread.sleep(10000);
}
catch (InterruptedException e)
{
}
myThread.stop();
```

The `stop()` method throws a ThreadDeath object at the thread to kill it. Thus when a thread is killed in this manner it dies asynchronously. The thread will die when it actually receives the **ThreadDeath** exception.

The `stop()` method causes a sudden termination of a Thread's `run()` method. If the `run()` method performs critical or sensitive calculations, `stop()` may leave the program in an inconsistent or awkward state. Normally, you should not call Thread's `stop()` method but arrange for a gentler termination such as setting a flag to indicate to the `run()` method that it should exit.

IllegalThreadStateException

The runtime system throws an IllegalThreadStateException when you call a method on a thread and that thread's state does not allow for that method call. For example, **IllegalThreadStateException** is thrown when you invoke `suspend()` on a thread that is not "Runnable".

As shown in the various examples of threads so far in this chapter, when you call a thread method that can throw an exception, you must either catch and handle the exception, or specify that the calling method throws the uncaught exception.

The isAlive() Method

A final word about thread state: the programming interface for the Thread class includes a method called `isAlive()`. The `isAlive()` method returns true if the thread has been started and not stopped. Thus, if the `isAlive()` method returns **false** you know that the thread is either a "New Thread" or "Dead". If the `isAlive()` method returns **true**, you know that the thread is either "Runnable" or "Not Runnable". You cannot differentiate between a "New Thread" and a "Dead" thread; nor can you differentiate between a "Runnable" thread and a "Not Runnable" thread.

12.2.7. Thread Priorities

The scheduler determines which thread should be running based on a priority number assigned to each thread. The range of priorities is from 1 to 10. The default priority of a thread is Thread.NORM_PRIORITY, which is assigned the value of 5. Two other static variables are made available, they are Thread.MIN_PRIORITY, which is set to 1, and Thread.MAX_PRIORITY, which is set to 10. The getPriority() method can be used to find the current value of the priority of a thread.

Example 12.11 Program to implement the Multithreading.

```
// Multi Thread
class MyThread extends Thread
{
    MyThread(String s)
    {
        super(s); // thread create with s name.
        start();
    }
    public void run()
    {
        for(int i=1;i<=5;i++)
        {
            System.out.println(getName()+" Thread here");
            try { sleep(1000); } catch(Exception ex){}
        }
    }
}
class MainThread
{
    public static void main(String aa[])
    {
        MyThread t1 = new MyThread("Mihir");
        MyThread t2 = new MyThread("Nitin");
        MyThread t3 = new MyThread("Dharam");
        MyThread t4 = new MyThread("Harsh");
        System.out.println("Main Thread exit..............");
    }
}
```

OUTPUT

```
Main Thread exit..............
Mihir Thread here
Nitin Thread here
Dharam Thread here
Harsh Thread here
Mihir Thread here
Nitin Thread here
```

```
Dharam Thread here
Harsh Thread here
Mihir Thread here
Nitin Thread here
Dharam Thread here
Harsh Thread here
Mihir Thread here
Nitin Thread here
Dharam Thread here
Harsh Thread here
Mihir Thread here
Nitin Thread here
Dharam Thread here
Harsh Thread here
```

SUMMARY

— In this chapter we have explained some additional features of java programming such as exceptional handling, Multithreading and Input/Output.

— Exceptional handling is a feature for handling the errors and exception at runtime using try and catch block.

— Multitasking is the ability of single processor to perform more than one operation at a time.

— Multhreading is the concept of multitasking.

— We have also explain the Input/Output console as well as file Input/Output as introduction.

SOLVED PROBLEMS

Q. 12.1. What is user defined exception?

Ans. There are many exception defined by java which are used to track the run time exceptions and act accordingly. User can also define his exceptions that can be thrown in the same way java exceptions.

Q. 12.2. What is the difference between Process and Threads?

Ans. Process is a program in execution. The execution of processes must be sequential. On the other hand thread is a light-weighted process which is executed in parallel with other threads. The theard shares some of the resources with other threads but ceratin things are unique to as particular thread like process id and program counter.

Q. 12.3. What are three ways in which a thread can enter the waiting state?

Ans. A thread can enter the waiting state by invoking its sleep() method, by blocking on I/O, by unsuccessfully attempting to acquire an object's lock, or by invoking an object's wait() method. It can also enter the waiting state by invoking its (deprecated) suspend() method.

Q. 12.4. What state does a thread enter when it terminates its processing?

Ans. When a thread terminates its processing, it enters the dead state.

Q. 12.5. What's the difference between Thread and Runnable types?

Ans. A Java Thread controls the main path of execution in an application. When you invoke the Java Virtual Machine with the java command, it creates an implicit thread in which to execute the main method. The Thread class provides a mechanism for the first thread to start-up other threads to run in parallel with it.

The Runnable interface defines a type of class that can be run by a thread. The only method it requires is run(), which makes the interface very easy to fulfil by extending existing classes. A runnable class may have custom constructors and any number of other methods for configuration and manipulation.

Q. 12.6. How does the run() method in Runnable work?

Ans. It may help to think of the run method like the main method in standard single threaded applications. The run method is a standard entry point to run or execute a class. The run method is normally only executed in the context of an independent Thread, but is a normal method in all other respects.

Q. 12.7. What's the difference between a thread created as a subclass of java.lang.Thread and one created as an implementation of Runnable?

Ans. As threads, both behave the same. However, a thread that implements runnable can also extend another class.

Q. 12.8. Why should synchronized blocks of code call wait()?

Ans. To give other waiting threads time to execute and prevent deadlock. Thread.yield() also works for this purpose

Q. 12.9. Can there be a try block without a matching catch block?

Ans. Yes, if there is a finally block. try without catch or finally is a compile-time error.

Q. 12.10. Can there be a catch block without a matching try block?

Ans. No. This is a compile-time error.

Q. 12.11. Can there be a finally statement that is not attached to a try-catch block?

Ans. No. This is a compile-time error.

Q. 12.12. When are exception objects created?

Ans. In general when an unexpected condition occurs. In specific, when you use throw to send a new exception, when the Java library uses throw to send a new exception, or when the Java runtime signals a RuntimeException.

Q. 12.13. Does an exception have to be handled by the same try-catch block that threw it?

Ans. Not necessarily. If an exception isn't caught by the nearest try-catch block, it can be caught higher up. If it's a runtime exception it may not be caught at all.

Q. 12.14. How can you make sure you catch every exception that a statement may throw?

Ans. Use this catch block as the last one in your hierarchy:

```
catch (Exception e) {

}
```

Q. 12.15. If an exception isn't caught in a user-defined class, where does it go and what does it do?

Ans. If you're in an applet it may be caught in the applet viewer or web browser. However, the most likely result is that the thread in which the exception was thrown will be killed.

Q. 12.16. Suppose you've defined an exception called NotANumberException(). Now suppose that NotAPositiveNumberException extends NotANumberException. Will

```
catch (NotANumberException) {
System.out.println("Caught a NotANumberException");
}
```

catch a NotAPositiveNumberException?

Ans. Yes. Superclasses always catch their subclasses.

Now suppose you have the following catch clause and throw a NotAPositiveNumberException. What happens?

```
catch (NotANumberException) {
    System.out.println("Caught a NotANumberException");
}
catch (NotAPositiveNumberException) {
    System.out.println("Caught a NotAPositiveNumberException");
}
```

"Caught a NotANumberException" is printed on System.out.

Q. 12.17. In general which catch statement should come first? The one that catches the subclass or the one that catches the superclass? Why?

Ans. The one that catches the subclass. If the catch block that catches a superclass comes first, then the subclass will never be seen. You should move from the most specific (the subclass) to the most general (the superclass).

Q. 12.18. What do you have to do to have an exception handled by both its own catch statement and by its superclass's catch statement?

Ans. Nest try statements and rethrow the exception inside the first catch block, like this

```
try {
    try {

        // statements that may fail
     }
    catch (SubclassException e) {
    // handle subclass
    throw(e);
    }
}
catch (superclass Exception e) {
// handle superclass

}
```

Q. 12.19. What's the difference between throw and throws?

Ans. The throw keyword throws a new exception. The throws keyword declares that a method may throw a particular exception.

Q. 12.20. Write a class that keeps a running total of all characters passed to it (one at a time) and throws an exception if it is passed a non-alphabetic character.

Ans.

```
public class NotALetterException extends Exception {

    static int count=0;
```

```
        String msg;

        public NotALetterException(String s) {
        msg = s;
    }

    public static void countLetter(char c) throws NotALetterException {
    if (Character.isJavaLetter(c)) {
    count++;
  }
  else {
        throw new NotALetterException(c + "is not a letter");
  }
  }
}
```

EXERCISES

Q. 12.1. What is meant of exceptions?
Q. 12.2. Explain use of try and catch with examples.
Q. 12.3. What do you mean by user defined exceptions?
Q. 12.4. Explain the Java defined exceptions?
Q. 12.5. Differentiate throw and throws clause.
Q. 12.6. What is the role of finally? When it is beneficial?
Q. 12.7. Give an example of multiple catch clause.
Q. 12.8. What do you mean by multithreading? Compare with multitasking.
Q. 12.9. What is the difference between Thread and process?
Q. 12.10. What do you mean by the life cycle of Thread?
Q. 12.11. How the processor gives the priorities to the Threads?
Q. 12.12. What is the role of is Alive () method?
Q. 12.13. Give the details about the console output and input.
Q. 12.14. Why we needed Stream objects?
Q. 12.15. Explain the concept of string Tokenizer.

Chapter 13

HTML and JAVA Applet

13.1. HTML

Hyper Text Markup Language, the coding language used to create hypertext documents for the World Wide Web. In HTML, a block of text can be surrounded with tags that indicate how it should appear (for example, in bold face or italics). Also, in HTML a word, a block of text, or an image can be linked to another file on the Web. HTML files are viewed with a World Wide Web browser.

HTML is an abbreviation of "HyperText Mark-up Language" which is already more than you need to know at this stage. However, for the sake of good order, let us explain in greater detail.

- **Hyper** is the opposite of linear. In the good old days—when a mouse was something the cat chased—computer programs ran linearly: when the program had executed one action it went to the next line and after that, the next line and so on. But HTML is different—you can go wherever you want and whenever you want. For example, it is not necessary to visit MSN.com before you visit HTML.net.
- **Text** is self-explanatory.
- **Mark-up** is what you do with the text. You are marking up the text the same way you do in a text editing program with headings, bullets and bold text and so on.
- **Language** is what HTML is. It uses many English words.

CHAPTER OUTLINE

- HTML
  - Simple Rules of HTML
  - HTML Tags
- Java Applets
  - Java Applet Vs Applications
  - Restrictions on a Java Applet
  - The Structure of Java Applet
  - Applet Life Cycle
- Applet Classes
- Some Important Classes
  - Color Class
  - Font Class
  - Graphic Class

Summary

Solved Problems

Exercises

> **Concept**
>
> *HTML is a language to create web pages.*

13.1.1. Simple rules of HTML

- Element names and attribute names are not case sensitive: attribute *values* **are**;
- Documents start with a <!doctype...> statement, followed by a header and a text body all enclosed in <html>...</html>
- The header is enclosed in <head>...</head>
- The text body is enclosed in <body>...</body>
- Comments are written as <!— A comment —>

To create HTML file Type in the following text in notepad:

```
<html>
<head>
<title>Title of page</title>
</head>
<body>
This is my first homepage. <b>This text is bold</b>
</body>
</html>
```

Save the file as "mypage.htm".

Start your Internet browser. Select "Open" (or "Open Page") in the File menu of your browser. A dialog box will appear. Select "Browse" (or "Choose File") and locate the HTML file you just created—"mypage.htm"—select it and click "Open". Now you should see an address in the dialog box, for example "C:\MyDocuments\mypage.htm". Click OK, and the browser will display the page.

The first tag in your HTML document is <html>. This tag tells your browser that this is the start of an HTML document. The last tag in your document is </html>. This tag tells your browser that this is the end of the HTML document.

The text between the <head> tag and the </head> tag is header information. Header information is not displayed in the browser window.

The text between the <title> tags is the title of your document. The title is displayed in your browser's caption.

The text between the <body> tags is the text that will be displayed in your browser.

The text between the <b> and </b> tags will be displayed in a bold font.

13.1.2. HTML Tags

- HTML tags are used to mark-up HTML **elements**
- HTML tags are surrounded by the **two characters < and >**
- The surrounding characters are called **angle brackets**
- HTML tags normally **come in pairs** like <b> and </b>
- The first tag in a pair is the **start tag,** the second tag is the **end tag**
- The text between the start and end tags is the **element content**

- HTML tags are **not case sensitive,** <b> means the same as <B>

HTML Elements

Remember the HTML example:

```
<html>
<head>
<title>Title of page</title>
</head>
<body>
This is my first homepage. <b>This text is bold</b>
</body>
</html>
```

This is an HTML element:

```
<b>This text is bold</b>
```

The HTML element starts with a **start tag**: <b>

The **content** of the HTML element is: This text is bold

The HTML element ends with an **end tag**: </b>

The purpose of the <b> tag is to define an HTML element that should be displayed as bold.

This is also an HTML element:

```
<body>
This is my first homepage. <b>This text is bold</b>
</body>
```

This HTML element starts with the start tag <body>, and ends with the end tag </body>.

The purpose of the <body> tag is to define the HTML element that contains the body of the HTML document.

Tag Attributes

Tags can have attributes. Attributes can provide additional information about the HTML elements on your page.

This tag defines the body element of your HTML page: <body>. With an added bgcolor attribute, you can tell the browser that the background color of your page should be red, like this: <body bgcolor="red">.

This tag defines an HTML table: <table>. With an added border attribute, you can tell the browser that the table should have no borders: <table border="0">

Attributes always come in name/value pairs like this: name="value".

Attributes are always added to the start tag of an HTML element.

Quote Styles, "red" or 'red'?

Attribute values should always be enclosed in quotes. Double style quotes are the most common, but single style quotes are also allowed.

In some rare situations, like when the attribute value itself contains quotes, it is necessary to use single quotes:

name='John "ShotGun" Nelson'

The most important tags in HTML are tags that define headings, paragraphs and line breaks.

Headings

Headings are defined with the <h1> to <h6> tags. <h1> defines the largest heading. <h6> defines the smallest heading.

```
<h1>This is a heading</h1>
<h2>This is a heading</h2>
<h3>This is a heading</h3>
<h4 >This is a heading</h4>
<h5>This is a heading</h5>
<h6>This is a heading</h6>
```

HTML automatically adds an extra blank line before and after a heading.

Paragraphs

Paragraphs are defined with the <p> tag.

```
<p>This is a paragraph</p>
<p>This is another paragraph</p>
```

HTML automatically adds an extra blank line before and after a paragraph.

Line Breaks

The <br> tag is used when you want to end a line, but don't want to start a new paragraph. The <br> tag forces a line break wherever you place it.

```
<p>This <br> is a para<br>graph with line breaks</p>
```

The <br> tag is an empty tag. It has no closing tag.

Comments in HTML

The comment tag is used to insert a comment in the HTML source code. A comment will be ignored by the browser. You can use comments to explain your code, which can help you when you edit the source code at a later date.

```
<!— This is a comment —>
```

Note that you need an exclamation point after the opening bracket, but not before the closing bracket.

Basic Notes - Useful Tips

When you write HTML text, you can never be sure how the text is displayed in another browser. Some people have large computer displays, some have small. The text will be reformatted every time the user resizes his window. Never try to format the text in your editor by adding empty lines and spaces to the text.

HTML will truncate the spaces in your text. Any number of spaces count as one. Some extra information: In HTML a new line counts as one space.

Using empty paragraphs <p> to insert blank lines is a bad habit. Use the <br> tag instead. (But don't use the <br> tag to create lists. Wait until you have learned about HTML lists.)

You might have noticed that paragraphs can be written without the closing tag </p>. Don't rely on it. The next version of HTML will not allow you to skip ANY closing tags.

HTML automatically adds an extra blank line before and after some elements, like before and after a paragraph, and before and after a heading.

13.2. JAVA APPLETS

A web page designed using HTML is a static page. Thus, the information presented in a web page is constant. As with any visual medium, the dynamic behavior applied to a web page will make it more attractive. The dynamic behavior in web page can b in the form of:

- Scrolling text
- Messages displayed in the status bar
- Changing images

To be able to apply such behavior to a web page, one solution is to use a java applet.

An applet is a little application program. Prior to the World Wide Web, the built-in writing and drawing programs that came with Windows were sometimes called "applets." On the Web, using Java, the object-oriented programming language, an applet is a small program that can be sent along with a Web page to a user. Java applets can perform interactive animations, immediate calculations, or other simple tasks without having to send a user request back to the server.

> **Concept**
>
> *An applet is a Java program used to create dynamic webpages on the browser with help of HTML.*

A java applet is a program that can be embedded in a HTML page. Thus, we can write an Applet that will display scrolling text for example, and then embed it into the HTML page, giving it the ability to display scrolling text.

13.2.1. Java Applet Vs Applications

There are some of the differences between Applet and an application in Java:

- A Java application is a stand-alone program. This means that the application can be run by itself.

  A Java applet is a program that has to be run from within a web browser or a similar java enabled application such as an Applet Viewer.
- The Java applet is run from within a browser and so can make use of all the infrastructure that the browser can provide. For example, the event handling capability, user interface etc. These may also be specifically built into the Applet by the programmer.
- Java applet are included in HTML pages using the special tag <APPLET>. When the user loads a page containing a Java applet in his browser, the browser downloads the applet on to the local machine and executes the bytecode on the local system.
- Since the applet is executed on the local computer, the browser limit the actions that an applet can perform. This is done so that the applet does not cause any damage to the client system.

13.2.2. Restrictions on a Java Applet

Some of the most popular browser in the world today are the Netscape Navigator and the Internet Explorer.

The Following are some of the restrictions placed by these browser on the Java Applets:

- Applets are not allowed to read from or write to files on the local system.
- Applets can communicate only with the server on they are originally resident and with no other server either on the Interet or on the client's local network.
- Applets are not allowed to execute any application on the local computer neither are they allowed to load any DLLs.

13.2.3. The structure of Java Applet

Consider the following code for Java Applet. The code given below is used to write a applet code which displays *Hello, World!.*

Example 13.1: Hello world

```
import java.applet.*;
import java.awt.*;

public class Our_Applet extends Applet
{
    public void paint(Graphics g)
    {
        g.drawString(""Hello,World!),50,25;
    }
}
```

The applet is a Java program that can be run only in a Browser or other similar Java enabled applications like an Appletviewer.

Embedding an applet in an HTML page is very simple and is done as shown below:

Create an html file in notepad with following code:

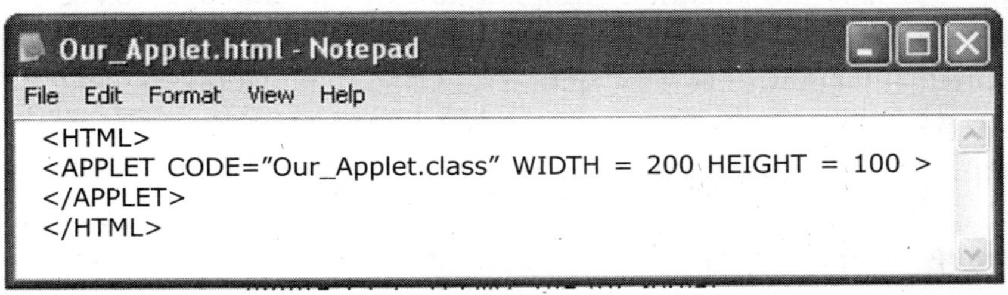

Fig. 13.1. HTML file for applet

```
<APPLET CODE="Our_Applet.class" WIDTH = 200 HEIGHT = 100 >
</APPLET>
```

The tag '<APPLET> </APPLET>' are used to signify that an Applet is being added to the HTML page. The attribute 'Code' is used to specify the name of the Applet's class file. The attributes 'Width' and 'Height' are used to define the applet's display area.

This HTML code should be saved as "Our_Apple.html".

To view the applet in an Appletviewer, HTML file created for the applet has to be run as shown below.

```
C:\JavaProgs> appletviewer <appletfilename>.html
```

For the applet created above, the applet program can be viewed as shown below

```
C:\JavaProgs> appletviewer Our_Applet.html
```

The output will appear as like the figure:

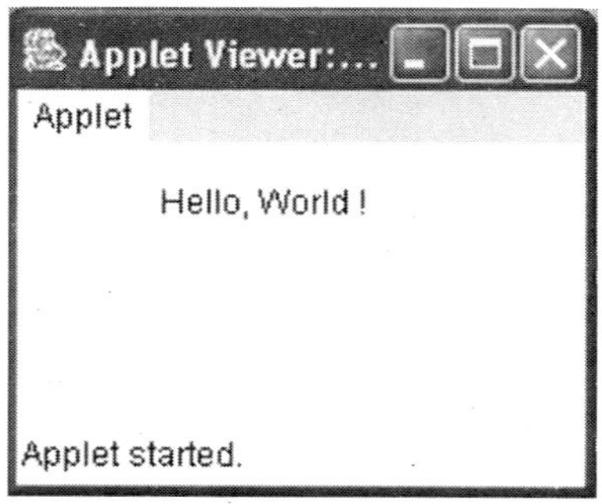

Fig. 13.2. Applet output

Applets can also be viewed with the help of Browser by opening the required HTML files in that browser.

13.2.4. Applet Life Cycle

To create your own applet, you derive a class from Java's Applet class and then add the desired functionality. The Applet class contains methods for actions such as interacting with the browser environment in which it is running, for retrieving images, for playing sound files, and for resizing its host windows. The path for execution or way to execute all these methods is referred as applet life cycle. Now question is that what are these functions and what is the way of execution. Let see the following:

- **init()** – for initialization, first method to be called by JVM.
- **start()** – ready for running.
- **repaint()** – clear the Applet screen
- **update()** – update the Applet Screen.
- **paint()** – drawing on the applet screen
- **stop()** – to suspend the execution of Applet
- **destroy()** – final cleanup like destructor in C++.

Now question arises, suppose we are using only paint() method then where are the remaining methods of life cycle ?

Actually all methods are concrete methods in Applet class, if user override any method then user's method to be called else Applet class methods called.

Let see an example where we demonstrate the use of all methods of life cycle.

Example 13.2 Applet program to use the methods of Applet life cycle.

```
import java.applet.*;
```

```
import java.awt.*;

public class example extends Applet
{
      private String history = "";

      public example()
      {
      }

      public void init()
      {
            history = history + "init ";
            resize(320, 240);
      }

      public void start()
      {
            history = history + "(start ";
      }

      public void stop()
      {
            history = history + "stop) ";
      }

      public void destroy()
      {
            history = history + "destroy ";
      }

      public void paint(Graphics g)
      {
            g.drawString(history, 10, 10);
      }
}

//<applet code= "example.java" width =300 hieght=300></applet>
```

To execute the above example, compile the program and use following command to execute:

```
C:\> appletviewer example.java
```

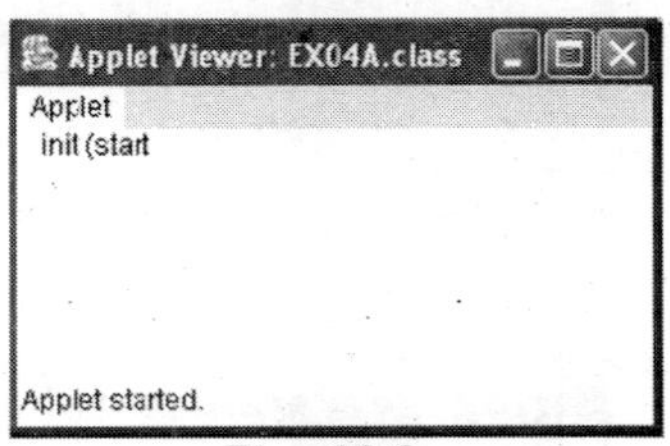

Fig. 13.3.

Even though the code in the destroy method of above example will execute, the results of adding this text will not be apparent. Because the history string is displayed only by the paint method and paint is not called after destroy, you will not be able to see the destroy message.

To see the full message of above example write the execution command in the command prompt and press restart option of Applet menu in appletviewer:

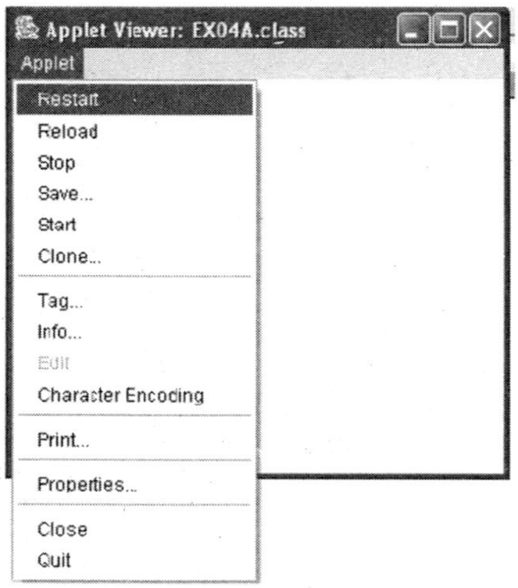

Fig. 13.4.

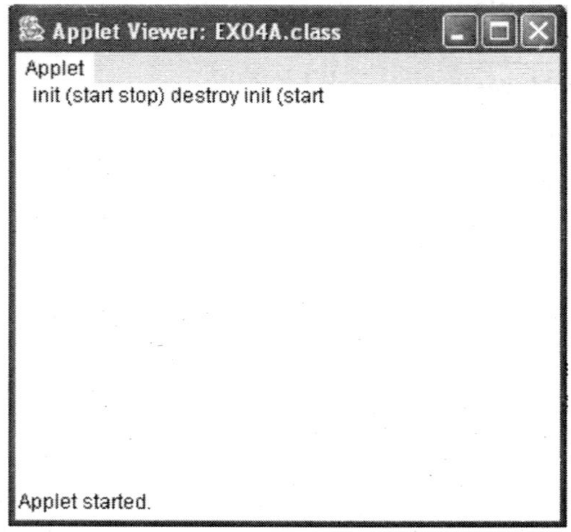

Fig. 13.5.

After the applet was loaded, the Restart button was pressed repeatedly. This had the effect of stopping, destroying and starting the applet each time, as you can see in Figure 13.4 Closing of appletviewer will stop the applet and then destroy it.

13.3. APPLET CLASSES

The Applet class basically exist in the java.applet package, which is also sub-package of java.awt package. See the class hierarchy of Applet class:

```
java.lang.Object
      |
      +--java.awt.Component
            |
            +--java.awt.Container
                  |
                  +--java.awt.Panel
                        |
                        +--java.applet.Applet
```

The Applet class contain three interfaces and some methods for easy functionality. The most commonly used methods of Applet Class are:

Methods	***Description***
void **destroy**()	Called by the browser or applet viewer to inform this applet that it is being reclaimed and that it should destroy any resources that it has allocated.
String getParameter(String name)	Returns the value of the named parameter in the HTML tag.

void **init**()	Called by the browser or applet viewer to inform this applet that it has been loaded into the system.
boolean **isActive**()	Determines if this applet is active.
void **play**(URL url)	Plays the audio clip at the specified absolute URL.
void **play**(URL url, String name)	Plays the audio clip given the URL and a specifier that is relative to it.
void **resize**(Dimension d)	Requests that this applet be resized.
Void **resize**(int width, int height)	Requests that this applet be resized.
void **showStatus**(String msg)	Requests that the argument string be displayed in the "status window".
void **start**()	Called by the browser or applet viewer to inform this applet that it should start its execution.
void **stop**()	Called by the browser or applet viewer to inform this applet that it should stop its execution.

The applet class contains following Interfaces:

1. interface java.applet.**AppletContext**
2. interface java.applet.**AppletStub**
3. interface java.applet.**AudioClip**

1. **AppletContext:** This interface corresponds to an applet's environment: the document containing the applet and the other applets in the same document. The methods in this interface can be used by an applet to obtain information about its environment.
2. **AppletStub:** When an applet is first created, an applet stub is attached to it using the applet's setStub method. This stub serves as the interface between the applet and the browser environment or applet viewer environment in which the application is running.
3. **AudioClip:** The AudioClip interface is a simple abstraction for playing a sound clip. Multiple AudioClip items can be playing at the same time, and the resulting sound is mixed together to produce a composite.

13.4. SOME IMPORTANT CLASSES

The Applet is a class which exist in java.applet package. The java.applet package provides many classes to support the design an applet more interactive. Here are some classes we discussed for applets and their uses with examples.

13.4.1. Color Class

To draw anything in multiple colors an applet you must select a color, draw or fill, then select another color, and draw or fill again.

You can define colors with the Color class. The java.awt.Color class offers predefined constants for the following 13 standard colors:

BLACK, BLUE, CYAN, DARK_GRAY, GRAY, GREEN, LIGHT_GRAY, MAGENTA, ORANGE, PINK, RED, WHITE, YELLOW

Example 13.3 Applet Program to set back ground color.

```
import java.applet.*;
import java.awt.*;
public class exampleApplet extends Applet
{

    public void paint(Graphics g)
    {
        setBackground(Color.YELLOW);
        g.drawString("Welcome", 100, 100);
    }
}
//<applet code= "exampleApplet.java" width =300 height=300> </applet>
```

COMPILE:

```
C:\>javac exampleApplet.java
```

EXECUTE:

```
C:\>appletviewer exampleApplet.java
```

OUTPUT

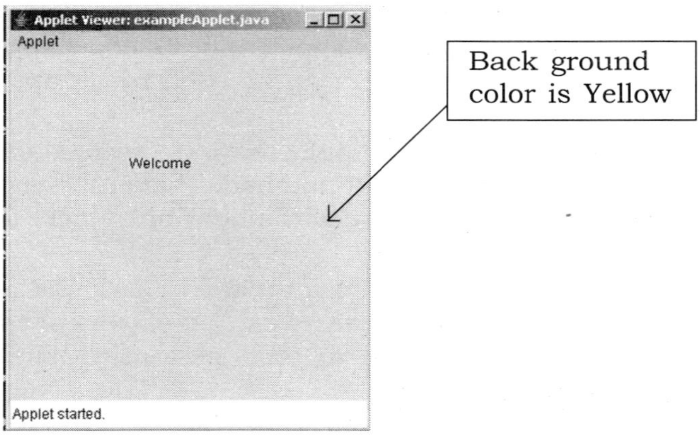

> **Note:** Before Java SE 1.4, color constant names were in lowercase, such as Color.red. You can now write the standard color names in uppercase or, in lowercase.

You can specify a custom color by creating a Color object by its red, green, and blue components. Using a scale of 0-255 (that is, one byte) for the redness, blueness, and greenness (also call RGB colors), call the Color constructor like this:

```
Color(int redness, int greenness, int blueness)
```

Here is an example of setting a custom color where we set back ground with custom color and you can change the values for changing the colors:

```
setBackgorund(new Color(0, 128, 128));        // a dull blue-green
```

Example 13.4 Applet Program to set a custom color on back ground.

```
import java.applet.*;
import java.awt.*;
public class exampleApplet extends Applet
{

    public void paint(Graphics g)
    {
        setBackground(new Color(0, 128, 128));
            g.drawString("Welcome", 100, 100);
    }
}
//<applet code= "exampleApplet.java" width =300 height=300> </applet>
```

COMPILE:

```
C:\>javac exampleApplet.java
```

EXECUTE:

```
C:\>appletviewer exampleApplet.java
```

OUTPUT

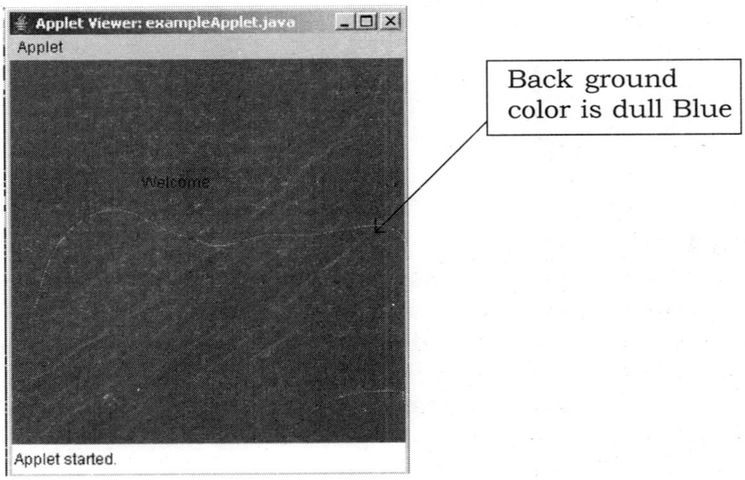

Example 13.5 Applet Program to set the fore ground & back ground colors.

```
import java.applet.*;
import java.awt.*;
public class exampleApplet extends Applet
{

    public void paint(Graphics g)
    {
        setBackground(Color.Yellow));
        g.setColor(Color.RED);
```

```
            g.drawString("Welcome", 100, 100);
        }
    }
    //<applet code= "exampleApplet.java" width =300 height=300> </applet>
```

COMPILE:

```
C:\>javac exampleApplet.java
```

EXECUTE:

```
C:\>appletviewer exampleApplet.java
```

OUTPUT

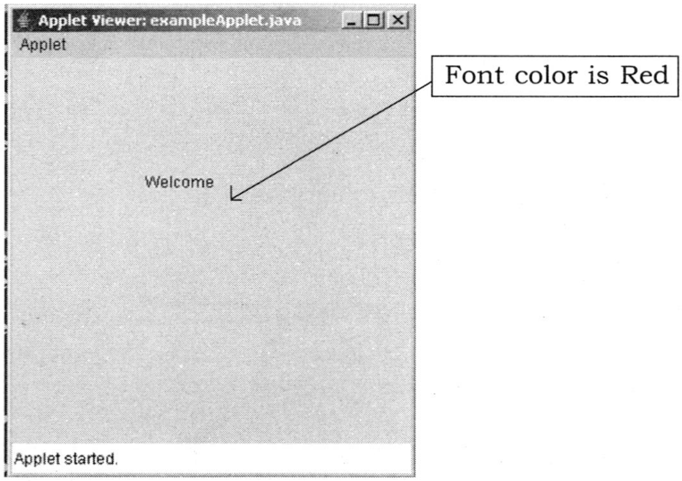

13.4.2. Font Class

In the previous sections, programs are displayed a string in the default font. Often, you want to show text in a different font. You specify a font by its font face name. A font face name is composed of a font family name, such as "Arial" and an optional suffix such as "Black." For example, the font faces "Arial" and "Arial Black" are both considered to be part of the family named "Arial".

To set the fonts in your program you have to use the given Font constructor:

```
Font(String fname, int fstyle, int fsize)
```

Where first parameter, fname is the font name available in your computer, and second parameter, fstyle which may be PLAIN, BOLD, ITALIC and BOLD + ITALIC. These are the constant and we can use constant integers for this using following table:

Style	Font Constant	Integer Constant
Plain	Font.PLAIN	0
Bold	Font.BOLD	1
Italic	Font.ITALIC	2
Bold and Italic	Font.BOLD + Font.ITALIC	3

Example 13.6 Applet Program to set the font.

```
import java.applet.*;
import java.awt.*;
public class exampleApplet extends Applet
{

    public void paint(Graphics g)
    {
        Font f = new Font("Courier New",1,45);
        g.setFont(f);
        g.drawString("Welcome", 100, 100);
    }
}

//<applet code= "exampleApplet.java" width =300 height=300> </applet>
```

COMPILE:

```
C:\>javac exampleApplet.java
```

EXECUTE:

```
C:\>appletviewer exampleApplet.java
```

OUTPUT

Example 13.7 Applet Program to set the font with back ground and fore ground colors.

```
import java.applet.*;
import java.awt.*;
public class exampleApplet extends Applet
{
    public void paint(Graphics g)
    {
        setBackground(Color.YELLOW);
```

```
                g.setColor(Color.RED);
                Font f = new Font("Courier New",1,45);
                g.setFont(f);
                g.drawString("Welcome", 100, 100);
            }
        }

        //<applet code= "exampleApplet.java" width =300 height=300> </applet>
```

COMPILE:

```
C:\>javac exampleApplet.java
```

EXECUTE:

```
C:\>appletviewer exampleApplet.java
```

OUTPUT

There are some methods of Font Class and their uses:

String getName()	- return the font name
int getStyle()	- return the font style
int getSize()	- return the font size
boolean isBold()	- if font is bold then true else false
boolean isItalic()	- if font is italic then true else false
boolean isPlain()	- if font is plain then true else false

13.4.3. Graphic Class

The Graphis class is most important class for the applet programming. The paint() method takes one parameter of type Graphics. A Graphics object provides a collection of settings for drawing images and text, such as the font you set or the current color. All drawing in Java must go through a Graphics object. It has methods that draw patterns, images, and text.

There are some methods of Graphics Class with their signatures and descriptions:

Methods	Descriptions
void drawString(String str,int x,int y)	To display the string at x, y coordinates
void drawLine(int x,int y,int x1,int y1)	To display the line from x, y to x1, y1 coordinates
void drawRect(int x,int y,int w,int h)	To draw the Rectangle at x, y coordinates with w width and h height
void fillRect(int x,int y,int w,int h);	To draw the filled Rectangle with fore ground color at x, y coordinates with w width and h height
void draw3DRect(int x,int y,int w,int h,true/false)	To draw the 3D-Rectangle at x, y coordinates with w width and h height where (true for 3D, false for 2D)
void fill3DRect(int x,int y,int w,int h,true/false)	To draw the filled 3D-Rectangle with fore ground color at x, y coordinates with w width and h height (true for 3D, false for 2D)
void drawRoundRect(int x,int y,int w,int h,int arcWidth,int arcHeight)	To draw the Rectangle with rounded corners at x, y coordinates with w width and h height where archeight is arc at corners.
void fillRoundRect(int x,int y,int w,int h, int arcWidth, int arcHeight)	To draw the filled Rectangle with rounded corners at x, y coordinates with w width and h height where archeight is arc at corners.
void drawOval(int x,int y,int w,int h)	To draw the Oval at x, y coordinates with w width and h height.
void fillOval(int x,int y,int w,int h)	To draw the filled Oval with fore ground color at x, y coordinates with w width and h height.
void drawArc(int x,int y,int w,int h,int start,int end)	To draw the Arc at x, y coordinates with w width and h height.
void fillArc(int x,int y,int w,int h,int start,int end)	To draw the filled Arc with fore ground color at x, y coordinates with w width and h height.
void drawImage(Image m,int x,int y,this)	To draw an Image in actual size at x, y coordinates.
void drawImage(Image m,int x, int y, int w, int h,this)	To draw an Image in user size at x, y coordinates with w width and h height.

In drawImage method this keyword used to draw image in current Applet and Image m is an object of Image class, which can be create like this:

```
Image m = getImage(getCodeBase(),"figure.jpg");
g.drawImage(m,100,100,this);
```

where getCodeBase() - return the current working directory.

Example 13.8 Program to display the Line, Rectangle and Round-Rectangle using colors.

```
import java.awt.*;
import java.applet.*;
public class Draw1 extends Applet
{
```

```
        public void paint(Graphics g)
        {
            setBackground(Color.YELLOW);
            g.setColor(Color.blue);
            g.drawLine(10,10,200,10);
            g.drawRect(10,20,200,100);
            g.fill3DRect(220,20,200,100,true);
            g.setColor(Color.red);
            g.drawRoundRect(10,130,200,100,50,40);
            g.fillRoundRect(220,130,200,100,50,40);
            g.setColor(Color.yellow);
            g.drawOval(10,240,200,100);
            g.fillOval(220,240,100,100);
            g.drawRect(10,20,200,100);
        }
    }
    //<applet code=Draw1 height=500 width=700></applet>
```

COMPILE:

```
C:\>javac Draw1.java
```

EXECUTE:

```
C:\>appletviewer Draw1.java
```

OUTPUT

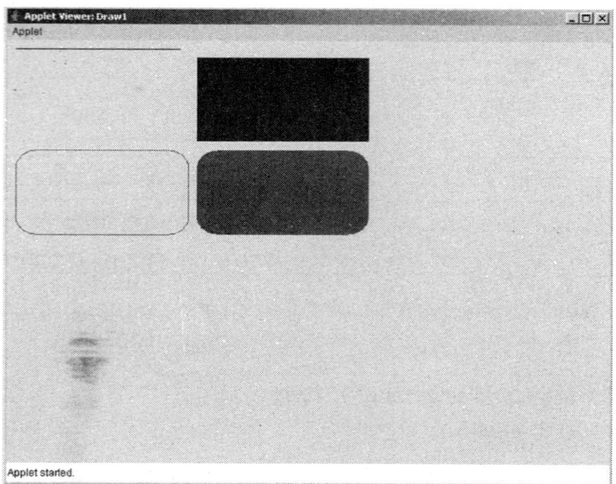

Example 13.9 **Program to display the use of all methods of Graphics class, using colors in a single applet.**

```
import java.awt.*;
import java.applet.*;
public class Draw1 extends Applet
{
```

```
    public void paint(Graphics g)
    {
        setBackground(new Color(100,248,158));
        g.setColor(Color.blue);
        g.drawLine(10,10,200,10);
        g.drawRect(10,20,200,100);
        g.fill3DRect(220,20,200,100,true);
        g.setColor(Color.red);
        g.drawRoundRect(10,130,200,100,50,40);
        g.fillRoundRect(220,130,200,100,50,40);
        g.setColor(Color.yellow);
        g.drawOval(10,240,200,100);
        g.fillOval(220,240,100,100);
        g.setColor(Color.black);
        g.drawArc(430,20,200,100,30,300);
        g.fillArc(430,130,200,100,30,300);
        g.fillArc(430,240,200,100,150,-300);
    }
}
//<applet code=Draw1 height=500 width=700></applet>
```

COMPILE:

```
C:\>javac Draw1.java
```

EXECUTE:

```
C:\>appletviewer Draw1.java
```

OUTPUT

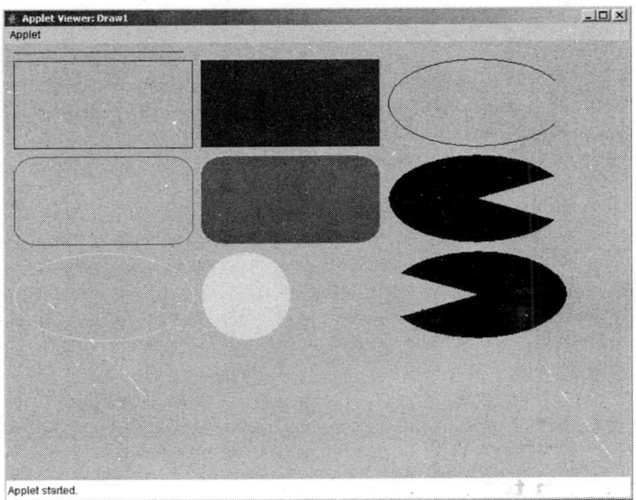

Example 13.10 Program to display an image on an applet.

(**Note:** We have used an image file abc.jpg existing in same directory, you can use any image file from your computer with file path.)

```
import java.awt.*;
```

```
import java.applet.*;
public class Draw1 extends Applet
{
     public void paint(Graphics g)
     {
          Image m = getImage(getCodeBase(),"abc.jpg");
          g.drawImage(m,10,10,680,480,this);
     }
}

//<applet code=Draw1 height=500 width=700></applet>
```

COMPILE:

```
C:\>javac Draw1.java
```

EXECUTE:

```
C:\>appletviewer Draw1.java
```

OUTPUT

Example 13.11 Applet Program to display a digital Clock on an applet.

```
import java.awt.*;
import java.applet.Applet;
public class clock extends Applet implements Runnable
```

```
{
      Thread t1;
      boolean flag=true;
      String nows;
      int h,m,s;
      public void init()
      {
      h=0;m=0;s=0;
      }
      public void start()
      {
      t1=new Thread(this);
      t1.start();
      }

      public void stop()
      {
      flag=false;
      }
      public void run()
      {
      while(flag)
            {

                  if(m>59)
                  {
                        h++;m=0;
                  }
                  else if(s>59)
                  {
                  m++; s=0;
                  }
                  else s++;

                  repaint();
                  try{
                        Thread.sleep(1000);
                        }
                  catch(InterruptedException e){}
            }
}
public void paint(Graphics g)
{

      setBackground(Color.yellow);
```

```
        g.setFont(new Font("Arial",1,30));
        String s1=String.valueOf(h)+": "+String.valueOf(m)+": "+String.valueOf(s);
        g.drawString(s1,100,200);

    }
    }
    //<applet code=clock.class width=300 height=300></applet>
```

COMPILE:

```
C:\>javac clock.java
```

EXECUTE:

```
C:\>appletviewer clock.java
```

OUTPUT

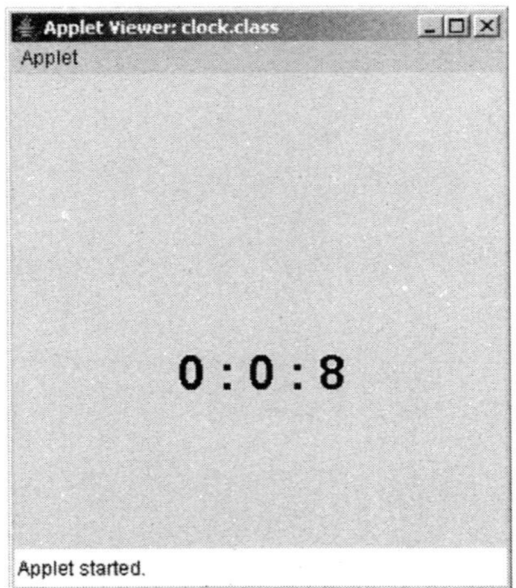

SUMMARY

- In this chapter we have explained the concept and style to design web pages using HTML and Java Applets.
- HTML is a language to create webpages.
- Java Applet is a Java program used to create dynamic web pages on the browser with the help of HTML.
- At last we have given the functionality of Applet Class of Java. We have provide most frequently used methods of this important class.

SOLVED PROBLEMS

Q. 13.1. What is the main difference between "APPLET" and "APPLICATION"?

Ans. An application is a standalone Java program characterized by the `public static void main ( String[] args){}` method. In contrast, an applet is a java program launched through a browser.

The main method is the entry point for a Java application program whereas init() method of the applet is its entry point for the program.

Q. 13.2. How will you initialize an Applet?

Ans. Using init() method.

Q. 13.3. What is the order of method invocation in an Applet?

Ans. The following is order:

1. init()
2. start()
3. paint()
4. stop()
5. destroy()

Q. 13.4. When is update method called?

Ans. update method is called every time whenever we call a repaint() method.

Q. 13.5. What are the guidelines for HTML formatted email?

Ans. The first (un-helpful) response is that email is a plain text medium and sending other MIME types is bound to be a problem for *some* users. One of the great advantages of email is that it is lightweight and usually takes a fraction of a second to download. If you include surplus HTML markup and perhaps image references you will significantly impair the user experience and turn subscribers away from your service, either by deleting the mail without reading or un-subscribing

Q. 13.6. What's the difference between reloading and restarting an applet?

Ans. Reloading clears all static variables and starts over again by constructing a new instance of the applet and calling its init() method. Depending on the platform reload may or may not reload the byte code from a .class file on the disk or on the network. Restarting an applet just calls its start() method. It does not call init(), the constructor, or reload the byte codes.

Q. 13.7. Assume that each letter occupies a box fifteen pixels wide by fifteen pixels high. Write an applet which takes a single letter as a parameter and fills the available space with that letter.

Ans.

```
import java.applet.Applet;
import java.awt.Graphics;

public class letterFillApplet extends Applet {
    String theLetter;

    public void init() {

    String s = getParameter("Letter");
    theLetter = s.substring(0, 1);

 }

        public void paint(Graphics g) {

        for (int i = 0; i < size().width; i += 15) {
        for (int j = 0; j < size().height; j += 15) {
        g.drawString(theLetter, i, j);
        }
```

```
    }

  }

 }
```

Q. 13.8. Create an applet for the PoetryApplet so that it includes a numlines parameter that will set the size of the poem array at runtime.

Ans.

```
    import java.applet.Applet;
    import java.awt.Graphics;

 public class PoetryApplet extends Applet {

    String[] poem;
    int numlines;

    public void init() {

    int i;
    String nextline;

  try {
    String s = getParameter("numlines");
    if (s == null) numlines = 101;
    else numlines = Integer.parseInt(s);
  }
    catch (NumberFormatException e) {
    numlines = 101;
  }

  poem = new String[numlines];

  for (i = 1; i <= numlines; i++) {
    nextline = getParameter("Line" + i);
    if (nextline == null) break;
    poem[i] = nextline;
  }
    numlines = i-1;

  }

    public void paint(Graphics g) {

    int y = 15;
```

```
    for (int i=1; i <= numlines; i++) {
    g.drawString(poem[i], 5, y);
    y += 15;
    }

  }

}
```

Q. 13.9. Change the PoetryApplet to use a Vector rather than an array so that it will truly handle an arbitrary number of parameters.

Ans.

```
    import java.applet.Applet;
    import java.awt.Graphics;
    import java.util.Vector;
    import java.util.Enumeration;

    public class PoetryApplet extends Applet {

    Vector poem;

     public void init() {

     int i = 1;
     String nextline;
    poem = new Vector();

     while ((nextline = getParameter("Line" + i++)) != null) {
         poem.addElement(nextline);
    }

  }

    public void paint(Graphics g) {

    int y = 15;

    Enumeration e = poem.elements();
    while (e.hasMoreElements()) {
    g.drawString((String) e.nextElement(), 5, y);
    y += 15;
    }

  }

}
```

Q. 13.10. Create a WrapText applet so that it optionally accepts FONT, SIZE and STYLE PARAMs which set the font face, point size and style in which the String is drawn. Be sure to gracefully handle the case when the APPLET tag does not include any or all of these PARAMs or includes invalid values (e.g. STYLE="UNDERLINE").

Ans.

```
import java.applet.Applet;
import java.awt.Graphics;
import java.awt.FontMetrics;
import java.util.StringTokenizer;
import java.awt.Font;

public class WrapTextApplet extends Applet
{
 String input_from_page;
 int margin = 5;
 Font theFont;
 public void init() {

 String s;
 int fontsize = 12;
 String fontname = "Courier";
 int fontstyle = Font.PLAIN;

 input_from_page = getParameter("Text");
 if ((s = getParameter("Size")) != null) {
 fontsize = Integer.parseInt(s);
 }
 if ((s = getParameter("Style")) != null) {
 if (s.equalsIgnoreCase("Italic")) fontstyle = Font.ITALIC;
 if (s.equalsIgnoreCase("Bold")) fontstyle = Font.BOLD;
 }
 if ((s = getParameter("Font")) != null) {
 fontname = s;
 }
 theFont = new Font(fontname, fontstyle, fontsize);
 }
 public void paint(Graphics g) {
 int i = 0;
 int linewidth = 0;
 StringBuffer sb = new StringBuffer();
 g.setFont(theFont);
    FontMetrics fm = g.getFontMetrics();
    StringTokenizer st = new StringTokenizer(input_from_page);
    while (st.hasMoreTokens()) {
     String nextword = st.nextToken();
     if (fm.stringWidth(sb.toString() + nextword) < size().width) {
```

```
    sb.append(nextword);
     sb.append(' ');
     }
     else if (sb.length() == 0) {
     g.drawString(nextword, margin, ++i*fm.getHeight());
     }
     else {
g.drawString(sb.toString(), margin, ++i*fm.getHeight());
sb = new StringBuffer(nextword + " ");
}

}
if (sb.length() > 0) {
     g.drawString(sb.toString(), margin, ++i*fm.getHeight());
     }
}

}
```

EXERCISES

Q. 13.1. What is the use of HTML?

Q. 13.2. List out the Basic rules of HTML?

Q. 13.3. What is the requirement of Java applets?

Q. 13.4. Differentiate between Applet and application.

Q. 13.5. Explain the life cycle of Java Applets.

Q. 13.6. What is the role of Applet class?

Q. 13.7. Can we use the constructors in Applets?

Q. 13.8. Create an applet in which one string moves from left to right.

Q. 13.9. Creat an applet to display the shape of triangle, Rectangle, circle with different colors.

Q. 13.10. Create an applet to display following figure.

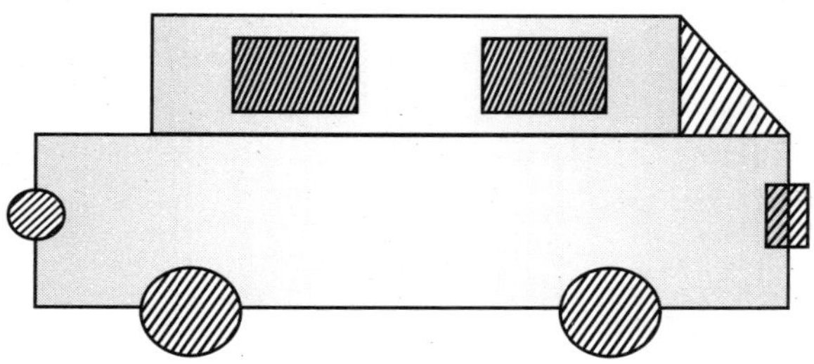

Q. 13.11. Explain the following tags:

(*a*) <head> (*b*) <h1>, <h2>

(*c*) <body> (*d*) <b>

(*e*) <title> (*f*) <p>

(*g*) <applet>

Chapter 14

JAVA Library

14.1. FILE HANDLING

When inputting and outputting to the screen we make use of the `System.in` and `System.out` objects as discussed in chapter 11, `System.in` and `System.out` are special *stream* objects which read data from the keyboard and write data to the screen. Just see example given below:

CHAPTER OUTLINE

- File Handling
  - File Input
  - File Output
  - The File Class
- Networking
  - Internet Basic
  - IP-Datagram
  - TCP-UDP
  - Ports
  - Java Networking
  - URL Class
  - Reading file from URL
  - InetAddress Class
  - Sockets

Summary
Solved Problems
Exercises

Example 14.1: HELLO WORLD PROGRAM

```
import java.io.*;
class HelloWorld2
{
    // Create BufferedReader class instance
    static InputStreamReader input = new
    InputStreamReader(System.in);
    static BufferedReader keyboardInput =
    new BufferedReader(input);

    /* Main method */
    public static void main(String[] args)
    throws IOException
    {
    String name;

    System.out.print("What is your name?");
    name = keyboardInput.readLine();

    System.out.print("\nHello " + name );
```

```
        System.out.println(" - Congratulations on writing your first" +
            " Java program which features some input!\n\n");
    }
}
```

If we wish to read and write to a file we need to replace the `in` and `out` objects with appropriate alternative stream objects.

14.1.1. File Input

Before we can read from or write to a file we must first *open* the file. To open a file for reading we use the constructor contained `FileReader` class:

```
FileReader file = new FileReader("HelloWorld2.java");
```

This creates a stream object, called `file`, to read from a file called `HelloWorld2.java` (Example 14.1) contained in the current directory. If the named file cannot be opened a `FileNotFoundException` is generated. Consequently, whereas previously we declared instances of the `InputStreamReader` class as class instances we must now declare them as local instances within a method. Having done this we can now use the `file` object to create an instance of the `BufferedReader` class as before:

```
FileReader file = new FileReader("HelloWorld2.java");
BufferedReader fileInput = new BufferedReader(file);
```

Note the similarity between this and the declarations used in the code given in Example 2:

```
InputStreamReader input = new InputStreamReader(System.in);
BufferedReader keyboardInput = new BufferedReader(input);
```

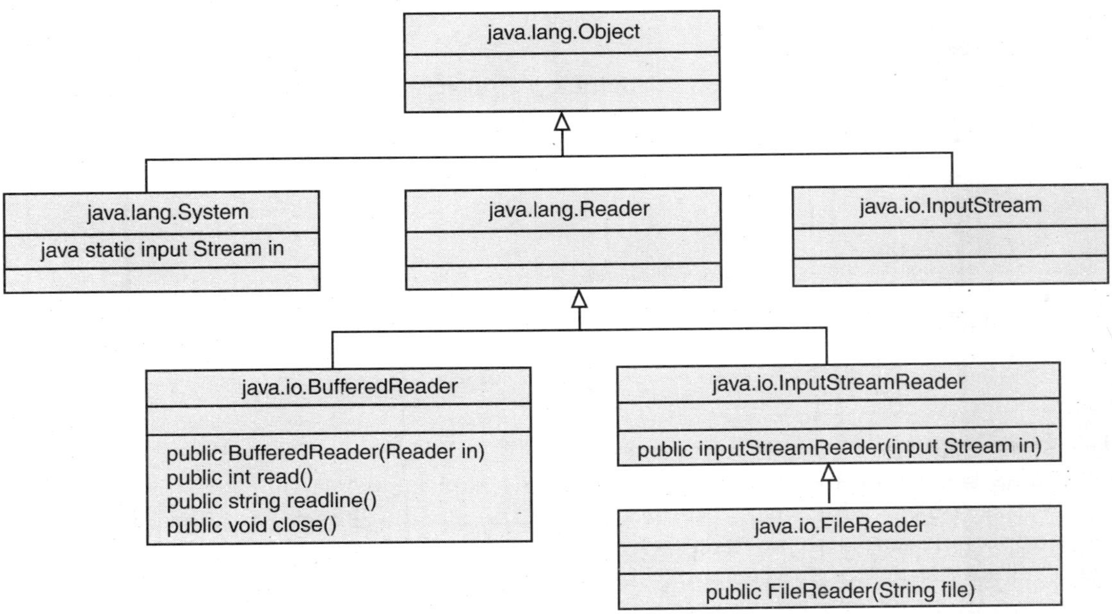

Fig. 14.1. Class diagram showing the connection between the different output classes

The BufferedReader constructor requires its argument to be an instance of the class Reader (or a sub-class of the class Reader), both the InputStreamReader and the FileReader classes are sub-classes of Reader and will therefore suffice. A class diagram illustrating the connections between these different classes is presented in figure 14.1.

Once we have finished reading from a file we should close it; this is achieved using the close instance method contained in the BufferedReader class. Because this is an instance method we must, of course, link it to an instance of the class BufferedReader:

```
fileInput.close();
```

The code presented in Example 14.2 reads the contents of the file HelloWorld2.java (presented in Example 14.1) and outputs the contents to the screen. Note that, in the Example 14.2, we keep reading from the file until we have read 31 lines.

Example 14.2: File Input Example

```
import java.io.*;

class FileIOExample1
{
    /* Main method */

    public static void main(String[] args) throws IOException
    {
        FileReader file = new FileReader("HelloWorld2.java");
        BufferedReader fileInput = new BufferedReader(file);
        String text;
        final int NUMBER_OF_LINES_IN_FILE = 31;

        // Read file and output
        for(int counter=0;counter < NUMBER_OF_LINES_IN_FILE;counter++) {
            text = fileInput.readLine();
            System.out.println(text);
            }

        // Close file

        fileInput.close();
        }
}
```

14.1.2. File Output

We have seen how we output to the screen (Example 14.1) using the system.out screen output object which is contained within the class System. An alteranative more generic approach can also be used where we create an instance of the PrintWriter "stream" class using the system.out object as shown in Example 14.3. Here we create an instance of the class PrintWriter, called screenOutput, which can be used in conjunction with the `print` and `println`methods contained in the `printWriter` class (as oposed to those methods with the same name contained in the `PrintStream` class and used in the example code in Example 14.1). The inter-relationship between the various output classes is illustrated by the class diagram presented in figure 14.2.

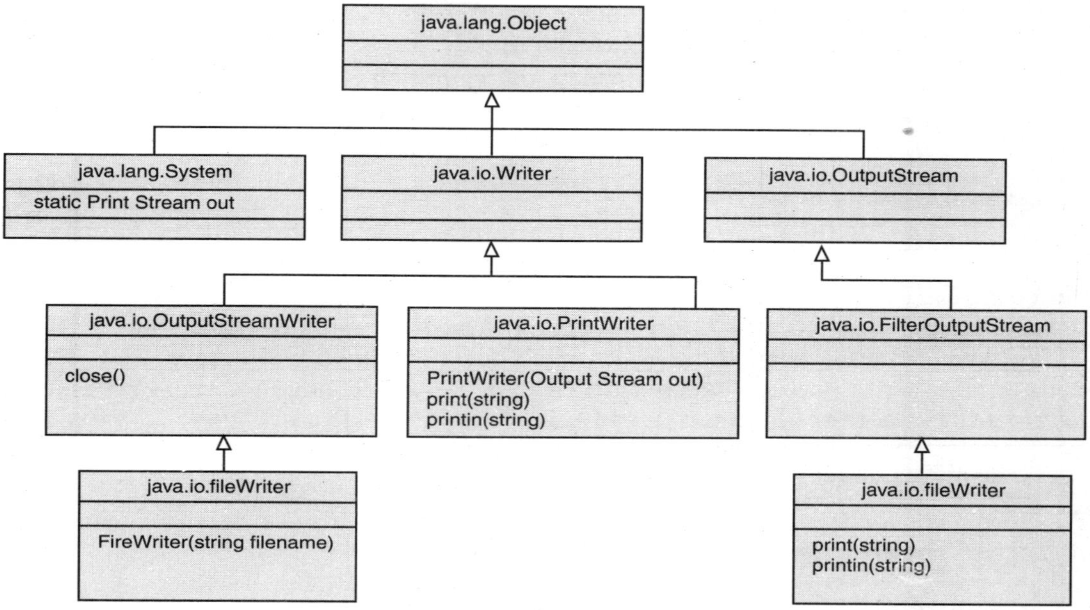

Fig. 14.2. Class diagram showing the connection between the different output classes

Example 14.3: Standard input/output

```
import java.io.*;

class HelloWorld4
{

    // Create BufferedReader class instance

    public static InputStreamReader input= new InputStreamReader (System.in);
    public static BufferedReader keyboardInput = new BufferedReader(input);

    // Crerste PrintWriter class instance

    static PrintWriter screenOutput = new PrintWriter(System.out,true);

    /* Main method */

    public static void main(String[] args) throws IOException
    {
```

```
        String name;
        screenOutput.print("What is your name? ");
        screenOutput.flush();
        name = keyboardInput.readLine();

        screenOutput.print("\nHello " + name );
        screenOutput.println(" - Congratulations on writing a more" +
            " sophisticated Java program that inovles output!\n\n");
    }
}
```

We can use exactly the same approach to output to a file by replacing the `System.out` object with an alternative "output to file" object. The Java class `FileWriter` class contained in the `java.io` package includes a constructor (with a single argument that must be a string) that creates such an object. The argument is the name of the file which we wish to write to (if there is no such file then Java will create one). Thus if we wish to (say) write to a file called `myFile` then we would have to create an instance of the class `FileWriter` (lets call it `file`) as follows:

```
FileWriter file = new FileWriter("myFile");
```

We now have an "output to file" object called `file` which we can use to create a `PrintWriter` object:

```
PrintWriter fileOutput = new PrintWriter(file);
```

which can be used as follows:

```
fileOutput.print("Halo World")
```

or:

```
fileOutput.println("Halo World Again")
```

Thus to write to a file we replace the `System.out` stream object with an appropriate "file output" stream object. We create such an object using the `FileWriter` class constructor as shown above. However, the class `FileWriter` does not contain the instance methods `print` and `println` which we would like to use. These are contained in the `PrintWriter` class, thus we must also create an instance of this class.

We can now adapt the code given in Example 14.2 so that the contents of the file `HelloWorld2` are output to a file called `testFile`; this is done in Example 14.4.

Example 14.4: File input and output example

```
import java.io.*;

class FileIOExample2
{

    /* Main method */
```

```
    public static void main(String[] args) throws IOException
    {
        FileReader file1 = new FileReader("HelloWorld2.java");
        BufferedReader fileInput = new BufferedReader(file1);
        FileWriter file2 = new FileWriter("testFile");
        PrintWriter fileOutput = new PrintWriter(file2);
        String text;
        final int NUMBER_OF_LINES_IN_FILE = 42;

    // Read file and output

    for(int counter=0;counter < NUMBER_OF_LINES_IN_FILE;counter++) {
        text = fileInput.readLine();
        System.out.println(text);
        fileOutput.println(text);
        }

        // Close file

        fileInput.close();
        fileOutput.close();
    }
}
```

14.1.3. The File Class

We express the location of a file using a *path name* (absolute or relative). The `File` class contains (amongst other things) a number of methods that can be used to check whether a named file exists or not. In Example 14.5, the code presented in Example 14.2 is repeated but with the inclusion of a method to check a file name, this time presented as a command line argument, using some of the methods contained in the `File` class.

Example 14.5: File input example with "file check"

```
import java.io.*;

class FileIOExample3
{

    /* Main method */

    public static void main(String[] args) throws IOException
    {
        String fileName = new String(args[0]);

        // CheckFile
```

```
        if (!checkFile(fileName)) System.exit(1);

        // Process

        FileReader file = new FileReader(fileName);
        BufferedReader fileInput = new BufferedReader(file);
        String text;
        final int NUMBER_OF_LINES_IN_FILE = 31;
        // Read file and output
        for(int counter=0;counter < NUMBER_OF_LINES_IN_FILE;counter++) {
            text = fileInput.readLine();
            System.out.println(text);
            }
        // Close file
        fileInput.close();
        }
    /* Check File */
    private static boolean checkFile(String fileName) {
        File src = new File(fileName);
        if (src.exists()) {
        if (src.canRead()) {
            if (src.isFile()) return(true);
            else System.out.println("ERROR 3: File is a directory");
            }
        else System.out.println("ERROR 2: Access denied");
        }
    else System.out.println("ERROR 1: No such file");
    return(false);
    }
}
```

14.2. NETWORKING

Networking architecture is based on the concept of layers. From Java Network Programming comes this layer model.

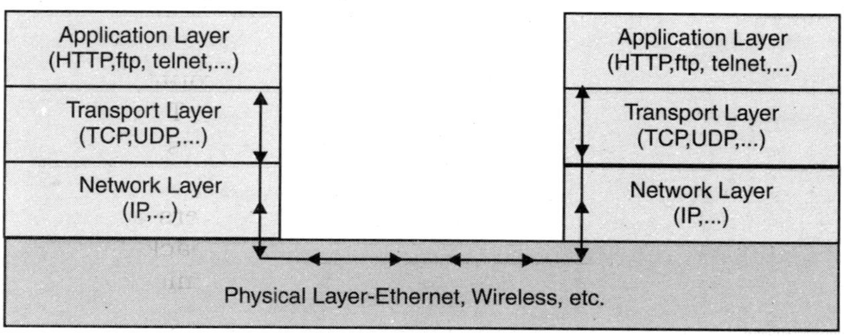

Fig. 14.3. Layer architecture

(A more formal OSI-Open System Interconnection - model has 7 layers but this one shows the essential layer definitions.)

Each layer has its own standardized protocols and applications programming interface (API), which refers to the functions, and their arguments and return values, called by the next higher layer.

Internally, the layers can be implemented in different ways as long as externally they obey the standard API.

For example, the Network Layer does not know if the Physical Layer is Ethernet or a wireless system because the device drivers respond to the function calls the same way.

The *Internet* refers primarily to the Network Layer that implements the Internet Protocol (IP) and the Transport Layer that implements the Transmission Control Protocol (TCP). In fact, you often here people refer to the "TCP/IP" network rather than calling it the Internet.

The application layer also includes various protocols, such as FTP (File Transport Protocol) and HTTP (Hypertext Transfer Protocol) for the Web, that rely on the TCP/IP layers.

Most users never look below the application layer. Most application programmers never work below the TCP/IP layers.

14.2.1. Internet Basic

As mentioned in the above, the Internet refers to the TCP & IP protocols used in the Transport and Network layers, respectively.

The diagram below shows how the TCP and IP layers split an email into groups of bytes called *packets*. A packet contains a *header* with the destination and source addresses and miscellaneous other information such as error correction bytes. (We will discuss the IP address format later.) The body of the packet consists of the *data payload*.

These packets are sent through the network towards there destination. At the *nodes* of the network, there are devices called routers. The routers read the destination addresses on the packets and send the packet to the next node closest to the final destination, just as a letter in the postal delivery system is passed along to the next node nearest the final address. If a node doesn't respond or if the traffic load needs balancing, the transmitting node looks for alternate routes.

When the packets reach their final destination, the original message is rebuilt. The packets may arrive out of order so the rebuilding must wait for all packets to arrive.

The above figure sketches the packet's travels through the network. A node in the local area network (LAN) could be another computer or a dedicated *router*, which is essentially a computer dedicated to packet transmission. A router receives packets from different nodes and sends them to the proper destination. Typically the local router sends the packet to an internet service provider (ISP), such as a DSL or cable internet system, which is connected to the *Internet backbone.*

The backbone uses very high speed, high capacity routers and fibers to send the packet to the ISP closest to the destination. From there the packet works its way to the destination computer, which might lie inside a LAN, which might in turn connect to a WAN (wide area network) such as on a computer campus.

The header consists of a destination address, source address, and other miscellaneous information such as error correction bytes. Routers read the destination address and send the packet to the next link determined to be the shortest path.

This path can vary if, for example, the router determines that a previously chosen link has too much traffic or the router at the other end of the link is not responding.

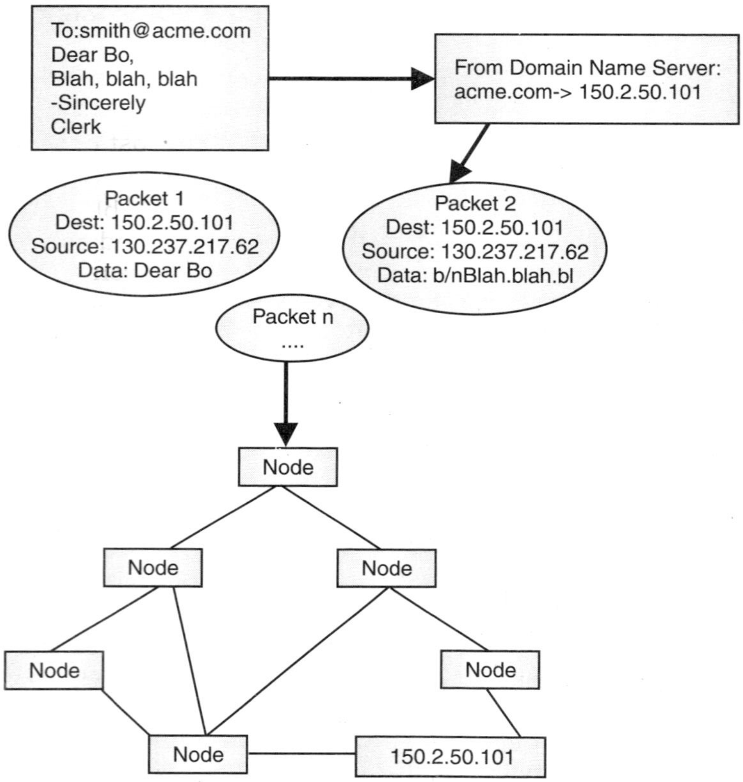

Fig. 14.4.

14.2.2. IP-Datagram

The IP layer communicates using packets called *datagrams*. Datagrams contain headers of 20-60 bytes and data payloads of up to 65K bytes.

Datagrams contain the basic IP source and destination addresses, where the IP (Internet Protocol) address consists of four bytes as in `130.237.217.62`. The left most byte is the highest order addressing and so could represent a region or country. The lower order bytes narrow the address down until the final byte indicates, for example, a particular computer on a LAN (Local Area Network).

There are a number of special addresses. For example, any address beginning with 127 such as 127.0.0.1 will translate as a *loop back* address. This means that any packets sent with this destination address will automatically return to the source computer.

Users typically deal with text addresses called hostnames that are easier to remember than the numerical addresses. For example, java.sun.com is a hostname. Special nodes on the Internet called *Domain Name Servers* (DNS) translate hostnames into the corresponding numerical IP addresses. A hostname is composed of a top level domain such as `com`, `se`, or `edu`. These domains are then divided into second level subdomains such as `sun.com` , `kth.se`, or `ucr.edu`. The systems at these locations can divide a domain further such as `java.sun.com`, `gluon.particle.kth.se`, or `physics.ucr.edu`.

14.2.3. TCP-UDP

Transmission Control Protocol (TCP) and the User Datagram Protocol (UDP) are two separate Internet protocols within the Transport Layer.

Datagrams from the same message can travel completely different paths as the routers dynamically choose paths for the same destination address so as to avoid loading down any one link. Thus datagrams may become lost or arrive out of order from how they were sent.

The Transport Layer attempts to smooth over the problems of the IP layer. The Transport Layer can place packets into the proper order and request *retransmission* of missing packets (that is, packets not arriving within a given time).

TCP guarantees bytes in correct order. UDP does not guarantee correct order but has lower overhead than TCP.

For some applications, such as audio and video transmission, the loss of a few bytes is not noticeable and so the UDP serves very well since it is faster than TCP.

Application Layer

The Application Layer involves all the user related programs like web browsers and ftp. The application layer programs use various protocols:

- **HTTP** - Hyper Text Transfer Protocol
- **FTP** - File Transfer Protocol
- **Telnet** - the telnet protocol provides for console sessions.
- **SMTP** - Simple Mail Transport Protocol - email.

These protocols require TCP since data can not be randomly dropped from source files and web pages. Other application layer programs like `ping`, which sends test packets to a given IP address, can use the simpler UDP. Most Java network programs only deal with this application layer.

14.2.4. Ports

While the IP address will get the packet to the right computer, how does the packet get to the right program? A computer might have:

- Several programs running at the same time
- Several programs trying to communicate via the Internet
- All the programs communicating out the same physical Ethernet cable

The *Port* number is the key for organizing all of this. The packets are guided to the correct program by the operating system according to the port number as indicated in figure 14.5.

The port is a 16-bit number. It is occasionally seen on Web addresses:

```
http://www.someplace.com:80/
```

where 80 is the default for the HTTP server. Nowadays the port number is seldom included in the URL since the default is almost always used.

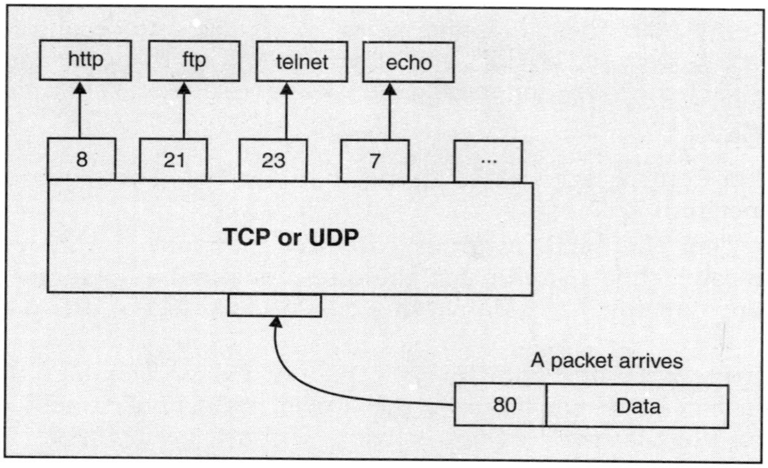

Fig. 14.5. Standard ports

Various applications using particular protocols use standard port values as shown in the above figure 14.5. UNIX machines reserve ports 1-1023 for standard services. Windows machines do not restrict the use of these ports but to make your Java programs portable, it is wise to use port values above 1023.

One type of firewall assigns port numbers to the machines behind its shield. Incoming packets all go to the same IP address but with different port numbers according to the machine they are destined for. This both reduces the exposure and accessibility to the machines and reduces the need for universally unique IP numbers. Behind the firewall, the local addresses can be the same as in other LANs since the LANs are isolated from each other.

This is one of the reasons that four bytes remain sufficient for the internet despite the explosion in the number of devices with IP addresses. (Nevertheless, a sixteen byte version of IP addressing called IPv6 will gradually take over.)

14.2.5. Java Networking

Java was designed from the start with networking in mind. Java became famous because of? applets, which were invented to provide dynamic content to web pages. Many other features of Java, however, extend Java network programming far beyond mere applets. Some of these include:

- `java.net` package in the core language includes numerous network related classes.
- ***Remote Method Invocation*** **(RMI)** packages allow a Java program running on one machine to invoke methods on another machine.
- ***Streaming I/O*** architecture provides a uniform structure in that I/O over a network is treated the same as I/O to a disk or the console.
- ***Serialization*** allows objects to be sent over a network and reconstructed on another machine.
- ***Threading*** allows easier creation of server programs where threads can be easily spun off to serve new clients as they connect.
- `java.security` and other packages allow for secure interactions over networks.

- ***Portability*** of java means that Java networking programs developed on one type of platform can also run and communicate with each other when used on other platforms.

14.2.6. URL Class

Browsers obtain resources from the Web by specifying the Web addresses, which are referred to officially as **Uniform Resource Locators** (URL). For example,

```
http://www.ibm.com/
```

The formal description of the URL components goes as follows:

```
Protocol_ID://Host_IP_address:Port/directory/Filename#Target
```

where

Protocol_ID = HTTP, FTP, etc

Host_IP_address = Host name in either numerical IP or hostname format

Port = default is 80

Filename = Name of a hypertext web page or other type of file. (Default index.htm[l])

Target = optional reference address within a Web page

In Java we can create an instance of the `java.net.URL` class. These specify the location of a file on the network. We have, for example, used `URL` objects to obtain the location of image files in applets as shown here:

```
getImage( getCodeBase(), filename);
```

The applet method `getCodeBase()` returns a `URL` object for the location of the applet file. Here it is assumed the image file resides in the same directory

The `URL` class provides several methods for obtaining URL information, For Example:

Example 14.6: Parse a URL address

```
import java.net.*;
import java.io.*;
/** Parse a URL address into its components. **/
public class ParseAddress
{
    public static void main (String[1 args) {
    if (arg.length !=1) {
          System.out.println ("Error: missing the url argument");
          System.out.println ("Usage: java ParseURL <url>");
          System.exit (0);
 }

    try {
          URL url = new URL (arg[0]);
          System.out.println ("Protocol = " + url.getProtocol ());
          System.out.println ("Host = " + url.getHost ());
          System.out.println ("File name = " + url.getFile ());
          System.out.println ("Port = " + url.getPort ());
```

```
        System.out.println ("Target = " + url.getRef ());
    }
    catch (MalformedURLException e) {
    System.out.println ("Bad URL = " + arg[0]);
    }

  }
} 7
```

You run the program as follows:

```
c:\> java ParseAddress http://www.myschool.edu:80/data/x.html#d
```

This program outputs:

```
Protocol = http
Host = www.myschool.edu
File name = /data/x.html
Port = 80
Target = d
```

Also, URL's can be constructed from a URL string

```
URL kth = new URL ("http//www.kth.se/index.html");
```

or from separate protocol, host, file strings:

```
URL kth = new URL ("http","www.kth.se","/index.html");
```

When you attempt to create a `URL` object, the constructor will validate the components. If they are in error, it will throw a `MalformedURLException`. So you must catch this exception from the `URL` constructors.

14.2.7. Reading file from URL

Often one of the first applets that new Java programmers want to write for their own work involves reading a file of some sort. This could be simple database information such as a list or table of values and labels in text format.

We have learned to read image files and audio files from an applet but have not yet read a text file. This is actually fairly easy to do. However, keep in mind that for applets the `SecurityManager` restricts access to only those files accessible from the location of the web page on which the applet code resides.

Below we show the applet `ReadFileViaURL` that reads a text file. Note that we use the `URL` class that we discussed in the previous section. In this case, the `URL` class provides a method to obtain file access via an `InputStream` object.

```
// Open a stream to the file using the URL.
try {
      InputStream in=url.openStream ();
      BufferedReader dis =
      new BufferedReader (new InputStreamReader (in));
      fBuf = new StringBuffer () ;

      while ( (line = dis.readLine ()) != null) {
```

```
      fBuf.append (line + "\n");
      }
      in.close ();
}
catch (IOException e) {
      fTextArea.append ("IO Exception = "+e);
      System.out.println ("IO Exception = "+e);
      return;
}
```

We wrap this stream with an `InputStreamReader` to provide proper character handling. We in turn wrap this stream with an `BufferedReader`, which provides buffering to smooth out the stream flow and also contains the convenient `readLine()` method that grabs an entire line and returns it as a string.

The program can also run as a standalone application.

14.2.8. InetAddress Class

The `java.net.InetAddress` class represents IP addresses. It works with either the hostname (e.g., `www.kth.se`) or the numerical IP address (e.g. 64.124.81.56) of an Internet host. `InetAddress` class offers a number of useful methods for dealing with hostnames and IP addresses.

The Example 14.9 program below illustrate some of the capabilities of the `InetAddress` class.

The applet `LocalAddress` displays the local host and IP address. Note that the SecurityManager in the browser JVM may block access to this information. You can try it in the appletviewer tool or run it as an application.

Example 14.7: Show how InetAddress can provide the local IP address

```
import java.applet.*;
import java.awt.*;
import java.net.*;
public class LocalAddress extends Applet
{
    String fMessage = "";
    /**
    * Create an instance of InetAddress for the
    * local host and display the local IP address.
    **/
public void init () {
    try {
    InetAddress local_Address = InetAddress.getLocalHost ();
    fMessage ="Local IP address = " + local_Address.toString ();
    }
    catch (UnknownHostException e) {
    fMessage ="Unable to obtain local IP address";
    }
```

```
    System.out.println (fMessage);
    } // init

    /** Paint IP info in Applet window. **/
    public void paint (Graphics g) {
    g.drawString (fMessage,20,20);
}
    /** Print out the local address in app mode.**/
    public static void main (String [] args) {
        LocalAddress applet =
        new LocalAddress ();
        applet.init ();
    } // main

} // Class LocalAddress
```

The following Example 14.8, returns the IP address if given a host name, or returns the host name if given the IP address.

Example 14.8: IP argument to a host name or a host name to an IP address.

```
import java.net.*;
class TranslateAddress {
    public static void main (String args[])
    {
     // Look for command line argument
    if (args.length != 1) {
    System.out.println ("Error! No IP or host name address");
    System.out.println ("Usage: java TranslateAddress java.sun.com");
    System.out.println ("   or java TranslateAddress 209.249.116.143");
    System.exit (0);
}
    try {
    // When the argument passed is a host name (e.g. sun.com),
    // the corresponding IP address is returned. If passed
    // an IP address, then only the IP address is returned.
    InetAddress address = InetAddress.getByName (args[0]);
    System.out.println ("Address " + args[0] + " = " + address);

    // To get the hostname when passed an IP address use
    // getHostName (), which will return the host name string.
    System.out.println ("Name of " + args[0] + " = " +
            address.getHostName ());
    } catch (UnknownHostException e) {
   System.out.println ("Unable to translate the address.");
  }
```

```
  } // main
}
```

Sample runs of this Example program:

```
c:\> java TranslateAddress gluon.particle.kth.com
Address gluon.particle.kth.se = gluon.particle.kth.se/130.237.34.133
Name of gluon.particle.kth.se = gluon.particle.kth.se

c:\> java TranslateAddress 130.237.34.133
Address 130.237.34.133 = /130.237.34.133
Name of 130.237.34.133 = gluon.particle.kth.se
```

Note that `getByName(String str)` returns an instance of `InetAddress` when given a host name. Then when this objects converts to a string it displays the `HostName/IP_Address` pair.

However, when given an IP address the resulting `InetAddress` object does not show the host name. Instead you must use the the `getHostName()` method to obtain the hostname.

14.2.9. Sockets

Sockets provide connections between applications and allow streams of data to flow. Sockets are fairly straightforward to set up in Java. Java provides two kinds of sockets:

1. **Socket:** provides a connection-oriented protocol that behaves like telnet or ftp. The connection remains active, even with no communications occurring, until explicitly broken.
2. **DatagramSocket:**
   - a connectionless protocol
   - transfers datagram packets
   - no fixed connection
   - does not keep packets in order
   - no guarantee a packet will arrive at its destination

The class `ServerSocket` is not a socket , but it monitors a given port and returns a `Socket` object when a client attempts to connect to that port.

> **Concept**
>
> *Sockets provide connection between applications of clients/ server and allow streams of data to flow.*

The Example program 14.9 shown below, illustrates how to run the ***whois*** internet operation that returns information about a given domain name. The registry site `whois.internic.net` provides this service from anywhere on the internet.

The code segment shows how to create a socket that connects a stream to port 43 at the remote site. An output stream is obtained from the socket and a PrintWriter wrapped around it. This is then used to send a whois query to that socket.

Similarly, an input stream is obtained from the socket and wrapped first with an InputStreamReader and then a BufferedReader. The latter class provides the readLine()

routine or obtained a whole line of text at one time from the whois output. ?You can enter other domain names in the text field.

Though written in applet form, the security manager in most browser JVMs will block accesses to IP addresses other than the source of the applet. So we only show an image of a typical output. You can, however, run the program as an application or an appletviewer.

Example 14.9: The whois output in a text area

```
import javax.swing.*;
import java.awt.*;
import java.awt.event.*;
import java.util.*;
import java.io.*;
import java.net.*;

public class WhoisApplet extends JApplet  implements ActionListener
{
    // A Swing textarea for display of string info
    JTextArea fTextArea = null;
    JTextField fTextField = null;
    String fFileToRead="data.txt";

    //Buttons
    JButton fGoButton;
    JButton fClearButton;
    JButton fExitButton;

    // Flag for whether the applet is in a browser
    // or running via the main () below.
    boolean fInBrowser = true;
    public void init () {

    // Create a User Interface with a textarea with sroll bars
    // and a Go button to initiate processing and a Clear button
    // to clear the textarea.
    Container content_pane = getContentPane ();
    JPanel panel = new JPanel (new BorderLayout ());

    // Create an instance of DrawingPanel
    fTextArea = new JTextArea ();
    fTextArea.setEditable (false);
    fTextArea.setLineWrap (true);

    // Add to a scroll pane so that a long list of
    // computations can be seen.
    JScrollPane area_scroll_pane = new JScrollPane (fTextArea);
```

```
panel.add (area_scroll_pane,"Center");
fTextField = new JTextField ("sun.com", 10);
// If return hit after entering text, the
// actionPerformed will be invoked.
fTextField.addActionListener (this);

fGoButton = new JButton ("Go");
fGoButton.addActionListener (this);

fClearButton = new JButton ("Clear");
fClearButton.addActionListener (this);

fExitButton = new JButton ("Exit");
fExitButton.addActionListener (this);

JPanel control_panel = new JPanel ();

control_panel.add (fTextField);
control_panel.add (fGoButton);
control_panel.add (fClearButton);
control_panel.add (fExitButton);

panel.add (control_panel,"South");
// Add text area with scrolling to the contentPane.
content_pane.add (panel);

} // init
/** Connect to the whois service via a socket. Write the
* address with a PrintWriter object and read the output
* via a BufferedReader. Send the output to the text area.
**/
public void whoisConnect ( String queryAddress) {

int port = 43; // Standard whois port
String reply;

try {
Socket whois_socket = new Socket ("whois.internic.net", port);

PrintWriter print_writer =
new PrintWriter ( whois_socket.getOutputStream (),true);
print_writer.println (queryAddress );

InputStreamReader input_reader =
new InputStreamReader (whois_socket.getInputStream ());
```

```
    BufferedReader buf_reader =  new BufferedReader (input_reader);

    while ( (reply = buf_reader.readLine ()) != null) {
    // Write the whois info to the textarea.
    fTextArea.append (reply + '\n');

}
    fTextArea.append ("\n\n"); // Add space between queries
    } catch (IOException e ){
    fTextArea.append ("IO Exception = "+e);
    System.out.println ("IO Exception = "+e);
    return;
}
    } // whoisConnect
    /** Respond to the buttons **/
    public void actionPerformed (ActionEvent e) {

    Object source = e.getSource ();
    if (source == fGoButton || source == fTextField ) {
    String query = fTextField.getText ();
    if (query == null || query.length () <=1 )
    fTextArea.setText ("Error: Bad host entry");
    else
        try {
        whoisConnect (query);
        }
    catch (Exception ec) {
    fTextArea.append("\n\n Error attempting to connect to WhoIs!\n");
    fTextArea.append("Exception = \n" + ec);

    } else if (source == fClearButton)
        fTextArea.setText (null);
    else if (!fInBrowser)
        System.exit (0);
    } // actionPerformed
    /** Display the println string on the text area **/
    public void println (String str) {
        fTextArea.append (str + '\n');
    }
    /** Display the print string on the text area **/
    public void print (String str){
    fTextArea.append (str);
}
    public static void main (String[] args) {
    //
```

```
    int frame_width=450;
    int frame_height=300;

    //
    WhoisApplet applet = new WhoisApplet ();
    applet.fInBrowser = false;
    applet.init ();

    // Following anonymous class used to close window & exit program
    JFrame f = new JFrame ("whois Demo");
    f.setDefaultCloseOperation (JFrame.EXIT_ON_CLOSE);

    // Add applet to the frame
    f.getContentPane ().add ( applet);
    f.setSize (new Dimension (frame_width,frame_height));
    f.setVisible (true);
    } // main

}// class WhoisApplet
```

Note: WhoisApplet operations are blocked by the JVM security manager in most browsers but you can run it in standalone mode instead.

Below we show an image of an example of the program in action:

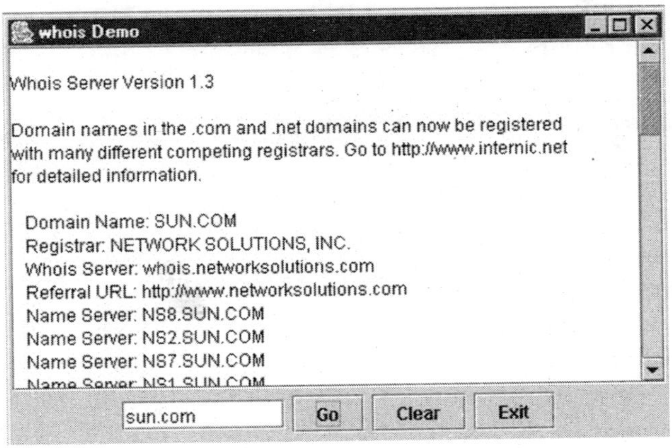

Fig. 14.6.

SUMMARY

— This chapter is dedicated to java library classes, we have explain major classes of java library which are used commonly.

— The two major classes of strings are **String** and String Buffer class, String Buffer classes used to handle the mutable strings, which can be change.

— File handling is performed by the input and output stream classes.

— Java was designed from the start with networking in mind thats why java provide many features for Networking via java programming.

— Event handling is the main advantage of Java for GUI programming we have explained main event handling examples.

SOLVED PROBLEMS

Q. 14.1. Why are streams used in Java?

Ans. In Java streams are used to process input and output data as a sequence of bytes, which does not assume a specific character content or encoding. Byte content can be read from network sources, files and other sources, and bytes copied between multiple inputs and outputs.

Streams types can be sub-classed to add filtering, to mark a point in the stream and re-read those bytes, or to skip a number of bytes. Streams also enable serializable objects to stored and re-constructed using ObjectInputStream and ObjectOutputStream types.

Q. 14.2. What's the difference between a file and a filename? Which is more closely related to the java.io.File class?

Ans. A file is the actual data located at a particular location on a disk. A filename is used to refer to the file. The name of the file can be changed without changing the data. The java.io.File class is closer to a filename than a file.

Q. 14.3. Explain RMI mechanism?

Ans. Sometimes it is necessary to invoke a method of an object that is being executed on a remote machine as part of distributed computing. This shares computation between 2 machines. Fortunately java has provided this mechanism. This mechanism is called RMI.

Q. 14.4. What is the difference between URLinstance and URLConnection instance?

Ans. A URLinstance represents the location of a resource, and a URLConnection instance represents a link for accessing or communicating with the resource at the location.

Q. 14.5. How do I make a connection to URL?

Ans. You obtain a URL instance and then invoke openConnection on it.

URLConnection is an abstract class, which means you can't directly create instances of it using a constructor. We have to invoke openConnection method on a URL instance, to get the right kind of connection for your URL.

Example:
```
URL url;
    URLConnection connection;
    try{ url = new URL("...");
    conection = url.openConnection();
    }catch (MalFormedURLException e) { }
```

Q. 14.6. What Is a Socket?

Ans. A socket is one end-point of a two-way communication link between two programs running on the network. A socket is bound to a port number so that the TCP layer can identify the application that data is destined to be sent. Socket classes are used to represent the connection between a client program and a server program. The java.net package provides two classes—Socket and ServerSocket—which implement the client side of the connection and the server side of the connection, respectively.

Q. 14.7. What information is needed to create a TCP Socket?

Ans. The Local System's IP Address and Port Number. And the Remote System's IPAddress and Port Number.

Q. 14.8. What are the two important TCP Socket classes?

Ans. Socket and ServerSocket.

ServerSocket is used for normal two-way socket communication. Socket class allows us to read and write through the sockets. getInputStream() and getOutputStream() are the two methods available in Socket class.

Q. 14.9. When MalformedURLException and UnknownHostException throws?

Ans. When the specified URL is not connected then the URL throw MalformedURLException and If InetAddress' methods getByName and getLocalHost are unabletoresolve the host name they throwan UnknownHostException.

Q. 14.10. The API doesn't list any constructors for InetAddress. How can we create an InetAddress instance?

Ans. In case of InetAddress the three methods getLocalHost, getByName, getByAllName can be used to create instances.

Example,

```
InetAddress add1;
InetAddress add2;
try{
add1 = InetAddress.getByName("java.sun.com");
add2 = InetAddress.getByName("199.22.22.22");
}catch(UnknownHostException e){}
```

Q. 14.11. Is it possible to get the Local host IP?

Ans. Yes. Use InetAddress's getLocalHost method.

Q. 14.12. What's the Factory Method?

Ans. Factory methods are merely a convention whereby static methods in a class return an instance of that class. The InetAddress class has no visible constructors. To create an InetAddress object, you have to use one of the available factory methods. In InetAddress the three methods getLocalHost, getByName, getByAllName can be used to create instances of InetAddress.

Q. 14.13. What's the difference between TCP and UDP?

Ans. These two protocols differ in the way they carry out the action of communicating. A TCP protocol establishes a two way connection between a pair of computers, while the UDP protocol is a one-way message sender. The common analogy is that TCP is like making a phone call and carrying on a two-way communication, while UDP is like mailing a letter.

Q. 14.14. What is the Proxy Server?

Ans. A proxy server speaks the client side of a protocol to another server. This is often required when clients have certain restrictions on which servers they can connect to. And when several users are hitting a popular web site, a proxy server can get the contents of the web server's popular pages once, saving expensive internetwork transfers while providing faster access to those pages to the clients.

Also, we can get multiple connections for a single server.

Q. 14.15. What are the seven layers of OSI model?

Ans. Application, Presentation, Session, Transport, Network, DataLink, Physical Layer.

Q. 14.16. What Transport Layer does?

Ans. It ensures that the mail gets to its destination. If a packet fails to get its destination, it handles the process of notifying the sender and requesting that another packet be sent.

Q. 14.17. What is DHCP?

Ans. Dynamic Host Configuration Protocol, a piece of the TCP/IP protocol suite that handles the automatic assignment of IP addresses to clients.

Q. 14.18. What is SMTP?

Ans. Simple Mail Transmission Protocol, the TCP/IP Standard for Internet mails. SMTP exchanges mail between servers; contrast this with POP, which transmits mail between a server and a client.

Q. 14.19. Write a program to recursively list all the files and directories contained in a directory given on the command line.

Ans.

```
import java.io.File;

public class rls {

    File dir;

 public static void main(String[] args) {

    for (int i = 0; i < args.length; i++) {
    rls r = new rls(args[i]);
    r.list();
    }

 }

 public rls(String s) {

    dir = new File(s);

 }

    // This method doesn't do anything if dir isn't a directory
    public void list() {

    if (dir.isDirectory()) {
    System.out.println("\n" + dir + ":\n");
    String[] files = dir.list();
    for (int i = 0; i < files.length; i++) {
    System.out.println(files[i]);
    }
    for (int i = 0; i < files.length; i++) {
    rls r = new rls(dir + File.separator + files[i]);
    r.list();
    }

 }
```

```
}

}
```

Note: This program can get stuck in an infinite loop if aliases or symbolic links point to a directory that's already been scanned. How would you fix this?

Q. 14.20. Difference between = = and equals method?

Ans. = = compares the references(hash codes) of the strings so it gives unreliable results. Equals compares the contents of the strings.

EXERCISES

Q. 14.1. How can you create a string in Java programming?

Q. 14.2. What do you mean by immutable strings?

Q. 14.3. How can we compare two strings? Give an example.

Q. 14.4. What are the difference between string class Vs String Buffer Class?

Q. 14.5. Explain the use of FileReader Class.

Q. 14.6. Explain the use of following Networking classes:

(*a*) URL (*b*) InetAddress

(*c*) ServerSocket

Q. 14.7. What do you mean by Events?

Q. 14.8. What are the events we can handle by the programming?

Q. 14.9. What are the information needed for TCP sockets?

Chapter 15

JAVA AWT

15.1. INTRODUCTION

The AWT stands for Abstract Window ToolKit. It is a portable GUI library between Solaris and Windows 95/NT and Mac System 7.X (soon) for stand-alone applications and/or applets. Since it can be used in applets it can be used on IRIX, SunOS, HP/UX, Linux which Netscape 2.0 supports.

The Abstract Window Toolkit provides many classes for programmers to use. It is your connection between your application and the native GUI. The AWT hides you from the underlying details of the GUI your application will be running on and thus is at very high level of abstraction. It takes the lowest common denominator approach to retain portability. No floating toolbars or Balloon help here.

The Abstract Window Toolkit is a Java package and can be used in any Java program by importing java.awt.* via the import keyword.

The Abstract Window Toolkit contains three kind of classes

- Containers
  - Frame, Dialog, Panel, Applet etc.
- Components
  - Label, TextField, Button, Checkbox, Scrollbar etc.
- Supporting classes
  - Color, Font, Dimesion etc.

CHAPTER OUTLINE

> **Concept**
>
> *AWT is a library of classes which provides GUI tools to develope GUI applications and applets.*

The Abstract Window Toolkit provides a class Component which an abstract class that provides all common methods to all the containers and components

- void setSize(int width, int height)
- void setLocation(int x, int y)
- void setBounds(int x, int y, int w, int h)
- void setVisible(boolean b)
- void setBackground(Color c)
- void setFont(Font f)
- void setCursor(Cursor c)

GUI programs can be developed in two ways

1. Java Applications
   - Loaded on every machine
   - To make a java program as GUI application, it must inherit from Frame class
2. Java Applets
   - Run using a web browser
   - Place only on one machine called web server
   - To make a program as applet, it must inherit from Applet class

15.2. WORKING WITH WINDOWS

The first thing you need to do when creating a GUI application is create the window. In Java AWT, top-level windows are represented by the Frame class.

Example 15.1 **Program to create a Window in Java Application.**

```
import java.awt.*;
class AWTexample extends Frame
{
    public AWTexample()
    {
        setTitle("My Window");
        setSize(300,200);
    }
    public static void main( String arg[])
    {
    AWTexample F=new AWTexample();
    F.setVisible(true);
    }
}
```

COMPILE:

```
C:\>javac AWTexample.java
```

EXECUTE:

```
C:\>java AWTexample.java
```

OUTPUT

> **Note:** The output window will not close using close button on title bar because it needs event handling, which will discuss in next chapter. Use Ctrl+C on command window to close.

Example 15.2 **Program to create a Window in Java Application using Frame object.**

```
import java.awt.*;
class AWTexample
{
    private Frame F;
    public AWTexample()
    {
        F=new Frame();
        F.setTitle("My Window");
        F.setSize(300,200);
        F.setBackground(Color.blue);
        F.setVisible(true);
    }
public static void main( String arg[])
    {
    new AWTexample();
    }
}
```

COMPILE:

```
C:\>javac AWTexample.java
```

EXECUTE:

C:\>java AWTexample.java

OUTPUT

A Panel, like a Frame, provides the space for you to attach any GUI components, including others panels. Each Panel, which inherits from Container, can have its own layout manager.

After you create a Panel, you must add it to a Window or Frame. This is done using the add method of the Container class. Then set the frame to visible so that the frame and panel are displayed.

Example 15.3 Program to create a small yellow Panel, and adds it to a Frame.

```
import java.awt.*;
class AWTexample
{
    private Frame F;
    private Panel pan;
    public AWTexample()
    {
        F = new Frame();
        pan = new Panel();
    }
    public void show()
    {
        F.setTitle("My Window");
        F.setSize(300,200);
        F.setBackground(Color.blue);
        F.setLayout(null);
        pan.setSize(100,100);
        pan.setBackground(Color.yellow);
        F.add(pan);
        F.setVisible(true);
    }
```

```
    public static void main( String arg[])
        {
            AWTexample A = new AWTexample();
            A.show();
        }
    }
```

COMPILE:

```
C:\>javac AWTexample.java
```

EXECUTE:

```
C:\>java AWTexample.java
```

OUTPUT

15.3. AWT CONTROLS

Controls are components that allow a user to interact with your application. The AWT supports the following types of controls:

- Buttons
- Checkbox
- CheckboxGroup
- Choice
- Label
- List
- Scrollbar
- TextField
- TextArea

Next, examine each of the AWT components below. Most, but not all, directly extend Component.

You've used most of the components should be familiar to you.

15.3.1. Button

A Button has a single line label and may be "pushed" with a mouse click. See an example how to add a button on the window.

Example 15.4 Program to add a Button on window.

```
import java.awt.*;
class AWTexample extends Frame
{
    public AWTexample()
    {
```

```
            setTitle("My Window");
            setSize(300,200);
            Button B=new Button("Press Me");
            setLayout(null);
            B.setBounds(50,50,90,30);
            add(B);
            setVisible(true);
        }

    public static void main( String arg[])
        {
            new AWTexample();
        }
    }
```

COMPILE:

```
C:\>javac AWTexample.java
```

EXECUTE:

```
C:\>java AWTexample.java
```

OUTPUT

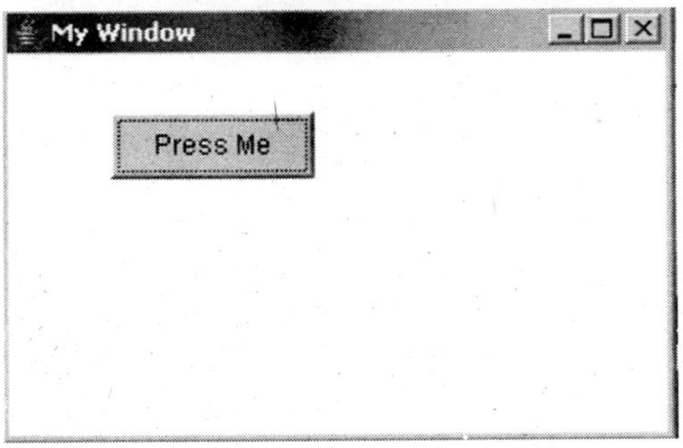

Note: In the above example there is no event handling added; pressing the button will not do anything.

The AWT button has no direct support for images as labels.

15.3.2. CheckBox

A Checkbox is a label with a small pushbutton. The state of a Checkbox is either true (button is checked) or false (button not checked). The default initial state is false. Clicking a Checkbox toggles its state. For example:

Example 15.5 Program to add a CheckBox on window.

```
import java.awt.*;
class AWTexample extends Frame
{
    public AWTexample()
```

```
        {
            setTitle("My Window");
            setSize(300,200);
            Checkbox C=new Checkbox("Check Me");
            setLayout(null);
            C.setBounds(50,50,90,30);
            add(C);
            setVisible(true);
        }
    public static void main( String arg[])
    {
        new AWTexample();
    }
```

COMPILE:

```
C:\>javac AWTexample.java
```

EXECUTE:

```
C:\>java AWTexample.java
```

OUTPUT

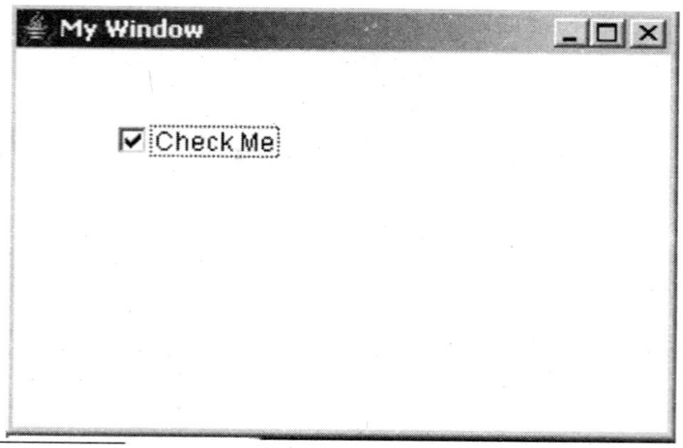

15.3.3. CheckBoxGroup

A CheckboxGroup is used to control the behavior of a group of Checkbox objects (each of which has a true or false state). Exactly one of the Checkbox objects is allowed to be true at one time. Checkbox objects controlled with a CheckboxGroup are usually referred to as "radio buttons".

The following example illustrates the basic idea behind radio buttons.

Example 15.6 Program to add two Radio Buttons on window.

```
import java.awt.*;
class AWTexample extends Frame
{
    public AWTexample()
    {
```

```
            setTitle("My Window");
            setSize(300,200);
            CheckboxGroup cbg=new CheckboxGroup();
            Checkbox C1=new Checkbox("I am Indian", cbg, ture);
            Checkbox C2=new Checkbox("I am American", cbg, false);
            setLayout(null);
            C1.setBounds(50,50,90,30);
            C2.setBounds(50,70,100,30);
            add(C1);
            add(C2);
            setVisible(true);
        }
        public static void main( String arg[])
        {
            new AWTexample(),
        }
    }
```

COMPILE:

```
C:\>javac AWTexample.java
```

EXECUTE:

```
C:\>java AWTexample
```

OUTPUT

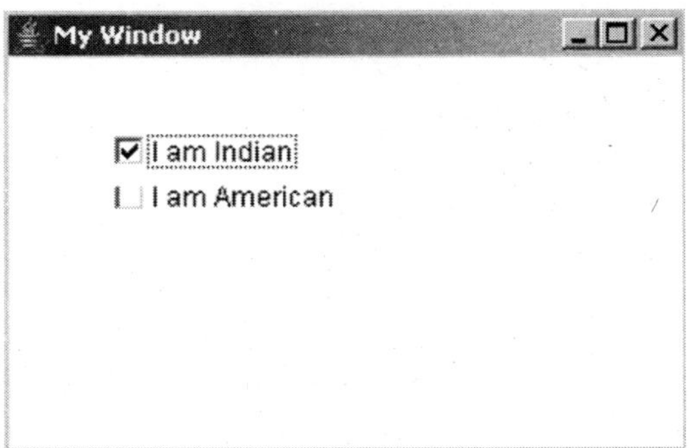

15.3.4. Choice

Choice objects are drop-down lists. The visible label of the Choice object is the currently selected entry of the Choice.

Example 15.7. Program to add a Combo-Box on window.

```
import java.awt.*;
class AWTexample extends Frame
{
    public AWTexample()
    {
        setTitle("My Window");
        setSize(300,200);
```

```
            Choice C=new Choice();
            setLayout(null);
            C.setBounds(50,50,90,30);
            C.add("Red");
            C.add("Green");
            C.add("Blue");
            add(C);
            setVisible(true);
        }
        public static void main( String arg[])
        {
            new AWTexample();
        }
    }
```

COMPILE:

```
C:\>javac AWTexample.java
```

EXECUTE:

```
C:\>java AWTexample
```

OUTPUT

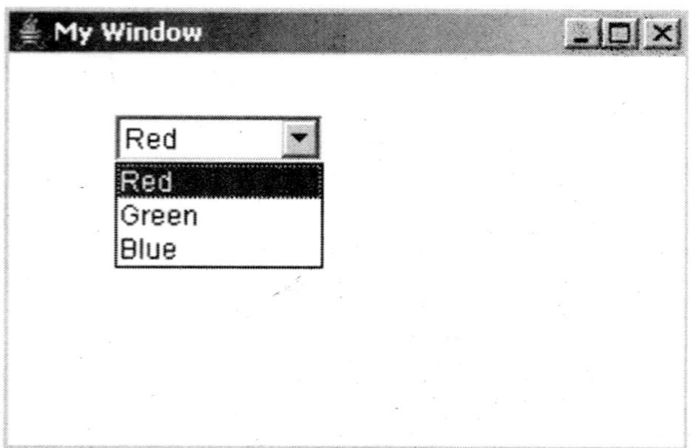

15.3.5. Label

A Label is a displayed Label object. It is usually used to help indicate what other parts of the GUI do, such as the purpose of a neighboring text field.

Example 15.8 Program to add two Labels on window.

```
import java.awt.*;
class AWTexample extends Frame
{
    public AWTexample()
    {
        setTitle("My Window");
        setSize(300,200);
```

```
            Label L1=new Label(" Label-1");
            Label L2=new Label(" Label-2");
            setLayout(null);
            L1.setBounds(50,50,90,30);
            L2.setBounds(50,70,100,30);
            add(L1);
            add(L2);
            setVisible(true);
        }

        public static void main( String arg[])
        {
            new AWTexample();
        }
    }
```

COMPILE:

```
C:\>javac AWTexample.java
```

EXECUTE:

```
C:\>java AWTexample
```

OUTPUT

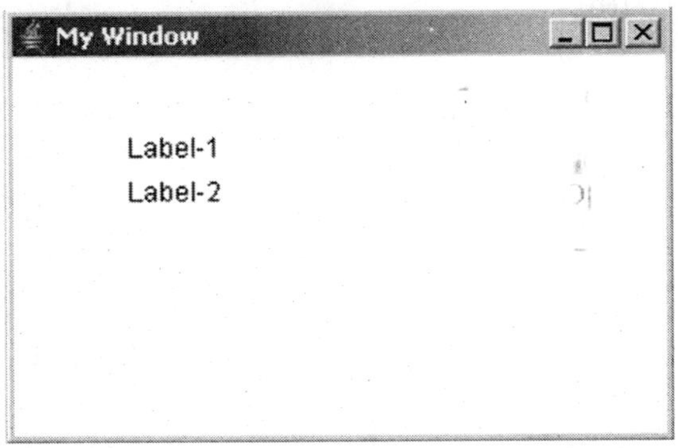

15.3.6. List

A List is a scrolling list box that allows you to select one or more items. Multiple selections may be used by passing true as the second argument to the constructor.

Example 15.9 Program to add a List-Box on window.

```
import java.awt.*;
class AWTexample extends Frame
{
    public AWTexample()
    {
        setTitle("My Window");
        setSize(300,200);
```

```
            List C=new List(4,false);
            setLayout(null);
            C.setBounds(50,50,90,50);
            C.add("India");
            C.add("Pakistan");
            C.add("Sri Lanka");
            C.add("Canada");
            add(C);
            setVisible(true);
        }
    public static void main( String arg[])
        {
            new AWTexample();
        }
    }
```

COMPILE:

```
C:\>javac AWTexample.java
```

EXECUTE:

```
C:\>java AWTexample
```

OUTPUT

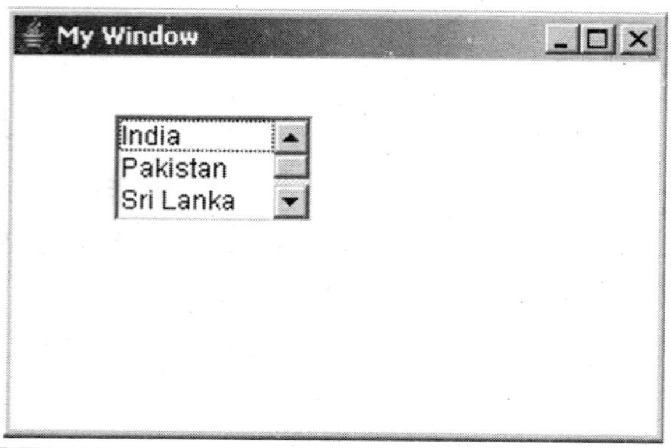

15.3.7. Scroll-Bar

A Scrollbar is a "slider" widget with characteristics specified by integer values that are set during Scrollbar construction. Both horizontal and vertical sliders are available.

Example 15.10 Program to add a Scroll-Bar on window.

```
import java.awt.*;
class AWTexample extends Frame
{
    public AWTexample()
    {
        setTitle("My Window");
```

```
            setSize(300,200);
            Scrollbar C=new Scrollbar(Scrollbar.HORIZONTAL,0,5,-100,105);
            //Scrollbar(HoR/VER, intial value, width of slider, min rage,
    max range)
            setLayout(null);
            C.setBounds(50,50,150,30);
            add(C);
            setVisible(true);
        }
    public static void main( String arg[])
        {
            new AWTexample();
        }
    }
```

COMPILE:

```
C:\>javac AWTexample.java
```

EXECUTE:

```
C:\>java AWTexample
```

OUTPUT

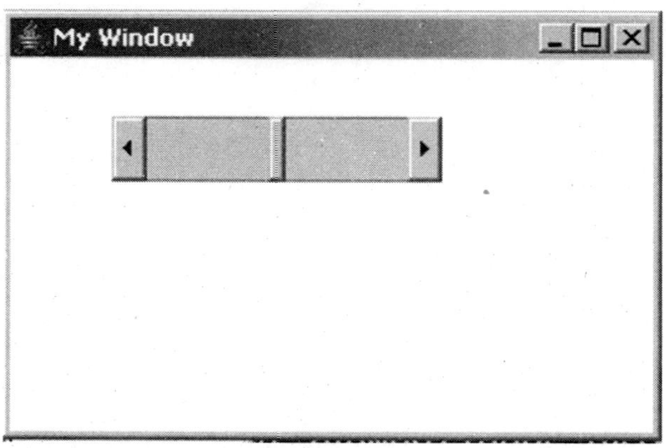

15.3.8. TextField

A TextField is a scrollable text display object with one row of characters. The preferred width of the field may be specified during construction and an initial string may be specified.

Example 15.11 Program to add a Text-Field on window.

```
import java.awt.*;
class AWTexample extends Frame
{
    public AWTexample()
    {
        setTitle("My Window");
```

```
            setSize(300,200);
            TextField C=new TextField("Type Some thing");
            setLayout(null);
            C.setBounds(50,50,100,30);
            add(C);
            setVisible(true);
        }
    public static void main( String arg[])
        {
            new AWTexample();
        }
    }
```

COMPILE:

```
C:\>javac AWTexample.java
```

EXECUTE:

```
C:\>java AWTexample
```

OUTPUT

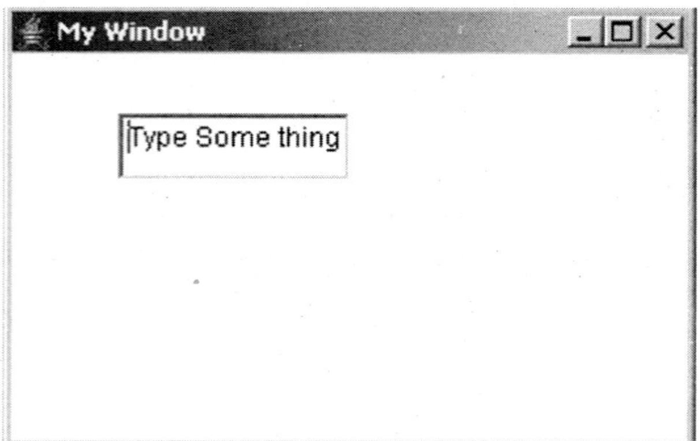

15.3.9. TextArea

A TextArea is a multi-row text field that displays a single string of characters, where newline ('\n' or '\n\r' or '\r', depending on platform) ends each row. The width and height of the field is set at construction, but the text can be scrolled up/down and left/right.

Example 15.12 Program to add a TextArea on window.

```
import java.awt.*;
class AWTexample extends Frame
{
    public AWTexample()
    {
        setTitle("My Window");
        setSize(300,200);
```

```
            TextArea C=new TextArea("Type code hare", 10,30);
            setLayout(null);
            C.setBounds(20,20,250,180);
            add(C);
            setVisible(true);
        }
    public static void main( String arg[])
        {
            new AWTexample();
        }
    }
```

COMPILE:

```
C:\>javac AWTexample.java
```

EXECUTE:

```
C:\>java AWTexample
```

OUTPUT

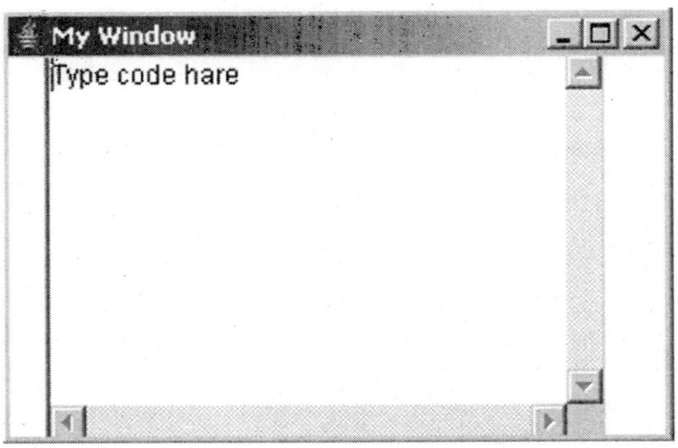

There is no way, for example, to put the cursor at beginning of row five, only to put the cursor at single dimension position 50.

There is a four-argument constructor that accepts a fourth parameter of a scrollbar policy. The different settings are the class constants: SCROLLBARS_BOTH, SCROLLBARS_HORIZONTAL_ONLY, SCROLLBARS_NONE, SCROLLBARS_VERTICAL_ONLY. When the horizontal (bottom) scrollbar is not present, the text will wrap.

15.4. LAYOUTMANAGER

How Components are "laid out" within a Container is described by the LayoutManager class. Since the LayoutManager class is abstract, we can not use it directly. You must sub-class it and provide your own functionality or use a *derived class of LayoutManager* (*i.e.*, BorderLayout, CardLayout, GridLayout, etc) already created for you.

There are many different layout schemes, but the ones pre-defined for us are

- **BorderLayout** - This scheme lays out the Component in 5 ways

1. North-Northern part of the Container
2. South-Southern part of the Container
3. East-Eastern part of the Container
4. West-Western part of the Container
5. Center-Centered in the Container

- **CardLayout** - Allows for what Windows programmers have called for years "tabbed dialogs" or dynamic dialogs (now available on all versions of Netscape).
- **GridLayout** - Allows for the layout of Components in a grid -like fashion rather than "North" or "Center".
- **GridBagLayout** - HTMLTable-ish style of layout.
- **FlowLayout** - Allows for Component to be laid out in a row (or flow) and aligned (left, right, center).
- **None**-No layout, the Container will not attempt to reposition the Components during a update.

To use a layout we must call setLayout() for the Container with an instance of a LayoutManager (since LayoutManager is abstract, a sub-class will do).

Using each layouts is different. For BorderLayout you specify the type of layout (North, South, etc.) by passing a string (i.e., "North", "South", etc to the method add() in a Container.

i.e.,

```
add("Center", myComponent);
```

adds *myComponent* to the Container and centers it. With other layouts, the first **String** parameter is different.

15.4.1. BorderLayout

The border layout is the default layout manager for all Window objects. It consists of five fixed areas: North, South, East, West, and Center. You do not have to put a component in every area of the border layout. The demonstration puts the label "Border-Layout" in the North area, the OK button in the South area, and the List of fruit trees in the Center area. The East (right) and West (left) areas are empty.

The five areas of a border layout are shown at left. Notice how the North and South areas run like a header and footer across the panel.

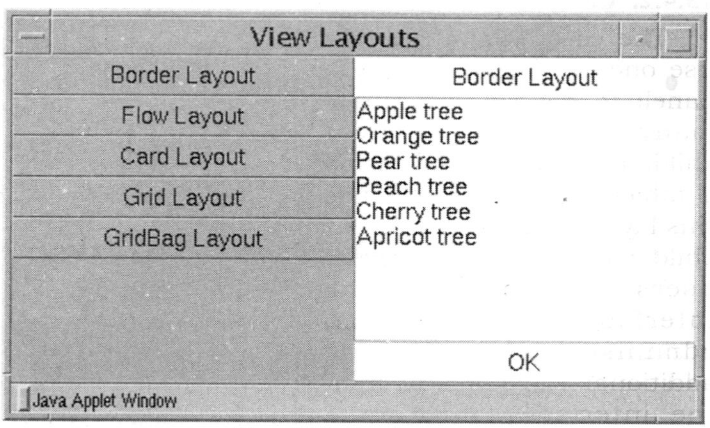

Fig. 15.1. Border layout

If any or all of the North, South, East, or West areas are left out, the Central area spreads into the missing area or areas. However, if the Central area is left out, the North, South, East, or West areas do not change.

You can indent the Central area by placing a label with one or more blank characters in the East or West areas. The following code shows two ways to do it:

```
Label label1 = new Label(" ");
p2.add("East, label1);

Label label2 = new Label();
label2.setText(" ");
p2.add("East", label2);
```

Here is the code for the borderlayout panel with the fruit tree list shown above:

```
p2.setLayout(new BorderLayout());
Label dialogLabel = new Label();
dialogLabel.setAlignment(Label.CENTER);
dialogLabel.setText("Border Layout");

List items = new List(6, true);
items.addItem("Apple tree");
items.addItem("Orange tree");
items.addItem("Pear tree");
items.addItem("Peach tree");
items.addItem("Cherry tree");
items.addItem("Apricot tree");

OK = new Button("OK");
p2.add("North", dialogLabel);
p2.add("South", OK);
p2.add("Center", items);
```

Note that when using the BorderLayout, you must add components with either the add (String, Component) or the add (Component, Object) method to ensure the components are all added correctly. A common mistake people make is to use add(Component), which can result in some of the components not being visible.

15.4.2. CardLayout

The card layout lets you use one container (usually a panel) to display one out of many possible component children (like flipping cards on a table). A program can use this layout to show a different child component to different users. For example, the interface shown to an administrator might have additional functionality from the interface shown to a regular user. With card layout,

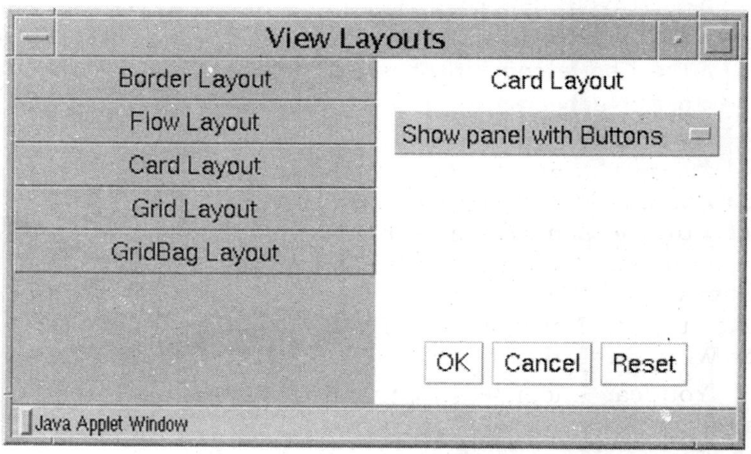

Fig. 15.2. Card layout

your program can show the appropriate interface depending on the type of user using the program.

You can also use card layout to let your end user toggle among different displays and choose the one they prefer. In this case, the program must provide a GUI for the user to make the selection. The Choice component is commonly used for this case and the following card layout example shows how to use a Choice component to set up a card layout that lets the end user toggle between two interfaces.

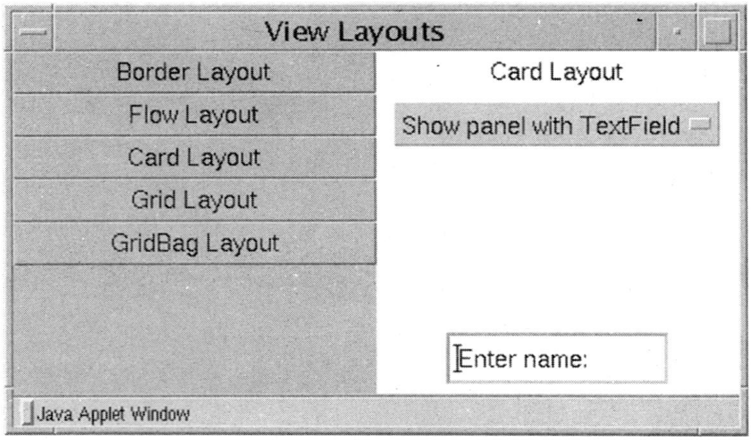

Fig. 15.3. Card layout

Here is the code for the card layout panel shown above:

```
  String BUTTONPANEL = "Show panel with Buttons";
String TEXTPANEL = "Show panel with TextField";

//Border layout is used to set the right panel up
//with the Choice component
//and "Card Layout" label

p2.setLayout(new BorderLayout());

//The Choice component is put in a panel for
//better appearance
Panel c2 = new Panel();
Choice c = new Choice();
c.addItem(BUTTONPANEL);
c.addItem(TEXTPANEL);
c2.add(c);

//The Label is centered and put in the North area
//of the right panel
Label cardLabel = new Label("Card Layout");
cardLabel.setAlignment(Label.CENTER);
p2.add("North", cardLabel);
```

```
//The choice is added to the
//Center area of the right panel
p2.add("Center", c2);

//The card panel is created and CardLayout used
cards = new Panel();
cards.setLayout(new CardLayout());

//The two panels controlled by the card panel
//are created.
p3 = new Panel();
p4 = new Panel();

//The button components are created
//and added to p3
OK = new Button("OK");
Button quit = new Button("Cancel");
reset = new Button("Reset");
p3.add(OK);
p3.add(quit);
p3.add(reset);

//The text field component is
//created and added to p4
tfield = new TextField("Enter name:", 20);
p4.add(tfield);

//Panels p3 and p3 are added to
//the card panel./c
cards.add(BUTTONPANEL, p3);
cards.add(TEXTPANEL, p4);

//The card panel is added to the
//south area of the right panel
p2.add("South", cards);
```

15.4.3. GridLayout

The Grid layout arranges components into a grid of rows and columns. You specify the number of rows and columns, the number of rows only and let the layout manager determine the numer of columns, or the number of columns only and let the layout manager determine the number of rows.

The cells in the grid are equal size based on the largest component in the grid. Resize the window to see how the components are resized to fit cells as they get larger and smaller.

The grid layout demonstration starts with a one column grid, and lets you choose to see the same components in two, three, and four column grids.

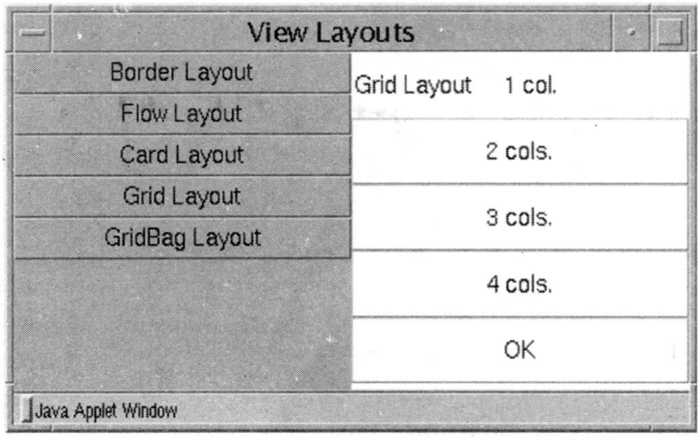

Fig. 15.4. Grid layout

Here is the code for the grid layout panel shown above.

```
   //Set grid to 5 rows and 1 column
p2.setLayout(new GridLayout(5,1));

two = new Button("2 cols.");
three = new Button("3 cols.");
four = new Button("4 cols.");
OK = new Button("OK");

p2.add(new Label("Grid Layout 1 col."));

p2.add(two);
p2.add(three);
p2.add(four);
p2.add(OK);

//Implement 2 column grid.
//Let layout manager determine number of rows.
p2.setLayout(new GridLayout(0,2));
two = new Button("2 cols.");
three = new Button("3 cols.");
four = new Button("4 cols.");
OK = new Button("OK");
p2.add(two);
p2.add(three);
p2.add(four);
p2.add(OK);
```

```
//Implement 3 column grid.
//Let layout manager determine number of rows.
p2.setLayout(new GridLayout(0,3));
two = new Button("2 cols.");
three = new Button("3 cols.");
four = new Button("4 cols.");
OK = new Button("OK");
p2.add(two);
p2.add(three);
p2.add(four);
p2.add(OK);

//Implement 4 column grid.
//Let layout manager determine number of rows.
p2.setLayout(new GridLayout(0,4));
two = new Button("2 cols.");
three = new Button("3 cols.");
four = new Button("4 cols.");
OK = new Button("OK");
p2.add(two);
p2.add(three);
p2.add(four);
p2.add(OK);
```

15.4.4. GridBagLayout

The Grid bag layout (like grid layout) arranges components into a grid of rows and columns, but lets you specify a number of settings to fine-tune how the components are sized and positioned within the cells. Unlike the grid layout, the rows and columns are not constrained to be a uniform size. For example, a component can be set to span multiple rows or columns, or you can change its position on the grid.

The layout manager uses the number of components in the longest row, the number of rows created, and the size of the components to determine the number and size of the cells in the grid. The set of cells a component spans can be referred to as its display area. A component's display area and the way in which it fills that display area are defined by a set of constraints that are represented by the GridBagConstraints object. These constraints are given below:

- gridx / gridy
- gridwidth/gridheight
- fill
- anchor
- ipadx / ipady
- insets

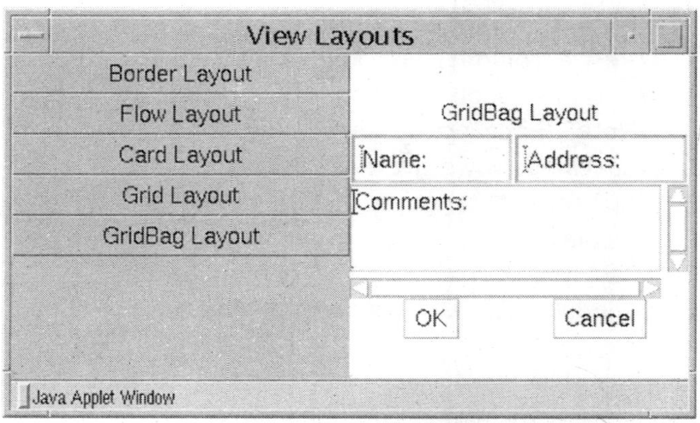

Fig. 15.5. Grid Bag Layout

The GridBagConstraints object is initialized with a set of default values, each of which can be customized to change the way the component will be layed out within the gridbag.

The following bulleted list outlines how to use the GridBagConstraints and GridBagLayout objects in code to achieve the look you want. This outline is for instructional purposes only and is used in the following code example. Once you understand how the objects work together, your code can, of course, do things in a different order or take another approach.

- Decide how you want the components sized and positioned.
- Create a GridBagLayout and GridBagConstraints object.
- Have the container use the GridBagLayout object.
- Customize the GridBagConstraints settings (or use defaults).
- Create Component A.
- Pass Component A and GridBagConstraints to the GridBagLayout object.
- Add Component A to the container. The previous bullet told the container's layout which settings to use for the component.
- Create Component B.
- Change the GridBagConstraints settings or leave as is.
- Pass Component B and GridBagConstraints to the GridBagLayout object.
- Add Component B to the container.
- and so on . . . .

Here is the code for the grid bag layout shown above.

```
//Create GridBagLayout and
//GridBagConstraints objects.
 //Have container p2 use grid bag layout.
 GridBagConstraints c = new GridBagConstraints();
 GridBagLayout gridbag = new GridBagLayout();
 p2.setLayout(gridbag);

 //Make components fill the available display area
 //in the horizontal and vertical directions
 c.fill = GridBagConstraints.BOTH;
 //Resize columns and keep them equal
 //during window resizing
 //Note: Since no weighty value is
 //specified, rows are not resized,
 //but remain together in the middle of the
 //display.
 //Test this by resizing the demonstration
 //window with GridBag Layout selected.
 c.weightx = 1.0;

 //Create the Label component.
 Label top = new Label("GridBag Layout");
```

```
top.setAlignment(Label.CENTER);

//Make the Label
component the last one in its row.
c.gridwidth = GridBagConstraints.REMAINDER;

//Pass the Label component and
//GridBagConstraints object to layout.
gridbag.setConstraints(top, c);

//Add the Label component to container p2.
//The GridBagConstraints
//passed to the GridBagLayout object for
//this container put the
//label in a row by itself, size the label
//to fill the row in the
//horizontal and vertical directions, and
//cause columns to resize
//equally when the window is resized.
p2.add(top);

//Create a Text Field component.
TextField name = new TextField("Name:", 25);

//Make the Text Field component the next
//to last one in its row.
//This means 1 or more components can be added
//to the row after
//the Text field component.
c.gridwidth = GridBagConstraints.RELATIVE;

//Pass the Text Field component and
//GridBagConstraints object to layout.
gridbag.setConstraints(name, c);

//Add the Text Field component to
//container p2.
p2.add(name);

//Create another Text Field component.
TextField addr = new TextField("Address:", 25);

//Make the Text field component the
//last one in its row.
c.gridwidth = GridBagConstraints.REMAINDER;
```

```
//Pass the Text Field component and
//GridBagConstraints object to layout.
gridbag.setConstraints(addr, c);

//Add the Text Field component to container p2.
p2.add(addr);

//Create a Text Area component.
TextArea comments = new TextArea(3, 25);
comments.setEditable(true);
comments.setText("Comments:");

//Pass the Text Area component and
//GridBagConstraints object to layout.
//Because GridBagConstraints are unchanged,
//the Text Area component
//is in a row by itself.
gridbag.setConstraints(comments, c);

//Add the Text Area component to container p2.
p2.add(comments);

//Create the OK and Cancel buttons.
OK = new Button("OK");
cancel = new Button("Cancel");

//Make the OK button the next to last
//one in its row.
//Components should not expand to fill
//the row in either
//the horizontal or vertical direction.
c.fill = GridBagConstraints.NONE;
c.gridwidth = GridBagConstraints.RELATIVE;

//Pass the OK button and GridBagConstraints
//object to layout.
gridbag.setConstraints(OK, c);

//Make the Cancel button the last one
//in its row.
c.gridwidth = GridBagConstraints.REMAINDER:

//Pass the Cancel button and
//GridBagConstraints object to layout.
```

```
gridbag.setConstraints(cancel, c);

//Add the OK and Cancel buttons to
//container p2.
p2.add(OK);
p2.add(cancel);
```

15.4.5. Flow Layout

The flow layout is the default layout manager for all Panel objects and applets. It places the panel components in rows according to the width of the panel and the number and size of the components. The best way to understand flow layout is to resize the demonstration window and notice how the components flow from one row to the other as you make the window wide and narrow.

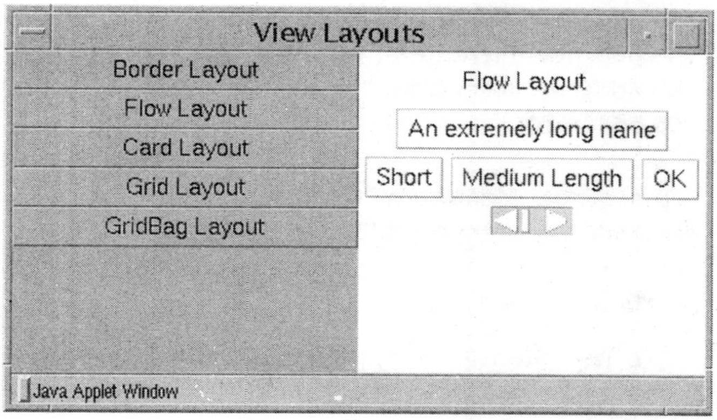

Fig. 15.6. Flow layout

Here is the code for the flow layout panel shown above:

```
    p2.setLayout(new FlowLayout());

p2.add(new Label("Flow Layout"));
OK = new Button("OK");
p2.add(new Button("An extremely long name"));
p2.add(new Button("Short"));
p2.add(new Button("Medium Length"));
p2.add(OK);
p2.add(new Scrollbar(Scrollbar.HORIZONTAL));
```

15.5. MENUS

Menus are at the center of your application. The difference between a usuable application and one that is absolutely frustrating lies in the organization of your menus. The user-interfaces rules for a good menu design are not clear.

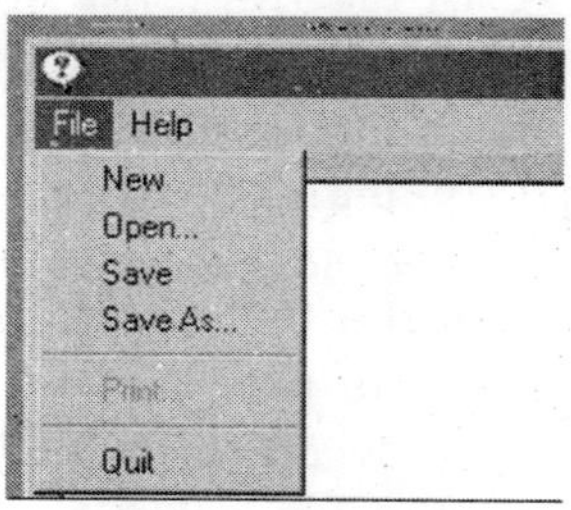

Fig. 15.7. Menus

The following classes are related to Menus

- MenuComponent
  - MenuBar
  - MenuItem
    - CheckboxMenuItem
    - Menu

Handling is done in the handleEvent() method which you will override.

To use menus you need to create a MenuBar and Menus with MenuItems, then attach the MenuBar to your application via setMenuBar. You can even embed, Menus in MenuItems..

```
MenuBar mb = new MenuBar();
Menu top = new Menu("File");

Menu sub = new Menu("New");
sub.add(new MenuItem("Document"));
sub.add(new MenuItem("Message"));
sub.add(new MenuItem("Image"));
top.add(sub);

top.add(new MenuItem("Open"));
top.add(new MenuItem("Save..));

setMenuBar(mb);
```

To use Menus in applet you'll need to create a new Frame and attach a menubar to that. You can not just change the browser's menubar. You are not dealing with OLE 2 here.

15.6. IMAGES

Have you ever wanted to do simple image manipulations like flipping, scaling, and cropping, but didn't want to bother with image filters? Starting with Java 1.1, you can do this very easily with a new drawImage method. The method signature is as follows:

```
public boolean drawImage (Image img, int dx1, int dy1, dx2, dy2,
    int sx1, int sy1, sx2, sy2,[Color backgroundColor,]
    ImageObserver observer)
```

The Image, optional background color, Color, and ImageObserver parameters are shared with the Java 1.0 varieties of the method, along with the return value. However, new and different to Java 1.1 are the four sets of points. They describe source and destination bounding rectangles for the image. Starting with the 's' parameters, the coordinates describe the source piece of the original image to draw. If the (x1, y1) corner is (0, 0), and the (x2, y2) corner is (img.getWidth(this), img.getHeight(this)), then the entire image will be drawn. On the other hand, if the coordinates specify a smaller size, the image is cropped to the area specified. The source corners must be the top-left and bottom-right corners of the area to draw, as shown in figure 15.8.

The following code in a Java 1.1 program demonstrates cropping to generate figure 15.8.

```
g.drawImage (im, 0, 0, 135, 90,65, 5, 200, 95, this);
```

Working with the 'd' parameters is where the real interesting things happen. These parameters describe the destination drawing area for the piece of the source image you selected. Since the points must be opposite corners of a rectangle, four options are available for these points:

- Keeping the coordinates in the same corners
- Swapping x-coordinates
- Swapping y-coordinates
- Swapping both coordinates

Keeping the coordinates in the same corners results in the scaling of the image. (No scaling will be done if the differences between the height and width of the two point sets is identical.) See figure 15.9 below.

Fig. 15.8. Image

The following source will compress the width to half size and double the height:

```
g.drawImage (im, 0, 0, im.getWidth(this)/2, im.getHeight(this)*2,
     0, 0, im.getWidth(this), im.getHeight(this), this);
```

Swapping x coordinates has the points in the top-right and bottom-left corners. This results in a flip along the vertical axis, as shown in figure 15.10. Here's the code for implementing this flip.

```
g.drawImage (im, im.getWidth(this), 0, 0, im.getHeight(this),
     0, 0, im.getWidth(this), im.getHeight(this), this);
```

Fig. 15.9. Scaled image to half width and twice height

Fig. 15.10. Flipping the image along the vertical axis

Swapping y coordinates, as shown in figure 15.11 and in the code below, has the points in the bottom-left and top-right corners. This results in a flip along the horizontal axis.

```
g.drawImage (im, 0, im.getHeight(this), im.getWidth(this), 0,
     0, 0, im.getWidth(this), im.getHeight(this), this);
```

Fig. 15.11. Flipping the image along the horizontal axis

Finally, figure 15.12 shows how swapping both coordinates results in a flip along both axis. The code is as follows:

```
g.drawImage (im, 0, im.getHeight(this), im.getWidth(this), 0,
     0, 0, im.getWidth(this), im.getHeight(this), this);
```

That's all there is to the new drawImage in Java 1.1. If the image is transparent, you can provide a background color that will seep through. Otherwise, the optional parameter is ignored. Unfortunately, you cannot use drawImage to rotate the image. If you do want to rotate the image, you have to do more work and subclass ImageFilter, just like with Java 1.0.

Fig. 15.12. Flipping the image along the vertical and horizontal axes

15.7. ADDITIONAL PACKAGES

15.7.1. Java Utility Package

The final Java package, java.util, contains a collection of utility classes.

- **Data Structure Classes**

  A variety of useful classes implementing standard computer science data structures: including BitSet, Dictionary, Hashtable, Stack and Vector. The java.util package also defines the Enumeration interface which provides a protocol for classes to count through a set of values.

- **Date**

  Use the Date class to create and manipulate calendar dates in a system-independent fashion.

- **StringTokenizer**

  This class converts a String of text into its tokens.

- **Properties**

  This class implements persistent properties. The properties table contains key/value pairs where both the key and the value are Strings. This class is used by the System class to implement System properties.

- **Observer and Observable**

  Classes that implement the Observer interface can "watch" Observable objects for state changes. When an Observable object changes it notifies all of its Observers of the change.

- **Random-Number Generator**

  The Random class implements a random-number generator.

- **Enumeration**

  The Enumeration interface defines a generic programming interface for iterating through a set of values.

15.7.2. Java Language Package

By default, each Java application/applet has access to the *java.lang* package. Inside *java.lang* are classes that represent primitive data types (such as int & char), as well as more complex classes. It contains classes pertaining to strings, string buffers, threads, and even the System class from which we obtain out input and output streams. *Java.lang* is quite extensive, and some aspects (such as threads) can be confusing for those new to the Java language. For this reason, we'll only present the more useful, less complex, parts of this package. This should allow you to experiment with some of the handy features of the Java language, while minimising the barrier to understanding. For more advanced Java programmers, you may wish to consult the Java API yourself, and experiment with other parts of *java.lang* not covered herein.

Interfaces in java.lang

- Cloneable
- Runnable

Classes in java.lang

- Boolean
- Byte
- Character
- Class
- ClassLoader
- Compiler
- Double
- Float
- Integer
- Long
- Math
- Number
- Object
- Process
- Runtime
- SecurityManager
- Short
- String
- StringBuffer
- System
- Thread
- ThreadGroup
- Throwable
- Void

SUMMARY

Abstract window took kit is a portable GUI library.

— It can work on Solaris, Window 95/NT and Mac system 7.x. for stand alone applications and /or applets.

— We have explained more of the features of AWT/GUI hierarchy.

— We have also explained addition packages for common use.

SOLVED PROBLEMS

Q. 15.1. Which containers use a border layout as their default layout?

Ans. The window, Frame and Dialog classes use a border layout as their default layout.

Q. 15.2. What method is used to specify a container's layout?

Ans. The setLayout() method is used to specify a container's layout.

Q. 15.3. Which containers use a FlowLayout as their default layout?

Ans. The Panel and Applet classes use the FlowLayout as their default layout.

Q. 15.4. What's the difference between a GridLayout and a GridBagLayout?

Ans. A GridLayout places each component into a cell of one size. Thus all components have the same size. A GridbagLayout allows components to span multiple cells. Therefore components can have different sizes. There are other differences, but that's the main one.

Q. 15.5. What's the difference between a Frame and a Dialog?

Ans. A Dialog is primarily for transitory and immediate responses. A Frame may last for a while.

Q. 15.6. What's the difference between a Menu and a MenuBar?

Ans. A MenuBar contains menus.

Q. 15.7. What's the difference between a Menu and a MenuItem?

Ans. Actually a Menu is a subclass of MenuItem. This allows you to make hierarchical menus by addinf menus to menus. However a MenuItem is generally thought of as a particular selection in a menu.

Q. 15.8. Why wouldn't we create a Menu inside the MenuBar's add method like you create MenuItem's inside a Menu's add method?

Ans. If you did this you would not have a convenient reference to the Menu, so it would be difficult if not impossible to add items to the menu. This isn't an issue for MenuItems since they generate events, but Menus don't generate events.

EXERCISES

Q. 15.1. What is the concept of AWT?

Q. 15.2. Give the code to create following controls.

(*a*) Button
(*b*) Canvas
(*c*) Checkbox
(*d*) CheckboxGroup
(*e*) Choice
(*f*) Label
(*g*) List
(*h*) Scroll bar
(*i*) Textfield
(*j*) TextArea

Q. 15.3. What are the difference between following

(*a*) Check box and CheckboxGroup
(*b*) Textfield and TextArea
(*c*) GridLayout and GridBagLayout
(*d*) Menu and MenuItem.

Q. 15.4. Give the code to flip the image as follows:

(*a*) flip an image alog the horizental axis
(*b*) flip an image alog the vertical axis
(*c*) flip an image alog the both horizental and vertical axis.

Q. 15.5. Explain the following classes of utility package.

(*a*) Hashtable
(*b*) Vector
(*c*) Random
(*e*) Enumeration
(*f*) Dictionary

Chapter 16

EVENT HANDLING

16.1. INTRODUCTION

CHAPTER OUTLINE

Your actions when you're using the GUI for a window-based program or an applet-clicking a menu item or a button, moving the mouse, and so on—are first recognized by the operating system. For each action, the operating system determines which of the programs currently running on your computer should know about it and passes the action on to that program. When you click a mouse button, it's the operating system that registers this and notes the position of the mouse cursor on the screen. It then decides which application controls the window where the cursor was when you

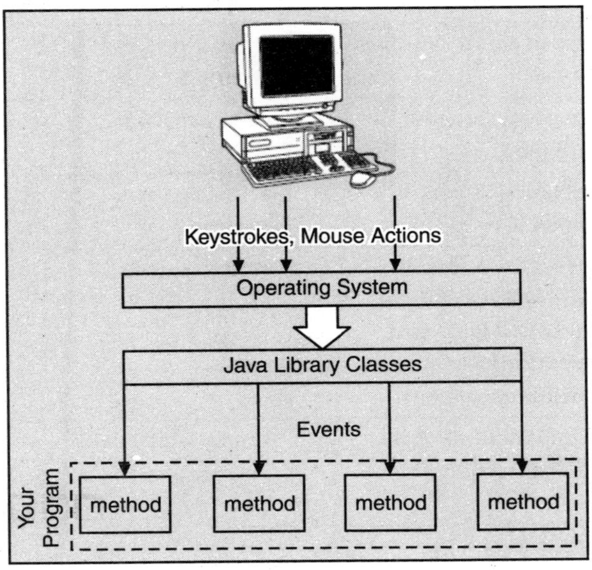

Fig. 16.1.

pressed the button, and communicates the mouse button-press to that program. The signals that a program receives from the operating system as a result of your actions are called events. The basic idea of how actions and events are communicated to your program code is illustrated in Figure 16.1.

> **Concept**
>
> *Any action performed by the user from input device, that is an event.*

Each event is some specific situation generated by keyboard, mouse or system itself.

For event handling there are some important concepts:

Event Source – that generate the events.

Event Listener – that listen the event.

Event handling – way to handling the events.

Java contains 11 Listener interfaces and 7 Adapter classes for event handling. Sun Micro systems added the Adapter classes in the java JDK 1.2. Java provides Adapter classes for only those interface, which we have more then one method. Using Adapter classes user can override methods according to his/her need may be one/more then one or may be all.

Here is the list of Listener interface and Adapter classes.

Category	**Listeners / Adapters**	**Methods**	**Controls to used**
Action	ActionListener	actionPerformed (ActionEvent)	Button Click / menu item / enter key / double click on any item
Item	ItemListener	itemStateChanged (ItemEvent)	Radio / List / Choice / CheckBox
Text	TextListener	textValueChanged (TextEvent)	when ever text value changed
Adjustment	AdjustmentListener	adjustmentValueChanged (AdjustmentEvent)	with Scrollbar
Focus	FocusListener and FocusAdapter	focusGained(FocusEvent) focusLost(FocusEvent)	Focus on any control
Mouse	MouseListener and MouseAdapter	mousePressed(MouseEvent) mouseReleased(MouseEvent) mouseEntered(MouseEvent) mouseExited(MouseEvent) mouseClicked(MouseEvent)	Mouse Clicks
Mouse Motion	MouseMotionListener and MouseMotion Adapter	mouseDragged(MouseEvent) mouseMoved(MouseEvent)	Mouse Movement
Key	KeyListener and KeyAdapter	keyPressed(KeyEvent) keyReleased(KeyEvent) keyTyped(KeyEvent)	Keyboard strikes

Window	WindowListener andWindowAdapter	windowClosing(WindowEvent) windowOpened(WindowEvent) windowIconified(WindowEvent) windowDeiconified(WindowEvent) windowClosed(WindowEvent) windowActivated(WindowEvent) windowDeactivated(WindowEvent)	Windows controls
Component	ComponentListener and ComponentAdapter	componentMoved(ConatinerEvent) componentHidden(ConatinerEvent) componentAdded(ConatinerEvent) componentResized(ConatinerEvent) componentShown(ConatinerEvent)	Components
Container	ContainerListener andContainerAdapter	componentAdded(ConatinerEvent) componentRemoved(ConatinerEvent)	Containers

16.2. ACTIONLISTENER EXAMPLE

An ActionListener interface is basically used for Button, Menu items, Enter Key and the Double click of mouse controls. The ActionListener interface contains only one method for handling the event, which is:

```
public void actionPerformed(ActionEvent e){}
```

Now how to use this method in different programs, see some examples:

Example 16.1 **Program to input two numbers using text boxes, create buttons for "+" ,"-", "*" , "/" and perform the action on them using event handling.**

```
import java.lang.*;
import java.awt.*;
import java.awt.event.*;
import java.applet.*;

public class Example extends Applet implements ActionListener
{
Label l1;
TextField t1,t2,t3;
Button b1,b2,b3,b4;
public void init()
{
l1=new Label("Enter value ");
b1=new Button ("+");
b2=new Button("-");
b3=new Button("*");
b4=new Button("/");
t1=new TextField(10);
t2=new TextField(10);
```

```
t3=new TextField(10);
setBackground(new Color(200,100,200));

add(l1);

add(t1);
add(t2);
add(t3);
add(b1);
add(b2);
add(b3);
add(b4);
b1.addActionListener(this);
b2.addActionListener(this);

b3.addActionListener(this);
b4.addActionListener(this);
}
public void actionPerformed(ActionEvent e)
{
int a=Integer.parseInt(t1.getText());
int b=Integer.parseInt(t2.getText());
int c=0;
if(e.getSource()==b1)
{
    c=a+b;
}
if(e.getSource()==b2)
{
c=a-b;
}
if(e.getSource()==b3)
{
c=a*b;
}
if(e.getSource()==b4)
{
    c=a/b;
}
t3.setText(String.valueOf(c));
}
}
//<applet code=Example height=500 width=300></applet>
```

COMPILE:

```
C:\>javac Example.java
```

EXECUTE:

```
C:\>appletviewer Example.java
```

OUTPUT

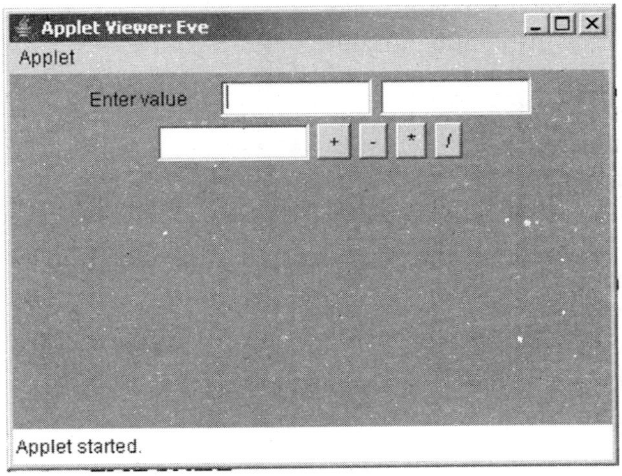

Example 16.2 **Program to input two numbers using text boxes, create buttons for Addition, Subtraction, Multiplication, Division and perform the action on them using event handling with proper settings of controls.**

```
import java.awt.event.*;
import java.applet.*;
import java.awt.*;

public class operation extends Applet implements ActionListener
{
      Label L1, L2,L3;
      TextField T1, T2, T3;
      Button b1,b2,b3,b4;
      public void init()
      {
            setLayout(null);
            L1=new Label("Enter first number:");
            L2=new Label("Enter second number:");
            L3=new Label("RESULT");

            T1=new TextField();
            T2=new TextField();
            T3=new TextField();

            b1=new Button("add");
            b2=new Button("Substract");
```

```
            b3=new Button("multiply");
            b4=new Button("divide");

            L1.setBounds(50,50,110,40);
            L2.setBounds(50,100,140,40);
            L3.setBounds(50,150,120,40);

            T1.setBounds(210,50,50,30);
            T2.setBounds(210,100,50,30);
            T3.setBounds(210,150,50,30);
            b1.setBounds(50,250,50,30);
            b2.setBounds(120,250,60,30);
            b3.setBounds(190,250,50,30);
            b4.setBounds(260,250,50,30);
            add (L1);
            add (L2);
            add (L3);
            add(T1);
            add(T2);
            add(T3);
            add(b1);
            add(b2);
            add(b3);
            add(b4);
            b1.addActionListener(this);
            b2.addActionListener(this);
            b3.addActionListener(this);
            b4.addActionListener(this);
        }
        public void actionPerformed(ActionEvent e)
        {
            int a=Integer.parseInt(T1.getText());
            int b=Integer.parseInt(T2.getText());
            int c=0;
            if(e.getSource()==b1)
                c =a + b;
            if(e.getSource()==b2)
                c=a-b;
            if(e.getSource()==b3) c=a*b;
            if(e.getSource()==b4)c=a/b;
            T3.setText(String.valueOf(c));
        }
    }
    //<applet code = operation.class width=500 height=500></applet>
```

COMPILE:

```
C:\>javac operation.java
```

EXECUTE:

```
C:\>appletviewer operation.java
```

OUTPUT

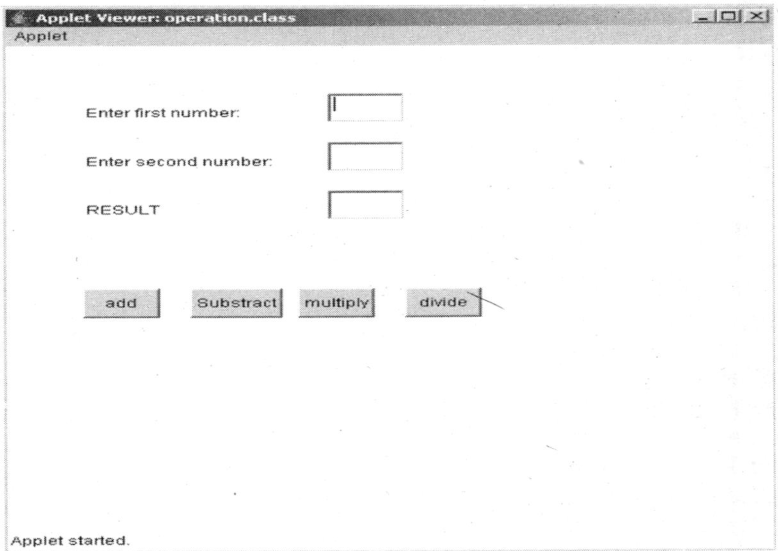

Example 16.3 **An AWT Program to create a Note-Pad with Menu items and perform the action on them using event handling.**

```
import java.awt.*;
import java.awt.event.*; // step - 1
class Notepad extends Frame implements ActionListener  // step - 2
{
    FileDialog fd; // for open & save Dialog
    TextArea ta = new TextArea();
    MenuBar mb = new MenuBar();
    Menu f1 = new Menu("F i l e ");
    Menu f2 = new Menu("E d i t ");
    MenuItem a1 = new MenuItem("N e w ");
    MenuItem a2 = new MenuItem("O p e n ");
    MenuItem a3 = new MenuItem("S a v e ....");
    MenuItem a4 = new MenuItem("E x i t ");
    MenuItem a5 = new MenuItem("C u t ");
    MenuItem a6 = new MenuItem("C o p y ");
    MenuItem a7 = new MenuItem("P a s t e ");
    MenuItem a8 = new MenuItem("S e l e c t  A l l ");
    Notepad()
    {
        setTitle("Notepad Example By Lalit Arora");
        setFont(new Font("Courier New",1,14));
```

```
        // add item in file menu
        f1.add(a1);
        f1.add(a2);
        f1.add(a3);
        f1.addSeparator();
        f1.add(a4);
        // add item in edit menu
        f2.add(a5);
        f2.add(a6);
        f2.add(a7);
        f2.addSeparator();
        f2.add(a8);
        // add menu in menu bar
        mb.add(f1);
        mb.add(f2);
        setMenuBar(mb); // set menubar in frame
        add(ta); // add ta in center
        setBounds(50,50,700,500); // set size of notepad
        setVisible(true);
        a1.addActionListener(this);
        a2.addActionListener(this);
        a3.addActionListener(this);
        a4.addActionListener(this);
        a5.addActionListener(this);
        a6.addActionListener(this);
        a7.addActionListener(this);
        a8.addActionListener(this);
        addWindowListener(new Abhi());
    }
    class Abhi extends WindowAdapter
    {
        public void windowClosing(WindowEvent e)
        {
            dispose();
        }
    }
    public void actionPerformed(ActionEvent e)
    {
        if(e.getSource()==a1) // for new
        {
            setTitle("Untited");
            ta.setText("");
        }
        if(e.getSource()==a2) // for open
        {
            fd = new FileDialog(this,"Open Me",FileDialog.LOAD);
            fd.setVisible(true);
            setTitle(fd.getFile());
        }
        if(e.getSource()==a3) // for save
        {
```

```
                fd = new FileDialog(this,"Save Me",FileDialog.SAVE);
                fd.setVisible(true);
                setTitle(fd.getFile());
            }
            if(e.getSource()==a4) // for exit
                dispose(); // setVisible(false);
            // mar gaya  / jinda hai
        }
        public static void main(String aa[])
        {
            new Notepad();
        }
    }
```

COMPILE:

```
C:\>javac Example.java
```

EXECUTE:

```
C:\>java Example
```

OUTPUT

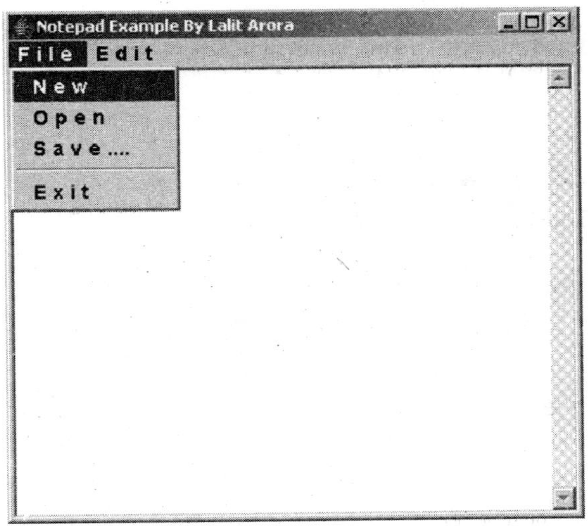

Example 16.4 Program to play Sounds files on applet.

```
import java.awt.*;
import java.awt.event.*;
import java.applet.*;
public class Sounds extends Applet implements ActionListener
{
    AudioClip c1,c2,c3;
    Button a1 = new Button("Hi !");
    Button a2 = new Button("Yahoo");
    Button a3 = new Button("Music");
    public void init()
```

```
        {
            c1 = getAudioClip(getCodeBase(),"hi.au");
            c2 = getAudioClip(getCodeBase(),"yahoo1.au");
            c3 = getAudioClip(getCodeBase(),"spacemusic.au");
            add(a1);
            add(a2);
            add(a3);
            a1.addActionListener(this);
            a2.addActionListener(this);
            a3.addActionListener(this);
        }
        public void actionPerformed(ActionEvent e)
        {
            if(e.getSource()==a1)
                c1.play(); // loop(); / stop();
            if(e.getSource()==a2)
                c2.play();
            if(e.getSource()==a3)
                c3.play();
        }
    }
    //<applet code=Sounds height=400 width=600></applet>
```

COMPILE:

```
C:\>javac Sounds.java
```

EXECUTE:

```
C:\>appletviewer Sounds.java
```

OUTPUT:

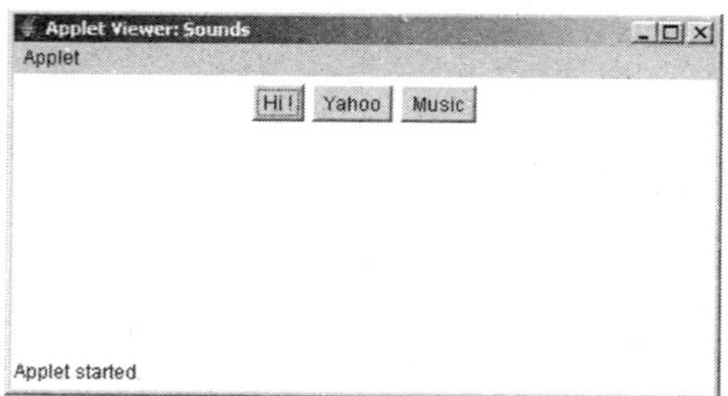

> **Note:** Here we used the sounds files "hi.au","yahoo1.au", "spacemusic.au" existing in the same directory of program. You can use your own file (*.au - audio files) or you can download from internet.

16.3. ITEMLISTENER EXAMPLE

An ItemListener interface is basically used for Radio, List, Choice and the Check Box controls. The ItemListener interface contains only one method for handling the event, which is:

```
public public void itemStateChanged(ItemEvent e){}
```

Now how to use this method in different programs, see some examples:

Example 16.5 **Program to create Radio buttons to select gender and perform the action on them using event handling and display gender on status-bar.**

```
import java.awt.*;
import java.awt.event.*;
import java.applet.*;
public class Radio extends Applet implements ItemListener
{
	CheckboxGroup cg = new CheckboxGroup(); // for grouping
	Checkbox c1 = new Checkbox("Male.",cg,true); // label ,group,state
	Checkbox c2 = new Checkbox("Female.",cg,false);
	public void init()
	{
		setFont(new Font("Courier New",1,14));
		setBackground(new Color(170,194,250));//Color.pink
		add(c1);
		add(c2);
		c1.addItemListener(this);
		c2.addItemListener(this);
	}
	public void itemStateChanged(ItemEvent e)
	{
		if(c1.getState()) // true if checked else return false
			showStatus("U r select - "+c1.getLabel());
		if(c2.getState()) // true if checked else return false
			showStatus("U r select - "+c2.getLabel());
	}
}
//<applet code=Radio height=400 width=600></applet>
```

COMPILE:

```
C:\>javac Radio.java
```

EXECUTE:

```
C:\>appletviewer Radio.java
```

OUTPUT

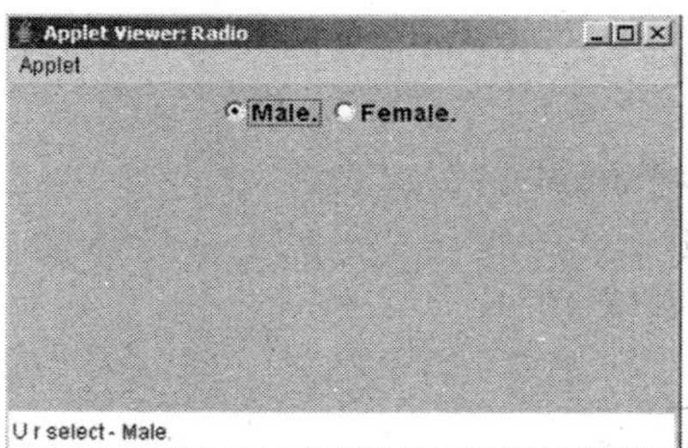

Example 16.6 Program to create Check Boxes for qualification and display the selected qualifications in Textbox.

```
    import java.awt.*;
    import java.awt.event.*;
    import java.applet.*;
    public class Check extends Applet implements ItemListener
    {
        Checkbox c1 = new Checkbox("BSc.",true); // true for checked ,
default is false
        Checkbox c2 = new Checkbox("MSc.");
        Checkbox c3 = new Checkbox("B.Tech.");
        Checkbox c4 = new Checkbox("MCA.");
        Checkbox c5 = new Checkbox("MBA.");
        TextField t1 = new TextField(50); // for 50 chars
        public void init()
        {
            setFont(new Font("Courier New",1,14));
            setBackground(new Color(170,194,250));
            add(c1);
            add(c2);
            add(c3);
            add(c4);
            add(c5);
            add(t1);
            c1.addItemListener(this);
            c2.addItemListener(this);
            c3.addItemListener(this);
            c4.addItemListener(this);
            c5.addItemListener(this);
        }
        public void itemStateChanged(ItemEvent e)
        {
            String s = "Qualification -> ";
            if(c1.getState())
                s+=c1.getLabel()+" , ";
            if(c2.getState()) // true if checked else return false
                s+=c2.getLabel()+" , ";
            if(c3.getState())
                s+=c3.getLabel()+" , ";
            if(c4.getState())
                s+=c4.getLabel()+" , ";
            if(c5.getState())
                s+=c5.getLabel();
            t1.setText(s);
        }
```

```
    }
    //<applet code=Check height=400 width=600></applet>
```

COMPILE:

```
C:\>javac Check.java
```

EXECUTE:

```
C:\>appletviewer Check.java
```

OUTPUT

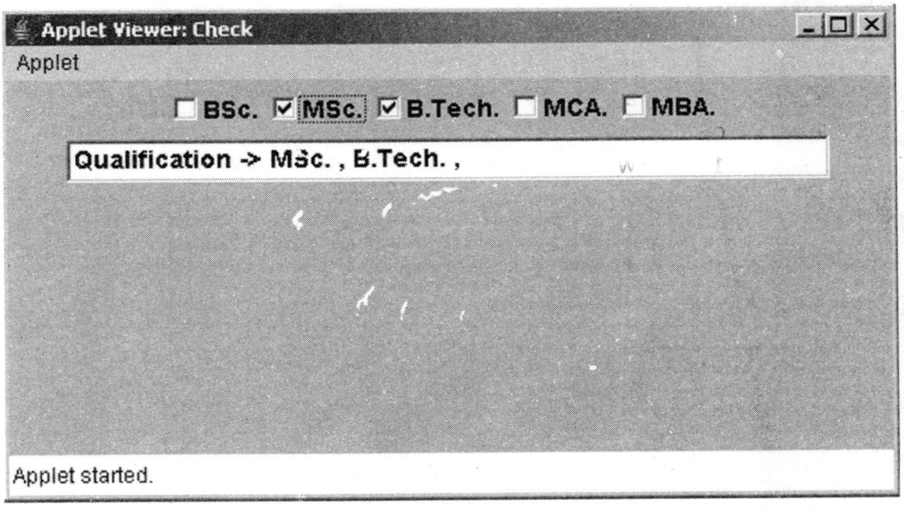

Example 16.7 **Program to create List Box and Combo Box and display the selected items in Text Box.**

```
import java.awt.*;
import java.awt.event.*;
import java.applet.*;
public class MyList extends Applet implements ItemListener
{
    List a = new List(3); // no of rows visible , default is 4
    Choice c = new Choice();
    TextField t1 = new TextField(50);
    String ar[] = new String[]{"Noida","Ghaziabad","Delhi","New
Delhi","Mathura","Gorakhpur","Lakhimpur","Basti","Nagpur"};
    public void init()
    {
        setFont(new Font("Courier New",1,14));
        setBackground(new Color(170,194,250));
        for(int i=0;i<ar.length;i++)
            a.add(ar[i]);
        for(int i=1970;i<=2008;i++)
            c.add(""+i);
        add(a);
```

```
            add(c);
            add(t1);
            c.select("1986");
            c.addItemListener(this);
            a.addItemListener(this);
        }
        public void itemStateChanged(ItemEvent e)
        {
            t1.setText("U r from - "+a.getSelectedItem()+" , selected year is - "
+c.getSelectedItem());
        }
    }
    //<applet code=MyList height=400 width=600></applet>
```

COMPILE:

```
C:\>javac MyList.java
```

EXECUTE:

```
C:\>appletviewer MyList.java
```

OUTPUT:

16.4. ADJUSTMENTLISTENER EXAMPLE

An AdjustmentListener interface is basically used only for Scrollbar control. The AdjustmentListener interface contains only one method for handling the event, which is:

```
public void adjustmentValueChanged(AdjusmentEvent e){}
```

Now how to use this method in different programs, see some examples:

Example 16.8 **Program to create three Scrollbars for Red, Green, Blue Colors and change the color of back ground according to adjustment of scrollbars.**

```
import java.awt.*;
import java.awt.event.*;
import java.applet.*;
public class Scroll extends Applet implements AdjustmentListener
{     //type(0-H/1-V),initial,thumb,min,max(scrollmax=max-thumb)
```

```
    Scrollbar r = new Scrollbar(0,0,20,0,275);
    Scrollbar g = new Scrollbar(0,0,20,0,275);
    Scrollbar b = new Scrollbar(0,0,20,0,275);
    int cr,cg,cb;
    public void init()
    {
        setLayout(null);
        r.setBounds(50,50,600,20);
        g.setBounds(50,75,600,20);
        b.setBounds(50,100,600,20);
        r.setBackground(Color.red);
        g.setBackground(Color.green);
        b.setBackground(Color.blue);
        add(r);
        add(g);
        add(b);
        r.addAdjustmentListener(this);
        g.addAdjustmentListener(this);
        b.addAdjustmentListener(this);
    }
    public void adjustmentValueChanged(AdjustmentEvent e)
    {
        cr = r.getValue();
        cg = g.getValue();
        cb = b.getValue();
        setBackground(new Color(cr,cg,cb));
        showStatus("Color is --> r = "+cr+" , g = "+cg+" , b = "+cb);
    }
}
//<applet code=Scroll height=400 width=600></applet>
```

COMPILE:

```
C:\>javac Scroll.java
```

EXECUTE:

```
C:\>appletviewer Scroll.java
```

OUTPUT

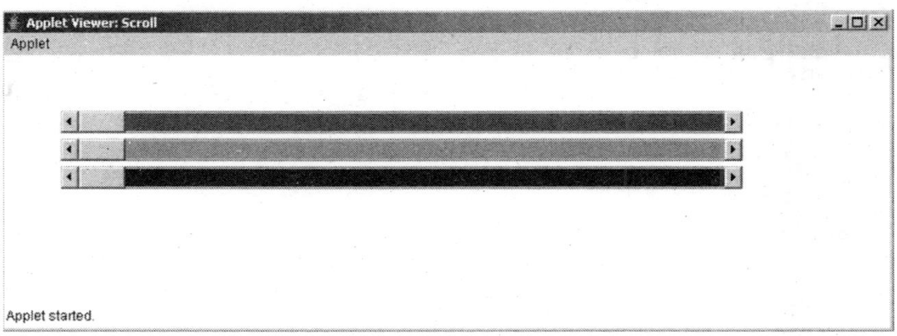

16.5. KEYLISTENER AND KEYADAPTER EXAMPLE

An KeyListener interface and KeyAdapter class is basically used only for keys of key board. The KeyListener interface and KeyAdapter class contains following methods for handling the event, which are:

```
public void keyPressed(KeyEvent e){}
public void keyTyped(KeyEvent e){}
public void keyReleased(KeyEvent e){}
```

Now how to use these methods in different programs, see some examples:

Example 16.9 Program to show the code of keys with their characters after pressing the individual key.

```
import java.awt.*;
import java.awt.event.*;
import java.applet.*;
import java.util.*;
public class Key extends Applet implements KeyListener
{
    int x;
    char c;
    public void init()
    {
        addKeyListener(this);
    }
    public void keyTyped(KeyEvent e){}
    public void keyReleased(KeyEvent e){}
    public void keyPressed(KeyEvent e)
    {
        x = e.getKeyCode(); // return key code not key ascii
        c = e.getKeyChar(); // return printable key char
        showStatus("Key Pressed , key code is --> "+x+" , key char is --> "+c);
    }
}
//<applet code=Key height=400 width=600></applet>
```

COMPILE:

```
C:\>javac Key.java
```

EXECUTE:

```
C:\>appletviewer Key.java
```

OUTPUT: (press any Key on the applet)

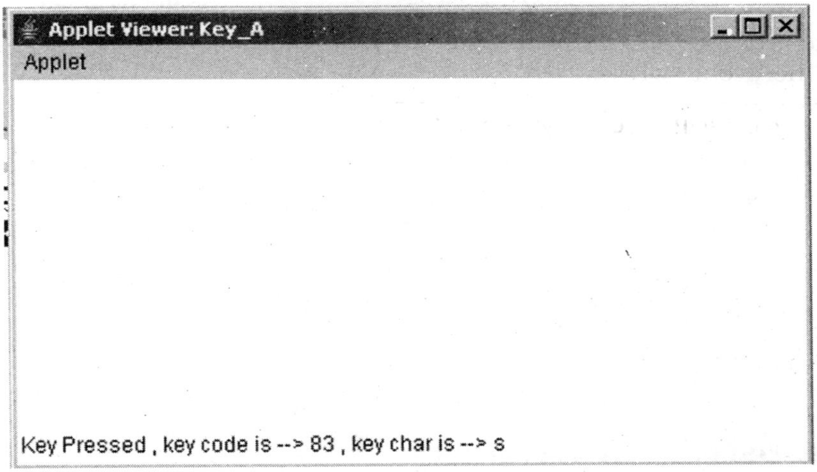

Example 16.10 **Program to show the code of keys with their characters after pressing the individual key.**

```
import java.awt.*;
import java.awt.event.*;
import java.applet.*;
public class Key_A extends Applet
{
    int x;
    char c;
    public void init()
    {
        addKeyListener(new Deepu());
    }
    class Deepu extends KeyAdapter
    {
        public void keyPressed(KeyEvent e)
        {
            x = e.getKeyCode(); // return key code not key ascii
            c = e.getKeyChar(); // return printable key char
            showStatus("Key Pressed , key code is --> "+x+" , key char is --> "+c);
        }
    }
}
//<applet code=Key_A height=400 width=600></applet>
```

COMPILE:

```
C:\>javac Key_A.java
```

EXECUTE:

```
C:\>appletviewer Key_A.java
```

OUTPUT: (press any Key on the applet)

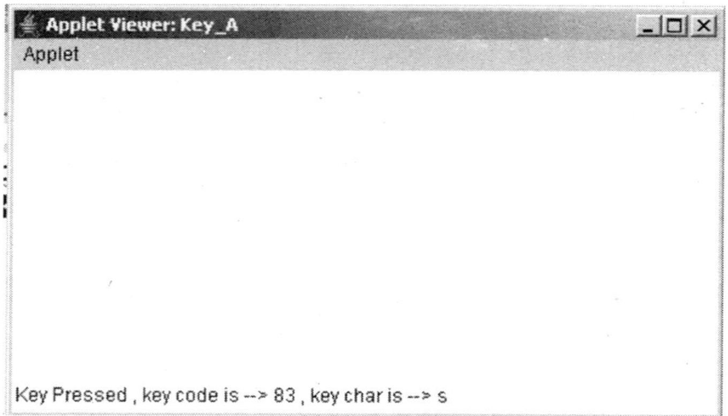

16.6. MOUSEMOTIONLISTENER

An MouseMotionListener interface and MouseMotionAdapter class is basically used for mouse movement. The MouseMotionListener interface and MouseMotionAdapter class contains only two methods for handling the event, which are:

```
public void mouseMoved(MouseEvent e){}
public void mouseDragged(MouseEvent e){}
```

Now how to use these methods in different programs, see some examples:

Example 16.11 Program to show the current status of Mouse Motion where mouse pointer will move around the applet.

Using MouseMotionListener

```
import java.awt.*;
import java.awt.event.*;
import java.applet.*;
public class Mouse extends Applet implements MouseMotionListener
{
      int x,y;
      public void init()
      {
            addMouseMotionListener(this);
      }
      public void mouseMoved(MouseEvent e)
      {
            x = e.getX(); // x hold the current mouse x position
            y = e.getY(); // y hold the current mouse y position
            showStatus("mouse Moved at --> x = "+x+" , y = "+y);
      }
      public void mouseDragged(MouseEvent e) { }
}
//<applet code=Mouse height=400 width=600></applet>
```

Using MouseMotionAdapter

```
import java.awt.*;
import java.awt.event.*;
import java.applet.*;
public class Mouse extends Applet
{
      int x,y;
      public void init()
      {
            addMouseMotionListener(new Deepa());
      }
      class Deepa extends MouseMotionAdapter
      {
            public void mouseMoved(MouseEvent e)
      {
            x = e.getX(); // x hold the current mouse x position
            y = e.getY(); // y hold the current mouse y position
            showStatus("mouse Moved at --> x = "+x+" , y = "+y);
      }
}
}
//<applet code=Mouse height=400 width=600></applet>
```

COMPILE:

```
C:\>javac Mouse.java
```

EXECUTE:

```
C:\>appletviewer Mouse.java
```

OUTPUT: (Move the mouse pointer on the applet)

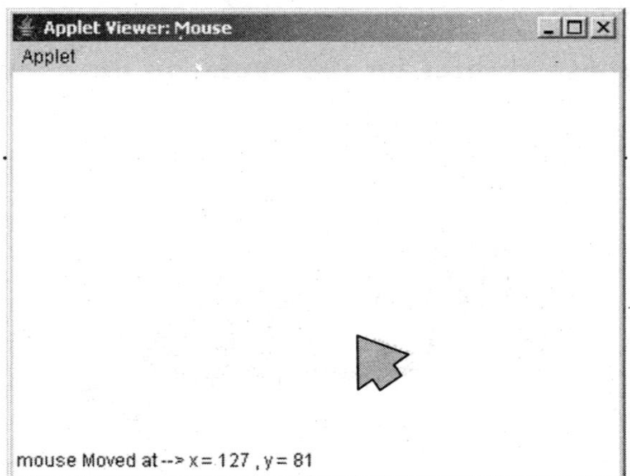

Example 16.12 **Program to design a Simple Paint Brush on an applet.**

```
import java.awt.*;
```

```
import java.awt.event.*;
import java.applet.*;
public class Mouse extends Applet implements MouseMotionListener
{
      Graphics g;
      int x,y;
      public void init()
      {
            g = getGraphics(); // g hold the current Applet graphics , now we can drawing without using paint method
            addMouseMotionListener(this);
      }
      public void mouseMoved(MouseEvent e)
      {
            x = e.getX(); // x hold the current mouse x position
            y = e.getY(); // y hold the current mouse y position
            showStatus("mouse Moved at --> x = "+x+" , y = "+y);
      }
      public void mouseDragged(MouseEvent e)
      {
            x = e.getX(); // x hold the current mouse x position
            y = e.getY(); // y hold the current mouse y position
            showStatus("mouse dragged at --> x = "+x+" , y = "+y);
            g.fillOval(x,y,10,10);
            //g.drawString("Arora",x,y);
      }
}
//<applet code=Mouse height=400 width=600></applet>
```

COMPILE:

```
C:\>javac Mouse.java
```

EXECUTE:

```
C:\>appletviewer Mouse.java
```

OUTPUT: (Drag the mouse pointer on the applet to draw any image)

16.7. MOUSELISTENER

An MouseListener interface and MouseAdapter class is basically used for mouse movement. The MouseListener interface and MouseAdapter class contains following methods for handling the event, which are:

```
public void mousePressed(MouseEvent e){}
public void mouseClicked(MouseEvent e){}
public void mouseReleased(MouseEvent e){}
public void mouseEntered(MouseEvent e){} - over
public void mouseExited(MouseEvent e){} - out
```

Now how to use these methods in different programs, see some examples:

Example 16.13 **Program to show all the methods of MouseListener on an applet.**

```
import java.awt.*;
import java.awt.event.*;
import java.applet.*;
public class Mouse extends Applet implements MouseListener
{
    int cg,cb,cr;
    public void init()
    {
        addMouseListener(this);
    }
    public void mousePressed(MouseEvent e)
    {
        showStatus("mouse Pressed");
    }
    public void mouseReleased(MouseEvent e)
    {
        showStatus("mouse Released");
    }
    public void mouseEntered(MouseEvent e)
    {
        showStatus("mouse over");
    }
    public void mouseExited(MouseEvent e)
    {
        showStatus("mouse out");
    }
    public void mouseClicked(MouseEvent e)
    {
        showStatus("mouse Clicked");
    }
}
//<applet code=Mouse height=400 width=600></applet>
```

COMPILE:

```
C:\>javac Mouse.java
```

EXECUTE:

```
C:\>appletviewer Mouse.java
```

OUTPUT: (Move the mouse pointer on the applet)

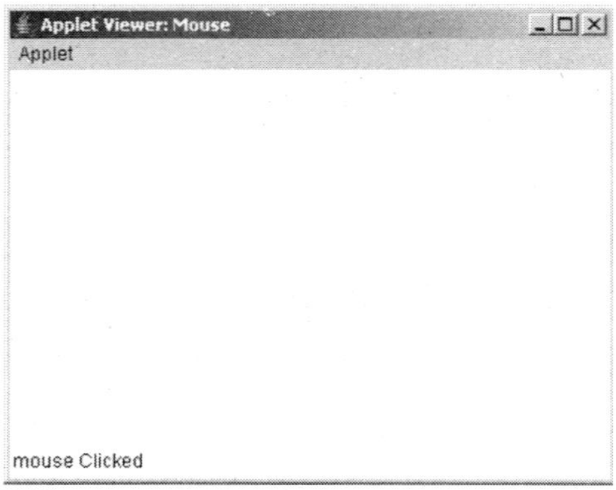

Example 16.14 Program to Change the back ground on Mouse click on an applet.

```
import java.awt.*;
import java.awt.event.*;
import java.applet.*;
public class Mouse extends Applet implements MouseListener
{
      int cg,cb,cr;
      public void init()
      {
            addMouseListener(this);
      }
      public void mousePressed(MouseEvent e)
      {
            showStatus("mouse Pressed");
      }
      public void mouseReleased(MouseEvent e)
      {
            showStatus("mouse Released");
      }
      public void mouseEntered(MouseEvent e)
      {
            showStatus("mouse over");
      }
      public void mouseExited(MouseEvent e)
      {
            showStatus("mouse out");
      }
      public void mouseClicked(MouseEvent e)
      {
```

```
            cr = (int)(Math.random()*255);
            cg = (int)(Math.random()*255);
            cb = (int)(Math.random()*255);
            setBackground(new Color(cr,cg,cb));
        }
    }
    //<applet code=Mouse height=400 width=600></applet>
```

COMPILE:

```
C:\>javac Mouse.java
```

EXECUTE:

```
C:\>appletviewer Mouse.java
```

OUTPUT: (Click the mouse on the applet)

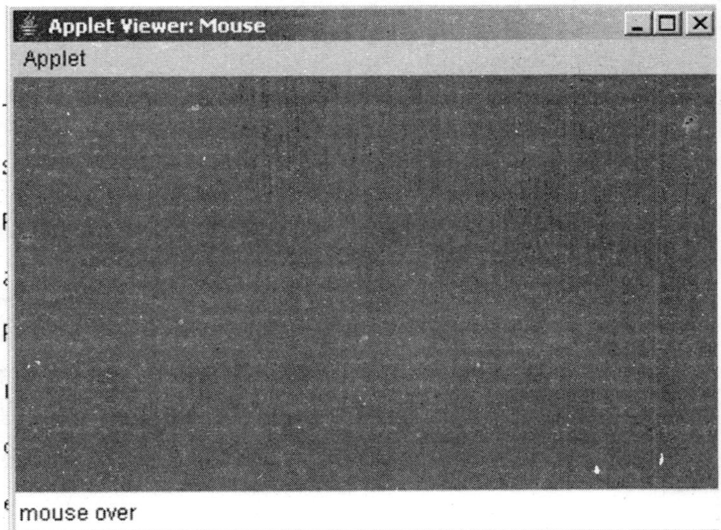

16.8. WINDOWLISTENER

An WindowListener interface and WindowAdapter class is basically used for windows operations movement. The WindowListener interface and WindowAdapter class contains following methods for handling the event, which are:

```
public void windowOpened(WindowEvent e){}
public void windowClosing(WindowEvent e){}
public void windowClosed(WindowEvent e){}
public void windowActivated(WindowEvent e){}
public void windowDeactivated(WindowEvent e){}
public void windowIconified(WindowEvent e){} - minimize
public void windowDeiconified(WindowEvent e){} - maximize
```

Now how to use these methods in different programs, see some examples:

Example 16.15 **An AWT Program to create a Note-Pad with Menu items and perform the action on them using event handling.**

```
import java.awt.*;
import java.awt.event.*;
class Notepad extends Frame implements ActionListener
{
    FileDialog fd; // for open & save Dialog
    TextArea ta = new TextArea();
    MenuBar mb = new MenuBar();
    Menu f1 = new Menu("F i l e ");
    Menu f2 = new Menu("E d i t ");
    MenuItem a1 = new MenuItem("N e w ");
    MenuItem a2 = new MenuItem("O p e n ");
    MenuItem a3 = new MenuItem("S a v e ....");
    MenuItem a4 = new MenuItem("E x i t ");
    MenuItem a5 = new MenuItem("C u t ");
    MenuItem a6 = new MenuItem("C o p y ");
    MenuItem a7 = new MenuItem("P a s t e ");
    MenuItem a8 = new MenuItem("S e l e c t  A l l ");
    Notepad()
    {
        setTitle("Notepad Example By Lalit Arora");
        setFont(new Font("Courier New",1,14));
        // add item in file menu
        f1.add(a1);
        f1.add(a2);
        f1.add(a3);
        f1.addSeparator();
        f1.add(a4);
        // add item in edit menu
        f2.add(a5);
        f2.add(a6);
        f2.add(a7);
        f2.addSeparator();
        f2.add(a8);
        // add menu in menu bar
        mb.add(f1);
        mb.add(f2);
        setMenuBar(mb); // set menubar in frame
        add(ta); // add ta in center
        setBounds(50,50,700,500); // set size of notepad
        setVisible(true);
        a1.addActionListener(this);
        a2.addActionListener(this);
        a3.addActionListener(this);
        a4.addActionListener(this);
        a5.addActionListener(this);
```

```
            a6.addActionListener(this);
            a7.addActionListener(this);
            a8.addActionListener(this);
            addWindowListener(new Abhi());
        }
        class Abhi extends WindowAdapter
        {
            public void windowClosing(WindowEvent e)
            {
                dispose();
            }
        }
        public void actionPerformed(ActionEvent e)
        {
            if(e.getSource()==a1) // for new
            {
                setTitle("Untited");
                ta.setText("");
            }
            if(e.getSource()==a2) // for open
            {
                fd = new FileDialog(this,"Open Me",FileDialog.LOAD);
                fd.setVisible(true);
                setTitle(fd.getFile());
            }
            if(e.getSource()==a3) // for save
            {
                fd = new FileDialog(this,"Save Me",FileDialog.SAVE);
                fd.setVisible(true);
                setTitle(fd.getFile());
            }
            if(e.getSource()==a4) // for exit
                dispose(); // setVisible(false);
            // mar gaya  / jinda hai
        }
        public static void main(String aa[])
        {
            new Notepad();
        }
    }
```

COMPILE:

```
C:\>javac Example.java
```

EXECUTE:

```
C:\>java Example
```

OUTPUT:

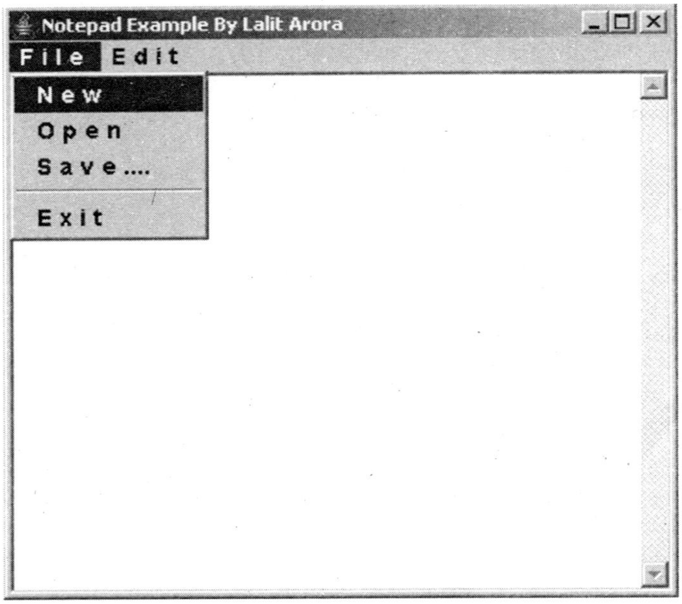

16.9. FOCUSLISTENER

An FocusListener interface and FocusAdapter class is basically used for Focus on Text box and Button controls. The FocusListener interface and FocusAdapter class contains only two methods for handling the event, which are:

```
public void focusGained(FocusEvent e){}
public void focusLost(FocusEvent e){}
```

Now how to use these methods in different programs, see some examples:

Example 16.16 Program to Add, Multiply, Subtract and Divide two numbers on applet.

```
import java.awt.*;
import java.awt.event.*;
import java.applet.*;
public class Focus extends Applet implements ActionListener,FocusListener
{
    String ar[] = new String[]{"Sum","Sub","Multi","Div","Clear"};
    Label a1 = new Label("Ist Num");
    Label a2 = new Label("IInd Num");
    Label a3 = new Label("Result");
    TextField t1 = new TextField();
    TextField t2 = new TextField();
    TextField t3 = new TextField();

    public void init()
```

```
{
    setLayout(null);
    setFont(new Font("Courier New",1,14));
    setBackground(new Color(129,164,250));
    a1.setBounds(100,100,100,30);
    a2.setBounds(100,160,100,30);
    a3.setBounds(100,220,100,30);
    t1.setBounds(210,100,100,30);
    t2.setBounds(210,160,100,30);
    t3.setBounds(210,220,100,30);
    add(a1);
    add(a2);
    add(a3);
    add(t1);
    add(t2);
    add(t3);
    for(int i=0,x=50;i<ar.length;i++,x+=100)
    {
        Button b = new Button(ar[i]);
        b.setBounds(x,300,90,30);
        b.addActionListener(this);
        add(b);
    }
    t3.setEditable(false); // user can't edit now
    t1.addFocusListener(this);
    t2.addFocusListener(this);
}
public void focusGained(FocusEvent e)
{
    if(e.getSource()==t1) // if focus gained on t1
    {
        t1.setText("");
        t2.setText("");
        t3.setText("");
    }
}
public void focusLost(FocusEvent e)
{
    String s,n="";
    if(e.getSource()==t1) // if focus lost from t1
    {
        s = t1.getText();
        for(int i=0;i<s.length();i++)
        {
            if(s.charAt(i) != ' ')
                n+=s.charAt(i);
        }
        if(n.equals("")) t1.setText("0");
```

```
            else t1.setText(n);
        }
        if(e.getSource()==t2) // if focus lost from t2
        {
            s = t2.getText();
            for(int i=0;i<s.length();i++)
            {
                if(s.charAt(i) != ' ')
                    n+=s.charAt(i);
            }
            if(n.equals("")) t2.setText("0");
            else t2.setText(n);
        }
    }
    public void actionPerformed(ActionEvent e)
    {
        int a,b,c;
        String cmd = e.getActionCommand(); // return the button caption
        try{ a = Integer.parseInt(t1.getText()); }catch(Exception ex){a = 0;}
        try{ b = Integer.parseInt(t2.getText()); }catch(Exception ex){b = 0;}
        if(cmd.equals("Sum"))
        {
            c = a + b;
            t3.setText(""+c);
        }
        if(cmd.equals("Sub"))
        {
            c = a - b;
            t3.setText(""+c);
        }
        if(cmd.equals("Multi"))
        {
            c = a * b;
            t3.setText(""+c);
        }
        if(cmd.equals("Div"))
        {
            try
            {
                c = a / b;
                t3.setText(""+c);
            }catch(Exception ex){ t3.setText("Infinity");}
        }
        if(cmd.equals("Clear"))
        {
            t1.setText("");
            t2.setText("");
            t3.setText("");
```

```
            }
        }
    }
    //<applet code=Focus height=400 width=600></applet> COMPILE:
```

COMPILE:

```
C:\>javac Focus.java
```

EXECUTE:

```
C:\>appletviewer Focus.java
```

OUTPUT:

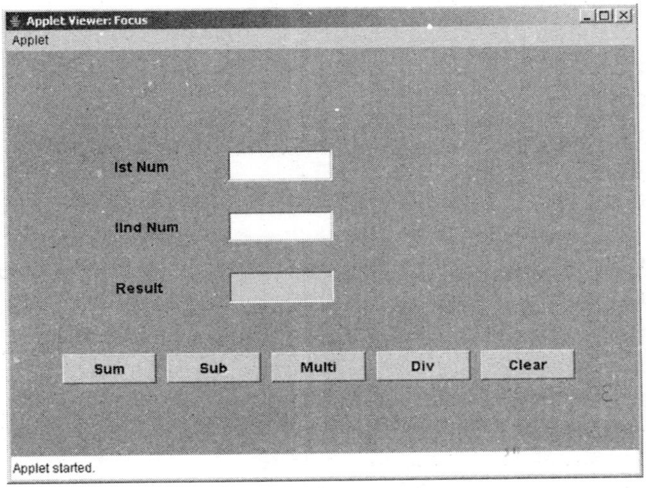

Example 16.17 Program of Employee Management System on applet.

```
import java.awt.event.*;
import java.awt.*;
import java.applet.*;
public class Focus extends Applet implements ActionListener,FocusListener
{
        String no[] = new String[50];
        String nm[] = new String[50];
        String sal[] = new String[50];
        int ctr=0,nor=0;
    String ar[] = new
String[]{"Exit","Cancel","Report","First","Prev","Next","Last","Delete","Update"};
        Label a1 = new Label("Empno");
        Label a2 = new Label("Ename");
        Label a3 = new Label("Salary");
        Label L  = new Label("System Message:");
        TextField t1 = new TextField();
        TextField t2 = new TextField();
        TextField t3 = new TextField();
```

```
Button b1 = new Button("Save");
public void init()
{
     setLayout(null);
     setFont(new Font("Courier New",1,14));
     setBackground(new Color(129,164,250));
     a1.setBounds(100,100,100,30);
     a2.setBounds(100,160,100,30);
     a3.setBounds(100,220,100,30);
     L.setBounds(100,360,500,30);
     t1.setBounds(210,100,100,30);
     t2.setBounds(210,160,150,30);
     t3.setBounds(210,220,100,30);
     b1.setBounds(50,300,90,30);
     add(a1);
     add(a2);
     add(a3);
     add(L);
     add(t1);
     add(t2);
     add(t3);
     add(b1);
     for(int i=0,x=150;i<3;i++,x+=100)
     {
          Button b = new Button(ar[i]);
          b.setBounds(x,300,90,30);
          b.addActionListener(this);
          add(b);
     }
     for(int i=3,x=50;i<ar.length;i++,x+=50)
     {
          Button b = new Button(ar[i]);
          b.setBounds(450,x,90,30);
          b.addActionListener(this);
          add(b);
     }
     t1.addFocusListener(this);
     b1.addActionListener(this);
     b1.setEnabled(false);
//   requestFocus(); // focus on frame
     }
     public void focusGained(FocusEvent e)
     {
          clear();
          b1.setEnabled(true);
     }
```

```
        public void focusLost(FocusEvent e)
        {
            if(t1.getText().equals(""))
                b1.setEnabled(false);
            for(int i=0;i<nor;i++)
            {
                if(t1.getText().equals(no[i]))
                {
                    disp(i);
            L.setText("System Message:Sorry U can't insert duplicate record , record
already exist");
            b1.setEnabled(false);
            break;
            }
        }
    }
    void save()
    {
        no[nor] = t1.getText();
        nm[nor] = t2.getText();
        sal[nor] = t3.getText();
        nor++; // ready for next record index.
    }
    void clear()
    {
        t1.setText("");
        t2.setText("");
        t3.setText("");
    }
    void disp(int a)
    {
        t1.setText(no[a]);
        t2.setText(nm[a]);
        t3.setText(sal[a]);
    }
    void updRec(int a)
    {
        nm[a] = t2.getText();
        sal[a] = t3.getText();
    }
    public void actionPerformed(ActionEvent e)
    {
        String cmd = e.getActionCommand();
        if(cmd.equals("Save"))
    {
        save();
```

```
            L.setText("System Message:O n e  R e c o r d  I n s e r t e d ");
            clear();
            b1.setEnabled(false);
        }
        if(cmd.equals("Update"))
        {
                updRec(ctr);
            L.setText("System Message:O n e  R e c o r d  U p d a t e d ");
        }
        if(cmd.equals("Cancel"))
            clear();
        if(cmd.equals("First"))
        {
            ctr=0;
            disp(ctr);
        }
        if(cmd.equals("Last"))
        {
            ctr=nor-1;
            disp(ctr);
        }
        if(cmd.equals("Prev"))
        {
            ctr--;
            if(ctr<0)
            {
                ctr = 0;
            L.setText("System Message:Sorry this is the First Record");

            }
            disp(ctr);
        }
        if(cmd.equals("Next"))
        {
            ctr++;
            if(ctr>=nor)
            {
                ctr = nor-1;
                L.setText("System Message:Sorry this is the Last Record");
            }
            disp(ctr);
        }
    }
    //<applet code=Focus height=400 width=600></applet>
```

COMPILE:

```
C:\>javac Focus.java
```

EXECUTE:

```
C:\>appletviewer Focus.java
```

OUTPUT:

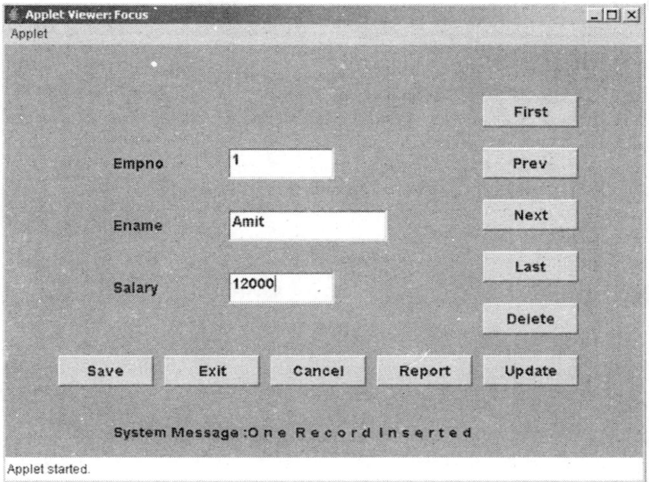

SUMMARY

— It this chapter we tried to explain all the event handling concept through examples.

— Action listener and after listeneris basically interface which are usedtop handle the events.

— In Java JDK 1, 2 also added some adapter classes, whichare subtitute ofthose interfaces which contain more methods to implement in derived classes.

EXERCISES

Q. 16.1. What is Event Source ?

Q. 16.2. What do you mean by Event Lintener ?

Q. 16.3. Write down the methods of Mons Listener.

Q. 16.4. What is the benefits of adapter classes ?

Q. 16.5. Write down the methods of window Listener ?

Q. 16.6. What is the difference between Mouse Listener and Mouse Motion Listener ?

Q. 16.7. Write a program to add a Button Control "Message" to display a welcome message on status bar.

Q. 16.8. Write a program to add some radio button controls, which input for category such as (Gen, OBC, SC/ST) and display the selected category on the lebel.

Q. 16.9. Write program to creater calculator with proper event handling.

Q. 16.10. Write a program to create a game of tick-tak-teo with Button control. (See Figure)

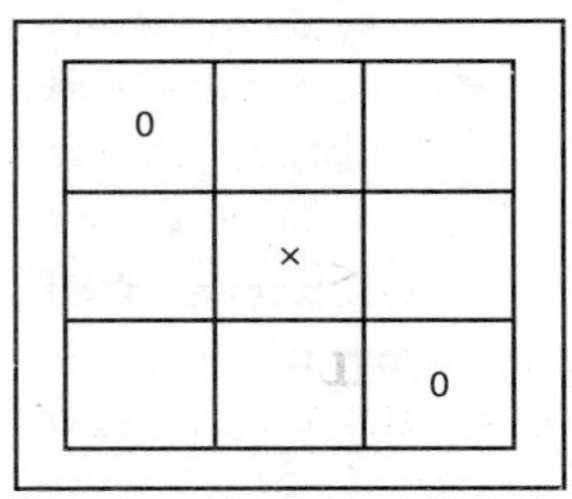

Chapter 17

Software Development Using JAVA

CHAPTER OUTLINE

17.1. JAVA BEANS

JavaBeans are reusable software components for Java that can be manipulated visually in a builder tool. Practically, they are classes written in the Java programming language conforming to a particular convention. They are used to encapsulate many objects into a single object (the bean), so that they can be passed around as a single bean object instead of as multiple individual objects. A JavaBean is a Java Object that is serializable, has a nullary constructor, and allows access to properties using getter and setter methods.

- JavaBeans is a portable, platform—independent component model written in the Java programming language, developed in collaboration with industry leaders.
- It enables developers to write reusable components once and run them anywhere — benefitting from the platform—independent power of Java technology.
- The goal of JavaBeans is to create a system whereby application developers can take a set of beans from a stock library and wire them together to make a full application.

Software Component: A software component is a unit of composition with contractually specified interfaces and explicit context dependencies only. A software component can be deployed independently and is subject to composition by third parties.

> **Concept**
>
> *"A Java bean is a reusable software component that can be visually manipulated in builder tools."*

JavaBeans are reusable software components written in Java. These components may be built into an application using an appropriate building environment. The Bean Development Kit (BDK) from Sun includes a simple example of a building environment which uses beans, called *the beanbox*, and some sample beans.

This note will tell you how to set up and use the BDK on the Computer Science and Informatics Fedora Linux workstations and how to add your own beans to the BeanBox environment.

17.1.1. The BDK

The BDK needs to create some directories and copy some files to your filespace before you can use it. So, the first time you want to use the BDK, give the following command:

```
% bdk-user-setup
```

This will create a directory called BDK1.1 and sub-directories BDK1.1/jars and BDK1.1/beanbox together with some links to files in the Bean Development Kit installation directory.

17.1.2. Running the BeanBox

Change directory to BDK1.1/beanbox and type the command ./run.sh:

```
% cd BDK1.1/beanbox % ./run.sh &
```

The BeanBox starts and loads beans from the JAR (Java Archive) files in the BDK1.1/jars directory. It then displays four windows on the screen:

- the ToolBox which lists all beans the BeanBox has found,
- the BeanBox which is a canvas where you will drop beans and connect them together,
- the Properties sheet which is used to set the properties of the current bean,
- the MethodTracer simple debugging facility.

17.1.3. JAR Files

In computing software, a JAR file (or Java ARchive) aggregates many files into one. Software developers generally use .jar files to distribute Java applications or libraries, in the form of Java class files and associated metadata and resources (text, images, etc.).

A JAR file allows Java runtimes to efficiently deploy a set of classes and their associated resources. The elements in a JAR file can be compressed, which, together with the ability to download an entire application in a single request, makes downloading a JAR file much faster than separately downloading the many uncompressed files which would form a single Java Application. The package java.util.zip contains classes that read and write JAR files.

Users can typically start such JAR files with a command similar to:

```
java -jar foo.jar
```

To unzip a JAR file users can use any standard unzip software. However, anyone with an installed Java Virtual Machine can use the jar command to expand the file:

```
jar -xf foo.jar
```

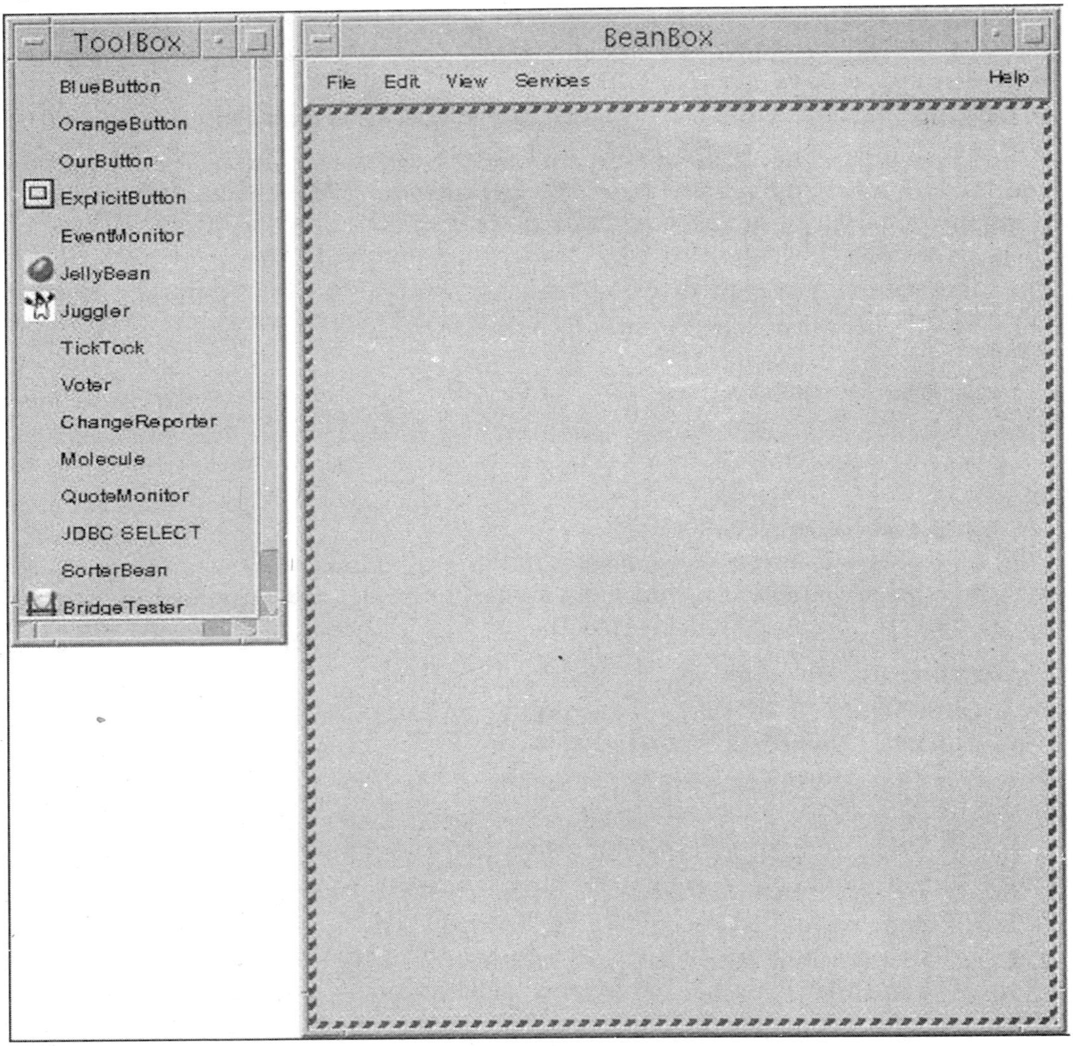

Fig. 17.1.

17.1.4. Features of Java Beans

- *Support for* ***introspection*** **:** so that a builder tool can analyze how a bean works
- *Support for* ***customization*** **:** so that when using an application builder a user can customize the appearance and behavior of a bean
- *Support for* ***events*** **:** as a simple communication metaphor than can be used to connect up beans
- *Support for* ***properties*** **:** both for customization and for programmatic use
- *Support for* ***persistence*** **:** so that a bean can be customized in an application builder and then have its customized state saved away and reloaded later

17.2. DEVELOPING SAMPLE BEAN

In this section you will learn more about beans by performing the following actions:

- Creating a simple bean
- Compiling the bean
- Generating a Java Archive (JAR) file
- Loading the bean into the GUI Builder of the NetBeans IDE
- Inspecting the bean's properties and events

Your bean will be named SimpleBean. Here are the steps to create it:

1. **Write the SimpleBean code**. Put it in a file named SimpleBean.java, in the directory of your choice. Here's the code:

```
import java.awt.Color;
import java.beans.XMLDecoder;
import javax.swing.JLabel;
import java.io.Serializable;

public class SimpleBean extends JLabel implements Serializable
{
     public SimpleBean() {
          setText( "Hello world!" );
          setOpaque( true );
          setBackground( Color.RED );
          setForeground( Color.YELLOW );
          setVerticalAlignment( CENTER );
          setHorizontalAlignment( CENTER );
     }
}
```

   SimpleBean extends the javax.swing.JLabel graphic component and inherits its properties, which makes the SimpleBean a visual component. SimpleBean also implements the java.io.Serializable interface. Your bean may implement either the Serializable or the Externalizable interface.

2. **Create a manifest, the JAR file, and the class file SimpleBean.class**. Use the Apache Ant tool to create these files. Apache Ant is a Java-based build tool that enables you to generate XML-based configurations files as follows:

```
<?xml version="1.0" encoding="ISO-8859-1"?>

<project default="build">

<dirname property="basedir" file="${ant.file}"/>

<property name="beanname" value="SimpleBean"/>
<property name="jarfile" value="${basedir}/${beanname}.jar"/>

<target name="build" depends="compile">
     <jar destfile="${jarfile}" basedir="${basedir}" includes="*.class">
```

```
            <manifest>
                <section name="${beanname}.class">
                    <attribute name="Java-Bean" value="true"/>
                </section>
            </manifest>
        </jar>
    </target>

    <target name="compile">
        <javac destdir="${basedir}">
            <src location="${basedir}"/>
        </javac>
    </target>

    <target name="clean">
        <delete file="${jarfile}">
            <fileset dir="${basedir}" includes="*.class"/>
        </delete>
    </target>

    </project>
```

It is recommended to save an XML script in the build.xml file, because Ant recognizes this file name automatically.

3. **Load the JAR file**. Use the NetBeans IDE GUI Builder to load the jar file as follows:
    1. Start NetBeans.
    2. From the File menu select "New Project" to create a new application for your bean. You can use "Open Project" to add your bean to an existing application.
    3. Create a new application using the New Project Wizard.
    4. Select a newly created project in the List of Projects, expand the Source Packages node, and select the Default Package element.
    5. Click the right mouse button and select New|JFrameForm from the pop-up menu.
    6. Select the newly created Form node in the Project Tree. A blank form opens in the GUI Builder view of an Editor tab.
    7. Open the Palette Manager for Swing/AWT components by selecting Palette Manager in the Tools menu.
    8. In the Palette Manager window select the beans components in the Palette tree and press the "Add from JAR" button.
    9. Specify a location for your SimpleBean JAR file and follow the Add from JAR Wizard instructions.
    10. Select the Palette and Properties options from the Windows menu.
    11. Expand the beans group in the Palette window. The SimpleBean object appears. Drag the SimpleBean object to the GUI Builder panel.

The following figure represents the SimpleBean object loaded in the GUI Builder panel:

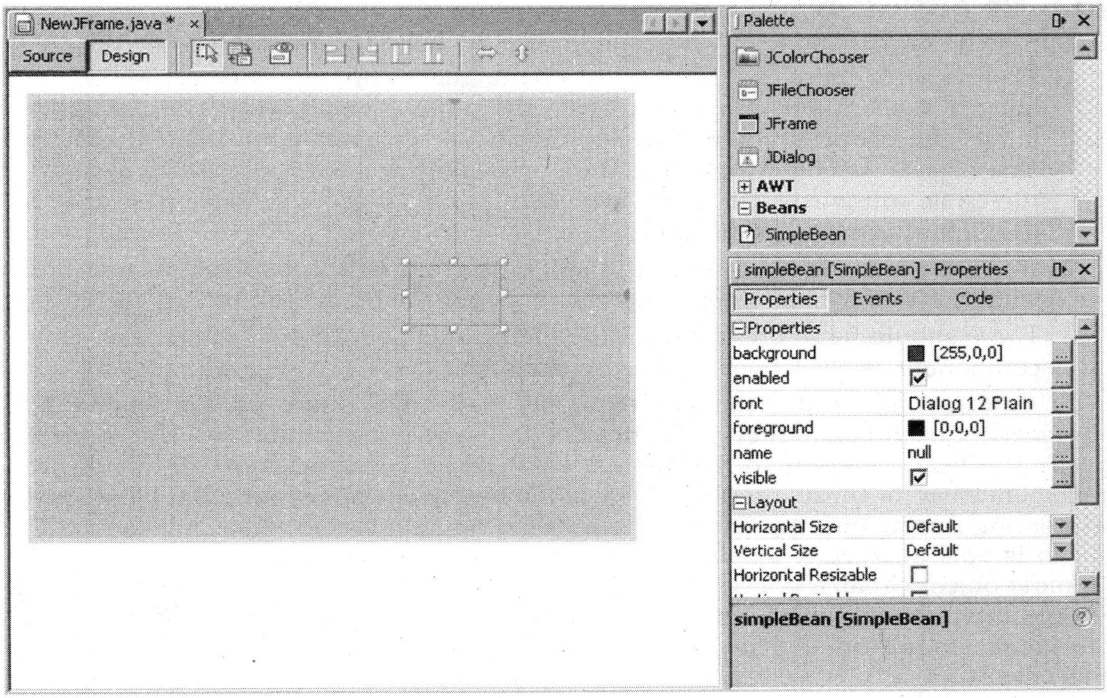

Fig. 17.2.

4. **Inspect Properties and Events**. The SimpleBean properties will appear in the Properties window. For example, you can change a background property by selecting another color. To preview your form, use the Preview Design button of the GUI Builder toolbar. To inspect events associated with the SimpleBean object, switch to the Events tab of the Properties window. You will learn more about bean properties and events in the lessons that follow.

17.3. ENTERPRISE JAVABEANS—AN INTRODUCTION

Enterprise JavaBeans (EJB) is a comprehensive technology that provides the infrastructure for building enterprise-level server-side distributed Java components. The EJB technology provides a distributed component architecture that integrates several enterprise-level requirements such as distribution, transactions, security, messaging, persistence, and connectivity to mainframes and Enterprise Resource Planning (ERP) systems. When compared with other distributed component technologies such as Java RMI and CORBA, the EJB architecture hides most the underlying system-level semantics that are typical of distributed component applications, such as instance management, object pooling, multiple threading, and connection pooling. Secondly, unlike other component models, EJB technology provides us with different types of components for business logic, persistence, and enterprise messages.

Thus, an Enterprise Java Bean is a remote object with semantics specified for creation, invocation and deletion. The EJB container is assigned the system-level tasks mentioned above. What a web container does for Java servlets and JSPs in a web server, the EJB container is for EJBs.

17.3.1. EJB Architecture

Any distributed component technology should have the following requirements:

1. There should be a mechanism to create the client-side and server-side proxy objects. A client-side proxy represents the server-side object on the client-side. As far as the client is concerned, the client-side proxy is equivalent to the server-side object. On the other hand, the purpose of the server-side proxy is to provide the basic infrastructure to receive client requests and delegate these request to the actual implementation object.
2. We need to obtain a reference to client-side proxy object. In order to communicate with the server-side object, the client needs to obtain a reference to the proxy.
3. There should be a way to inform the distributed component system that a specific component is no longer in use by the client.

In order to meet these requirements, the EJB architecture specifies two kinds of interfaces for each bean. They are home interface and remote interface. These interfaces specify the bean contract to the clients. However, a bean developer need not provide implementation for these interfaces. The home interface will contain methods to be used for creating remote objects. The remote interface should include business methods that a bean is able to serve to clients. One can consider using the home interface to specify a remote object capable of creating objects conforming to the remote interface. That is, a home interface is analogous to a factory of remote objects. These are regular Java interfaces extending the javax.ejb.EJBHome and javax.ejb.EJBObject interfaces respectively.

17.3.2. Types of EJBs

Enterprise JavaBeans server-side components come in two fundamentally different types:

- entity beans and
- session beans.

Basically **entity beans** model business concepts that can be expressed as nouns. For example, an entity bean might represent a customer, a piece of equipment, an item in inventory. Thus entity beans model real-world objects. These objects are usually persistent records in some kind of database.

Session beans are for managing processes or tasks. A session bean is mainly for coordinating particular kinds of activities. That is, session beans are plain remote objects meant for abstracting business logic. The activity that a session bean represents is fundamentally transient. A session bean does not represent anything in a database, but it can access the database.

17.4. JAVA SWING

Swing library is an official Java GUI toolkit released by Sun Microsystems. It is used to create Graphical user interfaces with Java.

The main characteristics of the Swing toolkit:

- platform independent
- customizable
- extensible
- configurable
- lightweight

The Swing API has 18 public packages:

- javax.accessibility
- javax.swing

- javax.swing.border
- javax.swing.event
- javax.swing.plaf
- javax.swing.plaf.metal
- javax.swing.plaf.synth
- javax.swing.text
- javax.swing.text.html.parser
- javax.swing.tree
- javax.swing.colorchooser
- javax.swing.filechooser
- javax.swing.plaf.basic
- javax.swing.plaf.multi
- javax.swing.table
- javax.swing.text.html
- javax.swing.text.rtf
- javax.swing.undo

Swing is an advanced GUI toolkit. It has a rich set of widgets. From basic widgets like buttons, labels, scrollbars to advanced widgets like trees and tables. Swing itself is written in Java.

Swing is a part of JFC, Java Foundation Classes. It is a collection of packages for creating full featured desktop applications. JFC consists of AWT, Swing, Accessibility, Java 2D, and Drag and Drop. Swing was released in 1997 with JDK 1.2. It is a mature toolkit.

The Java platform has Java2D library, which enables developers to create advanced 2D graphics and imaging.

There are basically two types of widget toolkits.

- Lightweight
- Heavyweight

A heavyweight toolkit uses OS's API to draw the widgets. For example, Borland's VCL is a heavyweight toolkit. It depends on WIN32 API, the built in Windows application programming interface. On UNIX systems, we have GTK+ toolkit, which is built on top of X11 library. Swing is a lightweight toolkit. It paints it's own widgets. Similarly does the Qt4 toolkit.

17.4.1. Java Swing Examples

Below is a java swing code for the traditional Hello World program.

Basically, the idea behind this Hello World program is to learn how to create a java program, compile and run it. To create your java source code you can use any editor (Text pad/Edit plus) or you can use an IDE like Eclipse.

```
import javax.swing.JFrame;
import javax.swing.JLabel;

//import statements
//Check if window closes automatically. Otherwise add suitable code
public class HelloWorldFrame extends JFrame {

    public static void main(String args[]) {
        new HelloWorldFrame();
    }
    HelloWorldFrame() {
        JLabel jlbHelloWorld = new JLabel("Hello World");
        add(jlbHelloWorld);
        this.setSize(100, 100);
        // pack();
        setVisible(true);
```

```
        }
    }
```

Output

17.5. JAVA SERVLETS

Servlets are Java classes that process the request dynamically and generate response independent of the protocol. Servlets are defined in Java Servlet API specification.

Servlets are server side Java programs which extends the functionality of web server. Servlet are protocol independent that means it can be used virtually with any protocol to process the request and generate the response. However in practice Servlets are used to process the HTTP requests and generate the HTML response.

Servlets are commonly used with web servers, where they can take the place of CGI scripts. A servlet is similar to a proprietary server extension, except that it runs inside a Java Virtual Machine (JVM) on the server (see Figure 17.3), so it is safe and portable. Servlets operate solely within the domain of the server: unlike applets, they do not require support for Java in the web browser.

Unlike CGI and FastCGI, which use multiple processes to handle separate programs and/or separate requests, servlets are all handled by separate threads within the web server process. This means that servlets are also efficient and scalable. Because servlets run within the web server, they can interact very closely with the server to do things that are not possible with CGI scripts.

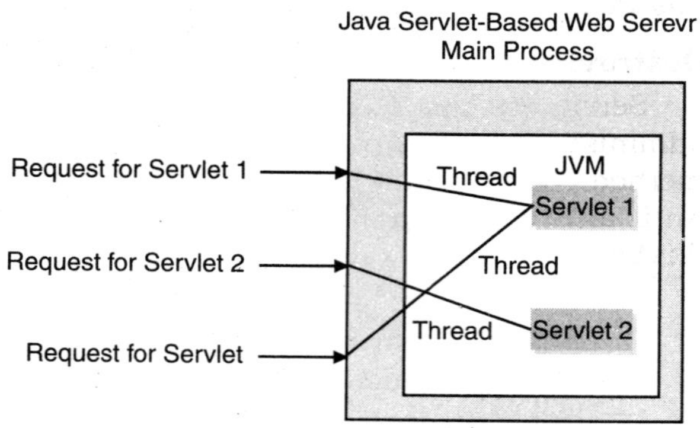

Fig. 17.3.

Another advantage of servlets is that they are portable: both across operating systems as we are used to with Java and also across web servers. As you'll see shortly, all of the major web servers support servlets. We believe that Java servlets offer the best possible platform for web application development, and we'll have much more to say about this later in the chapter.

Although servlets are most commonly used as a replacement for CGI scripts on a web server, they can extend any sort of server. Imagine, for example, a Java-based FTP server that handles each command with a separate servlet. New commands can be added by simply plugging in new servlets. Or, imagine a mail server that allows servlets to extend its functionality, perhaps by performing a virus scan on all attached documents or handling mail filtering tasks.

This book emphasizes the use of servlets as a replacement for CGI programs. We believe that, at least in the near term, most servlet developers will design and deploy servlets for use with HTTP servers. In the long term, however, other uses are likely to catch on, so this book takes pains to point out what functionality is applicable to generic servlets and what applies only to HTTP servlets. Whatever you hope to do with servlets, this book can help you with your task.

17.5.1. The Servlet Life Cycle

Each servlet has the same life cycle:

- A server loads and initializes the servlet
- The servlet handles zero or more client requests
- The server removes the servlet (some servers do this step only when they shut down)

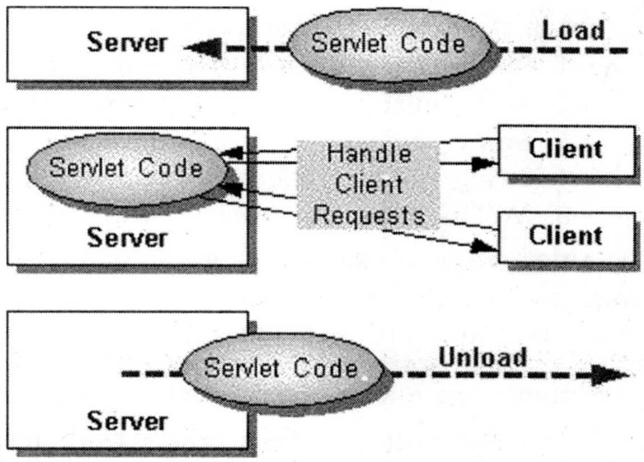

Fig. 17.4. Servlet life cycle

Initializing a Servlet

When a server loads a servlet, the server runs the servlet's `init` method. Initialization completes before client requests are handled and before the servlet is destroyed.

Even though most servlets are run in multi-threaded servers, servlets have no concurrency issues during servlet initialization. The server calls the `init` method once, when the server loads the servlet, and will not call the `init` method again unless the server is reloading the servlet. The server can not reload a servlet until after the server has destroyed the servlet by running the `destroy` method.

Destroying a Servlet

Servlets run until the server destroys them, for example, at the request of a system administrator. When a server destroys a servlet, the server runs the servlet's `destroy` method. The method is run once; the server will not run the `destroy` method again until after the server reloads and reinitializes the servlet. When the server calls the `destroy` method, another thread might be running a service request.

SUMMARY

— Java are used to create software components which can be use many places and we can reuse it.

— Java swings are basically based an AWT architecture and provides GUI features.

— Swings are the part of JFC (Java Foundation Classes)

— Java Servlet provides the mechanism of communicating among the client and servers.

SOLVED PROBLEMS

Q. 17.1. Are Swings a lightweight component? What is a h+eavyweight component?

Ans. A heavyweight component is one that is associated with its own native screen resource (commonly known as a peer). A lightweight component is one that "borrows" the screen resource of an ancestor (which means it has no native resource of its own — so it's "lighter").

As we know, AWT uses the underlying Window Tool Kit for UserInterface (i.e., native screen resource) where as Swing components does not use the native resources and each part of the Swing component is fully developed in Java and hence Swings are light weight component.

Swings are developed based on AWT architecture. A significant benefit of such layering is that it greatly facilitates porting AWT-component programs to Swing. In fact, because both the AWT and Swing component sets use the same AWT infrastructure, it's possible to mix both kinds of components in the same program.

Q. 17.2. Differentiate: AWT, Swing.

Ans. 1. Swing is based on AWT architecture.

2. Swing mixes light- and heavy-weight components.
3. Swing components are all Java beans.
4. Swing is faster and more complete.
5. AWT is older and used in Java's Micro Edition.

Q. 17.3. What is the use of Servlets?

Ans. Servlets are used as a middleware and contains the whole business logic and keep client and server free with their presentation and data processing parts respectively. They are mainly used in web projects where many clients are requesting the service from the server. The communication always takes place through servlets.

Q. 17.4. How will you pass values from HTML page to the Servlet?

Ans. HTML page calls Servlet using action attribute of form tag. In the servlet paramaters names and their values can be retreived using

request.getParameterNames() and request.getParameterValues(paramName) functions.

Q. 17.5. What is the servlet?

Ans. Servlets are modules that extend request/response-oriented servers, such as Java-enabled web servers. For example, a servlet might be responsible for taking data in an HTML order-entry form and applying the business logic used to update a company's order database.

Servlets are to servers what applets are to browsers. Unlike applets, however, servlets have no graphical user interface.

Q. 17.6. Whats the advantages using servlets than using CGI?

Ans. Servlets provide a way to generate dynamic documents that is both easier to write and faster to run. Servlets also address the problem of doing server-side programming with platform-specific APIs: they are developed with the Java Servlet API, a standard Java extension.

Q. 17.7. What are the uses of Servlets?

Ans. A servlet can handle multiple requests concurrently, and can synchronize requests. This allows servlets to support systems such as on-line conferencing.

Servlets can forward requests to other servers and servlets. Thus servlets can be used to balance load among several servers that mirror the same content, and to partition a single logical service over several servers, according to task type or organizational boundaries.

Q. 17.8. Which package provides interfaces and classes for writing servlets?

Ans. javax

Q. 17.9. Whats the Servlet Interface?

Ans. The central abstraction in the Servlet API is the Servlet interface. All servlets implement this interface, either directly or, more commonly, by extending a class that implements it such as HttpServlet.

Servlets—>Generic Servlet—>HttpServlet—>MyServlet.

The Servlet interface declares, but does not implement, methods that manage the servlet and its communications with clients. Servlet writers provide some or all of these methods when developing a servlet.

Q. 17.10. When a servlet accepts a call from a client, it receives two objects. What are they?

Ans. ServeltRequest: Which encapsulates the communication from the client to the server. ServletResponse: Whcih encapsulates the communication from the servlet back to the client. ServletRequest and ServletResponse are interfaces defined by the javax.servlet package.

Q. 17.11. What information that the ServletRequest interface allows the servlet access to?

Ans. Information such as the names of the parameters passed in by the client, the protocol (scheme) being used by the client, and the names of the remote host that made the request and the server that received it.

The input stream, ServletInputStream.Servlets use the input stream to get data from clients that use application protocols such as the HTTP POST and PUT methods.

Q. 17.12. What information that the ServletResponse interface gives the servlet methods for replying to the client?

Ans. It Allows the servlet to set the content length and MIME type of the reply.

Provides an output stream, ServletOutputStream and a Writer through which the servlet can send the reply data.

Q. 17.13. What is the servlet Lifecycle?

Ans. Each servlet has the same life cycle:

- A server loads and initializes the servlet (init())
- The servlet handles zero or more client requests (service())
- The server removes the servlet (destroy())
  (some servers do this step only when they shut down)

Q. 17.14. How HTTP Servlet handles client requests?

Ans. An HTTP Servlet handles client requests through its service method. The service method supports standard HTTP client requests by dispatching each request to a method designed to handle that request.

Q. 17.15. How to use C++ code in Java Program?

Ans. By using a native method. A native method can specify checked exceptions in a throws clause. JNI API allows java methods to invoke native functions implemented in C language.

```
class Native{
/*
* Loading the native library in a static block ensures that it's
loaded
* before the native method call.
*/
static{System.loadLibrary("LibraryOfNativeMethods");}
native void nativeMethod();
}
class Client{
public static void main(String[] args){
Native nativeOne = new Native();
nativeONe.nativeMethod();
}
}
```

EXERCISES

Q. 17.1. Explain the architecture of JavaBeans.

Q. 17.2. What are the features of JavaBeans?

Q. 17.3. Java swings are light-weight processes. What day on mean by this?

Q. 17.4. Explain about the JFC.

Q. 17.5. What is meant by component mode?

Q. 17.6. What is meant by Introspectional.

Q. 17.7. What are the features of Java Swings?

Q. 17.8. What is the difference between Swings and AWT?

Q. 17.9. What is the role of servlets?

Q. 17.10. What is the life cycle of servlet?

Q. 17.11. What is the ServletInterface?

Chapter 18

JDBC

18.1. INTRODUCTION

The JDBC API is a Java API that can access any kind of tabular data, especially data stored in a Relational Database.

JDBC helps you to write java applications that manage these three programming activities:

1. Connect to a data source, like a database
2. Send queries and update statements to the database
3. Retrieve and process the results received from the database in answer to your query. The following simple code fragment gives a simple example of these three steps:

```
Connection con = DriverManager.getConnection
        (      "jdbc:myDriver:wombat",
"myLogin","myPassword");

Statement stmt = con.createStatement();
ResultSet rs = stmt.executeQuery("SELECT a,
b, c FROM Table1");
while (rs.next()) {
        int x = rs.getInt("a");
        String s = rs.getString("b");
        float f = rs.getFloat("c");
        }
```

This short code fragment instantiates a `DriverManager` object to connect to a database driver and log into the database, instantiates a Statement object that carries

CHAPTER OUTLINE

your SQL language query to the database; instantiates a ResultSet object that retrieves the results of your query, and executes a simple while loop, which retrieves and displays those results. It's that simple.

> **Concept**
>
> *JDBC (Java Database Connectivity) is an application program Interface (API) specification for connecting programs written in Java to the data in popular databases.*

18.2. JDBC PRODUCT COMPONENTS

JDBC includes four components:

1. JDBC API
2. JDBC Driver Manager
3. JDBC Test Suite
4. JDBC-ODBC Bridge

18.2.1. The JDBC API

The JDBC™ API provides programmatic access to relational data from the Java™ programming language. Using the JDBC API, applications can execute SQL statements, retrieve results, and propagate changes back to an underlying data source. The JDBC API can also interact with multiple data sources in a distributed, heterogeneous environment.

The JDBC API is part of the Java platform, which includes the *Java™ Standard Edition* (Java™ SE) and the *Java™ Enterprise Edition* (Java™ EE). The JDBC 4.0 API is divided into two packages: java.sql and javax.sql. Both packages are included in the Java SE and Java EE platforms.

18.2.2. JDBC Driver Manager

The JDBC DriverManager class defines objects which can connect Java applications to a JDBC driver. DriverManager has traditionally been the backbone of the JDBC architecture. It is quite small and simple.

The Standard Extension packages javax.naming and javax.sql let you use a DataSource object registered with a *Java Naming and Directory Interface*™ (JNDI) naming service to establish a connection with a data source. You can use either connecting mechanism, but using a DataSource object is recommended whenever possible.

18.2.3. JDBC Test Suite

The JDBC driver test suite helps you to determine that JDBC drivers will run your program. These tests are not comprehensive or exhaustive, but they do exercise many of the important features in the JDBC API.

18.2.4. JDBC-ODBC Bridge

The Java Software bridge provides JDBC access via ODBC drivers. Note that you need to load ODBC binary code onto each client machine that uses this driver. As a result, the ODBC driver is most appropriate on a corporate network where client installations are not a major problem, or for application server code written in Java in a three-tier architecture.

This Trail uses the first two of these four JDBC components to connect to a database and then build a java program that uses SQL commands to communicate with a test Relational Database. The last two components are used in specialized environments to test web applications, or to communicate with ODBC-aware DBMSs.

18.3. JDBC ARCHITECTURE

The JDBC API supports both two-tier and three-tier processing models for database access.

18.3.1. Two-Tier Processing Models

In the two-tier model, a Java applet or application talks directly to the data source. This requires a JDBC driver that can communicate with the particular data source being accessed. A user's commands are delivered to the database or other data source, and the results of those statements are sent back to the user. The data source may be located on another machine to which the user is connected via a network. This is referred to as a client/server configuration, with the user's machine as the client, and the machine housing the data source as the server. The network can be an intranet, which, for example, connects employees within a corporation, or it can be the Internet.

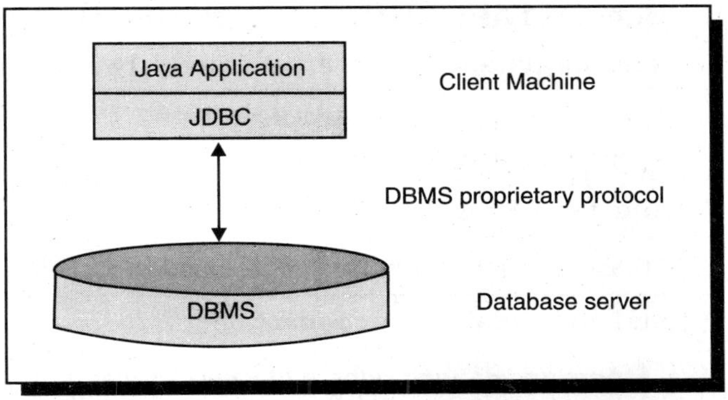

Fig. 18.1. Two-tier Architecture for Data Access

18.3.2. Three-Tier Processing Model

In the three-tier model, commands are sent to a "middle tier" of services, which then sends the commands to the data source. The data source processes the commands and sends the results back to the middle tier, which then sends them to the user. MIS directors find the three-tier model very attractive because the middle tier makes it possible to maintain control over access and the kinds of updates that can be made to corporate data. Another advantage is that it simplifies the deployment of applications. Finally, in many cases, the three-tier architecture can provide performance advantages.

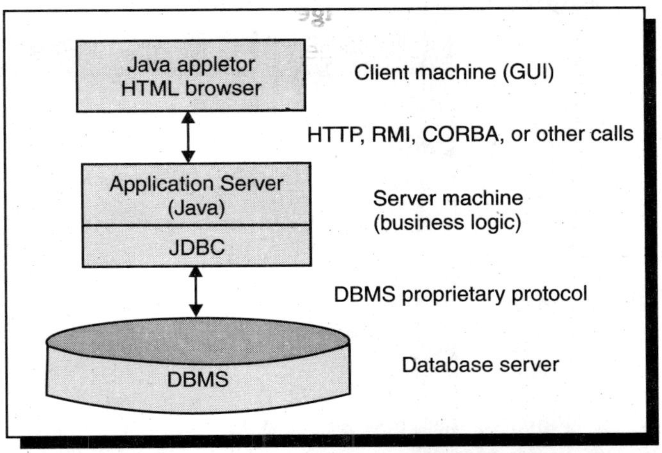

Fig. 18.2. Three-tier Architecture for Data Access

Until recently, the middle tier has often been written in languages such as C or C++, which offer fast performance. However, with the introduction of optimizing compilers that translate Java bytecode into efficient machine-specific code and technologies such as Enterprise JavaBeans™, the Java platform is fast becoming the standard platform for middle-tier development. This is a big plus, making it possible to take advantage of Java's robustness, multithreading, and security features.

With enterprises increasingly using the Java programming language for writing server code, the JDBC API is being used more and more in the middle tier of a three-tier architecture. Some of the features that make JDBC a server technology are its support for connection pooling, distributed transactions, and disconnected rowsets. The JDBC API is also what allows access to a data source from a Java middle tier.

18.4. ESTABLISHING A CONNECTION

To connect Database you must import two packages:

```
import java.net.URL;
import java.sql.*;\
```

And then load the JDBC driver:

```
Class.forName("sun.jdbc.odbc.JdbcOdbcDriver");
```

And then make the connection:

```
String url = "jdbc:odbc:b22access prices";
Connection con = DriverManager.getConnection(url);
```

You have to create a DNS for the data base which you have used with your program, here we given a method to create a DNS for MS-Access:

Step-1: Select Control Panel → Admin tools → ODBC

Step-2: Select System DSN → Add... → Select Microsoft Access Driver (*.mdb)

Step-3: Provide the DSN Name (b22access) and select the database filename.

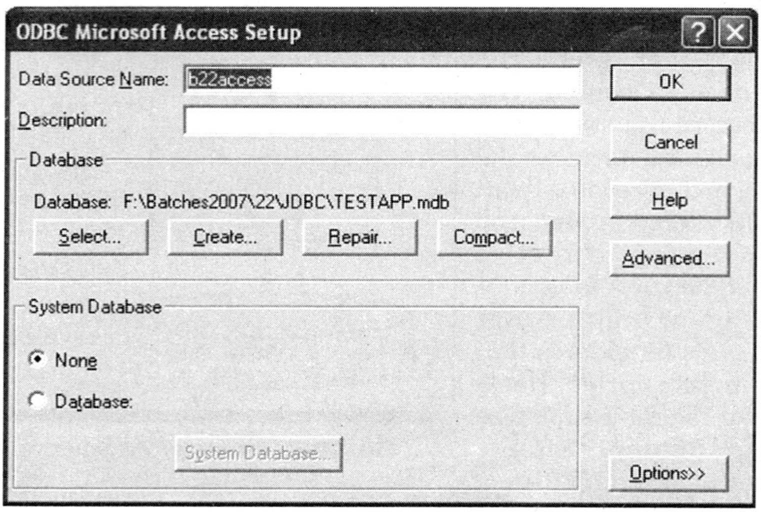

18.5. CREATING AND EXECUTING JDBC STATEMENTS

A Statement object is what sends your SQL statement to the DBMS. You simply create a Statement object and then execute it, supplying the appropriate execute method with the SQL statement you want to send. For a SELECT statement, the method to use is executeQuery. For statements that create or modify tables, the method to use is executeUpdate.

```
Statement stmt = con.createStatement();
```

18.5.1. Executing Statements (actually, SQL *queries*)

For Insertion:

```
sqlString = "insert into FoodPrice (StoreKey, FoodKey, Price) "
        + "values (3, 6, '$3.99')";
insertStmt.executeUpdate(sqlString);
```

For Selection:

```
sqlString = "select FoodKey, Price from FoodPrice " +
      "order by Price";
selectStmt.executeQuery(sqlString);
```

18.5.2. Retrieving Values from Result Sets

When executing a SELECT statement, JDBC returns results in a ResultSet object:

```
ResultSet prices = selectStmt.executeQuery(sqlString);
```

Using the Method `next`: works like the `hasNext` method of the `Iterator` type

```
while (prices.next())
{
      // process each row in the result set
}
```

Using the `getXXX` Methods: you may retrieve fields in any type using `getString`

```
// hard-coded for a resultset with two fields in each row
for (int i=1; i<=2; i++)
// be careful: index starting with 1, NOT 0!
System.out.print(prices.getString(i) + " ");
```

18.5.3. Updating Tables

For update:

```
String updateString = "UPDATE FoodPrice " +
                "SET Price = '$0.75' " +
                "WHERE FSKey = 3";
```

For delete: apply to a whole record in a table, not just a few fields in a record

```
String deleteString = "DELETE FROM FoodPrice " +
                "WHERE FSKey = 3";
```

18.5.4. Using Prepared Statements

A stored procedure is a group of SQL statements that can be called by name. In other words, it is executable code, a mini-program, that performs a particular task that can be invoked the same way one can call a function or method. Traditionally, stored procedures have been written in a DBMS-specific programming language. The latest generation of database products allows stored procedures to be written using the Java programming language and the JDBC API. Stored procedures written in the Java programming language are bytecode portable between DBMSs. Once a stored procedure is written, it can be used and reused because a DBMS that supports stored procedures will, as its name implies, store it in the database.

When to Use a `PreparedStatement` Object:

If you want to execute a Statement object many times, it will normally reduce execution time to use a `PreparedStatement` object instead.

Compare the following equivalent code segments:

Using *normal* statement:

```
String updateString = "UPDATE COFFEES SET SALES = 75 " +
            "WHERE COF_NAME LIKE 'Colombian'";
stmt.executeUpdate(updateString);
```

- Using prepared statement:

```
PreparedStatement updateSales = con.prepareStatement(
      "UPDATE COFFEES SET SALES = ? WHERE COF_NAME LIKE ?");
updateSales.setInt(1, 75);
updateSales.setString(2, "Colombian");
updateSales.executeUpdate();
```

The following code is an example of how to create a very simple stored procedure using the Java programming language. Note that the stored procedure is just a static Java method that contains normal JDBC code. It accepts two input parameters and uses them to change an employee's car number.

Do not worry if you do not understand the example at this point. The code example below is presented only to illustrate what a stored procedure looks like. You will learn how to write the code in this example.

Example 18.1: Prepared Statement

```
import java.sql.*;

public class UpdateCar {

    public static void UpdateCarNum(int carNo, int empNo)
            throws SQLException {
 Connection con = null;
 PreparedStatement pstmt = null;

 try {
```

```
    con = DriverManager.getConnection("jdbc:default:connection");

        pstmt = con.prepareStatement(
        "UPDATE EMPLOYEES SET CAR_NUMBER = ? " +
        "WHERE EMPLOYEE_NUMBER = ?");
        pstmt.setInt(1, carNo);
        pstmt.setInt(2, empNo);
        pstmt.executeUpdate();
    }
    finally {
    if (pstmt != null) pstmt.close();
    }
  }
}
```

18.6. JDBC SAMPLE APPLICATIONS

Example 18.2: A Simple JDBC Example

```
import java.sql.*;

public class JDBCSample {
 public static void main(java.lang.String[] args) {
 try {
    // This is where we load the driver
    Class.forName("sun.jdbc.odbc.JdbcOdbcDriver");
 }
 catch (ClassNotFoundException e) {
    System.out.println("Unable to load Driver Class");
    return;
 }

 try {
    // All database access is within a try/catch block. Connect to
    database,
 // specifying particular database, username, and password
    Connection con = DriverManager.getConnection("jdbc:odbc:
    companydb","", "");

    // Create and execute an SQL Statement
    Statement stmt = con.createStatement( );
    ResultSet rs = stmt.executeQuery("SELECT FIRST_NAME FROM EMPLOYEES");

    // Display the SQL Results
    while(rs.next( )) {
```

```
    System.out.println(rs.getString("FIRST_NAME"));
  }

    // Make sure our database resources are released
    rs.close( );
    stmt.close( );
    con.close( );

  }
    catch (SQLException se) {
    // Inform user of any SQL errors
    System.out.println("SQL Exception: " + se.getMessage( ));
    se.printStackTrace(System.out);
    }
  }
}
```

Example 18.2 starts out by loading a JDBC driver class (in this case, Sun's JDBC-ODBC Bridge). Then it creates a database connection, represented by a Connection object, using that driver. With the database connection, we can create a Statement object to represent an SQL statement. Executing an SQL statement produces a `ResultSet` that contains the results of a query. The program displays the results and then cleans up the resources it has used. If an error occurs, a `SQLException` is thrown, so our program traps that exception and displays some of the information it encapsulates.

Here is an another example for JDBC

Example 18.3: Simple JDBC program

```
import java.net.URL;
import java.sql.*;
import java.util.*;

public class InsertTest {
 public static void main(String[] args) {
 String driver = "sun.jdbc.odbc.JdbcOdbcDriver";
 String url = "jdbc:odbc:Grocery prices";

 try {
 Class.forName(driver);
 Connection con = DriverManager.getConnection(url);

 //get the metadata; not used in this sample
 DatabaseMetaData dma = con.getMetaData();

 // retrieving info from db
 Statement selectStmt = con.createStatement();
```

```
ResultSet prices =
selectStmt.executeQuery("select FoodKey, Price " +
"from FoodPrice " +
"order by Price");

while (prices.next()) {
for (int i=1; i<=2; i++)
System.out.print(prices.getString(i) + " ");
System.out.println();
}

selectStmt.close();

// inserting data into db
String insertSql = "insert into FoodPrice " +
"(StoreKey, FoodKey, Price) " +
"values (3, 6, '$3.99')";
Statement insertStmt = con.createStatement();
boolean failed = !(insertStmt.execute(insertSql //,
//Statement.RETURN_GENERATED_KEYS
));
System.out.println("Insertion failed: "+failed);

insertStmt.close();

con.close();
} //load the Bridge driver
catch (Exception e) {
e.printStackTrace();
}

}
}
```

Example 18.4 **Program to display the data of table emp of database empdata which contains name and age field. (Note database is created in MS-Access).**

```
import java.sql.*;

public class Database
{
      public static void main(String[] args)
      {
      try
      {
```

```
            Class.forName("sun.jdbc.odbc.JdbcOdbcDriver");
            Connection con =
            DriverManager.getConnection("jdbc:odbc:empdata");

            Statement stmt=con.createStatement();

            ResultSet rs=stmt.executeQuery("SELECT * FROM emp");

            while(rs.next())
            {
            System.out.print("Name:"+rs.getString("name"));
            System.out.println("Age:"+rs.getString("age"));
            }
                rs.close();
                //stmt.close();

            /* String qu="insert into emp(name,age) values('anshul',20)";
            stmt=con.createStatement();
            Boolean b=stmt.execute(qu);*/
            stmt.close();

                con.close();

        }
        catch(Exception e)
        {
            System.out.println(e);
        }
        }
    }
```

18.7. METADATA

Databases store user data, and they also store information about the database itself. Most DBMSs have a set of system tables, which list tables in the database, column names in each table, primary keys, foreign keys, stored procedures, and so forth. Each DBMS has its own functions for getting information about table layouts and database features. JDBC provides the interface DatabaseMetaData, which a driver writer must implement so that its methods return information about the driver and/or DBMS for which the driver is written. For example, a large number of methods return whether or not the driver supports a particular functionality. This interface gives users and tools a standardized way to get metadata. In general, developers writing tools and drivers are the ones most likely to be concerned with metadata.

JDBC provides two classes to access the Metadata:

- DatabaseMetaData
- ResultSetMetaData

DatabaseMetaData: It provides information about the database using getMetaData() method from Connection. The methods of DatabaseMetdaData are:

- getDatabaseProductName()
- getDatabaseProductVersion()
- getTables()

> **Concept**
>
> *Some information about the tables fields, its size and type is called as meta information.*

ResultSetMetaData: It returned from ResultSet with getMetaData() method. The methods of ResultSetMetaData are:

- int getColumnCount()
- String getColumnName(index)
- String getColumnTypeName(index)
- int getColumnDisplaySize(index)

SUMMARY

— JDBC helps you to write java applications that connect to a databases.

— JDBC includes four components JDBC API, JDBC Driver Manager, JDBC Test driver and JDBC-ODBC Bridge.

— JDBC API supports both two-tier and three tier processing model.

— We have given an over view of relational database to work with SQL Queries.

— We have also discussed the method of using prepared statements.

SOLVED PROBLEMS

Q. 18.1. What are the steps involved in establishing a connection?

Ans. This involves two steps: (1) loading the driver and (2) making the connection.

Q. 18.2. How can you load the drivers?

Ans. Loading the driver or drivers you want to use is very simple and involves just one line of code. If, for example, you want to use the JDBC-ODBC Bridge driver, the following code will load it:

Example,

```
Class.forName("sun.jdbc.odbc.JdbcOdbcDriver");
```

Your driver documentation will give you the class name to use. For instance, if the class name is jdbc.DriverXYZ, you would load the driver with the following line of code:

Example,

```
Class.forName("jdbc.DriverXYZ");
```

Q. 18.3. What Class.forName will do while loading drivers?

Ans. It is used to create an instance of a driver and register it with the DriverManager.

When you have loaded a driver, it is available for making a connection with a DBMS.

Q. 18.4. How can you make the connection?

Ans. In establishing a connection is to have the appropriate driver connect to the DBMS. The following line of code illustrates the general idea:

Example,

```
String url = "jdbc:odbc:Fred";
Connection con = DriverManager.getConnection(url, "Fernanda", "J8");
```

Q. 18.5. How can you create JDBC statements?

Ans. A Statement object is what sends your SQL statement to the DBMS. You simply create a Statement object and then execute it, supplying the appropriate execute method with the SQL statement you want to send. For a SELECT statement, the method to use is executeQuery. For statements that create or modify tables, the method to use is executeUpdate.

Example,

It takes an instance of an active connection to create a Statement object. In the following example, we use our Connection object con to create the Statement object stmt:

```
Statement stmt = con.createStatement();
```

Q. 18.6. How can you retrieve data from the ResultSet?

Ans. Step 1.

JDBC returns results in a ResultSet object, so we need to declare an instance of the class ResultSet to hold our results. The following code demonstrates declaring the ResultSet object rs.

Example,

```
ResultSet rs = stmt.executeQuery("SELECT COF_NAME, PRICE FROM COFFEES");
Step2.
String s = rs.getString("COF_NAME");
```

The method getString is invoked on the ResultSet object rs , so getString will retrieve (get) the value stored in the column COF_NAME in the current row of rs.

Q. 18.7. What are the different types of Statements?

Ans. There are three types of statements:

1. Statement (use createStatement method)
2. Prepared Statement (Use prepareStatement method) and
3. Callable Statement (Use prepareCall)

Q. 18.8. How can you use PreparedStatement?

Ans. This special type of statement is derived from the more general class, Statement.If you want to execute a Statement object many times, it will normally reduce execution time to use a PreparedStatement object instead.

The advantage to this is that in most cases, this SQL statement will be sent to the DBMS right away, where it will be compiled. As a result, the PreparedStatement object contains not just an SQL statement, but an SQL statement that has been precompiled. This means that when the PreparedStatement is executed, the DBMS can just run the PreparedStatement's SQL statement without having to compile it first.

Example,

```
PreparedStatement updateSales = con.prepareStatement("UPDATE COFFEES
SET SALES = ? WHERE COF_NAME LIKE ?");
```

Q. 18.9. What setAutoCommit does?

Ans. When a connection is created, it is in auto-commit mode. This means that each individual SQL statement is treated as a transaction and will be automatically committed right after it is executed. The way to allow two or more statements to be grouped into a transaction is to disable auto-commit mode

Example,

```
con.setAutoCommit(false);
```

Once auto-commit mode is disabled, no SQL statements will be committed until you call the method commit explicitly.

Example,

```
con.setAutoCommit(false);
PreparedStatement updateSales = con.prepareStatement(
```

```
"UPDATE COFFEES SET SALES = ? WHERE COF_NAME LIKE ?");
updateSales.setInt(1, 50);
updateSales.setString(2, "Colombian");
updateSales.executeUpdate();
PreparedStatement updateTotal = con.prepareStatement("UPDATE COFFEES SET TOTAL =
TOTAL + ? WHERE COF_NAME LIKE ?");
updateTotal.setInt(1, 50);
updateTotal.setString(2, "Colombian");
updateTotal.executeUpdate();
con.commit();
con.setAutoCommit(true);
```

Q. 18.10. How to call a Strored Procedure from JDBC?

Ans. The first step is to create a CallableStatement object. As with Statement an and PreparedStatement objects, this is done with an open Connection

object. A CallableStatement object contains a call to a stored procedure;

Example,

```
CallableStatement cs = con.prepareCall("{call SHOW_SUPPLIERS}");
ResultSet rs = cs.executeQuery();
```

Q. 18.11. How to Retrieve Warnings?

Ans. SQLWarning objects are a subclass of SQLException that deal with database access warnings. Warnings do not stop the execution of an application, as exceptions do; they simply alert the user that something did not happen as planned.

A warning can be reported on a Connection object, a Statement object (including PreparedStatement and CallableStatement objects), or a ResultSet object. Each of these classes has a getWarnings method, which you must invoke in order to see the first warning reported on the calling object

Example,

```
SQLWarning warning = stmt.getWarnings();
if (warning != null) {
System.out.println("\n—Warning—\n");
while (warning != null) {
System.out.println("Message: " + warning.getMessage());
System.out.println("SQLState: " + warning.getSQLState());
System.out.print("Vendor error code: ");
System.out.println(warning.getErrorCode());
System.out.println("");
warning = warning.getNextWarning();
}
}
```

Q. 18.12. How can you Move the Cursor in Scrollable Result Sets ?

Ans. One of the new features in the JDBC 2.0 API is the ability to move a result set's cursor backward as well as forward. There are also methods that let you move the cursor to a particular row and check the position of the cursor.

Example,

```
Statement stmt = con.createStatement(ResultSet.TYPE_SCROLL_SENSITIVE,
ResultSet.CONCUR_READ_ONLY);
```

```
ResultSet srs = stmt.executeQuery("SELECT COF_NAME, PRICE FROM COFFEES");
```

The first argument is one of three constants added to the ResultSet API to indicate the type of a ResultSet object: TYPE_FORWARD_ONLY, TYPE_SCROLL_INSENSITIVE, and TYPE_SCROLL_SENSITIVE.

The second argument is one of two ResultSet constants for specifying whether a result set is read-only or updatable: CONCUR_READ_ONLY and CONCUR_UPDATABLE . The point to remember here is that if you specify a type, you must also specify whether it is read-only or updatable. Also, you must specify the type first, and because both parameters are of type int, the compiler will not complain if you switch the order.

Specifying the constant TYPE_FORWARD_ONLY creates a nonscrollable result set, that is, one in which the cursor moves only forward. If you do not specify any constants for the type and updatability of a ResultSet object, you will automatically get one that is TYPE_FORWARD_ONLY and CONCUR_READ_ONLY.

Q. 18.13. What's the difference between TYPE_SCROLL_INSENSITIVE, and TYPE_SCROLL_SENSITIVE?

Ans. You will get a scrollable ResultSet object if you specify one of these ResultSet constants.The difference between the two has to do with whether a result set reflects changes that are made to it while it is open and whether certain methods can be called to detect these changes. Generally speaking, a result set that is TYPE_SCROLL_INSENSITIVE does not reflect changes made while it is still open and one that is TYPE_SCROLL_SENSITIVE does. All three types of result sets will make changes visible if they are closed and then reopened.

Example,

```
Statement stmt = con.createStatement(ResultSet.TYPE_SCROLL_INSENSITIVE,
ResultSet.CONCUR_READ_ONLY);
ResultSet srs = stmt.executeQuery("SELECT COF_NAME, PRICE FROM COFFEES");
srs.afterLast();
while (srs.previous()) {
String name = srs.getString("COF_NAME");
float price = srs.getFloat("PRICE");
System.out.println(name + " " + price);
}
```

Q. 18.14. How to Make Updates to Updatable Result Sets?

Ans. Another new feature in the JDBC 2.0 API is the ability to update rows in a result set using methods in the Java programming language rather than having to send an SQL command. But before you can take advantage of this capability, you need to create a ResultSet object that is updatable. In order to do this, you supply the ResultSet constant CONCUR_UPDATABLE to the createStatement method.

Example,

```
Connection con =
DriverManager.getConnection("jdbc:mySubprotocol:mySubName");
Statement stmt = con.createStatement(ResultSet.TYPE_SCROLL_SENSITIVE,
ResultSet.CONCUR_UPDATABLE);
ResultSet uprs = stmt.executeQuery("SELECT COF_NAME, PRICE FROM COFFEES");
```

EXERCISES

Q. 18.1. How can we store the data in the database?

Q. 18.2. What is the role of JDBC?

Q. 18.3. What are the products of JDBC?

Q. 18.4. Explain the Architecture of JDBC?

Q. 18.5. Explain the integrity rules of relational database.

Q. 18.6. Give the syntax of following SQL commands.

(*a*) Select (*b*) Insert

(*c*) Delete (*d*) Update

(*e*) Create table (*f*) Drop table

(*g*) Alter table

Q. 18.7. How to establish a connection with database in Java?

Q. 18.8. What is the role of Prepared Statement?

❑❑❑

Bibliography

1. Balaguruswamy, E. Objet Oriented Programming in Java, TMH.
2. Gredy Boach, Object Oriented Analysis and Design, Pearson Education.
3. Ram Baugh, James, BLAG, EDDY, Object-Oriented Modeling and Design, TMH.
4. Herbert Schildt, The Complete Reference Java 2, TMH.
5. George Ruse, Database Programming with JDBC and Java, O'Reilly.
6. Harold, Java Network programming, O'Reilly.
7. Horstmann CayS., Cornell, Care Java 2 Vol-I and II, Pearsen Education.
8. Downing Troy Brayan, Java RMI: Remake melthod Invocation, IDG Books.
9. Herbers schildt, Joseph O'Neil, JavaBeans programming, TMH.
10. www.Javaworld.com
11. www.sun.java.com

APPENDIX

Appendix A

PLACEMENT QUESTIONS

Q.1. Describe the principles of OOPS.

Ans. There are three main principals of oops which are called Polymorphism, Inheritance and Encapsulation.

Q.2. Explain the Encapsulation principle.

Ans. Encapsulation is a process of binding or wrapping the data and the codes that operates on the data into a single entity. This keeps the data safe from outside interface and misuse. One way to think about encapsulation is as a protective wrapper that prevents code and data from being arbitrarily accessed by other code defined outside the wrapper.

Q.3. Explain the Inheritance principle.

Ans. Inheritance is the process by which one object acquires the properties of another object.

Q.4. Explain the Polymorphism principle.

Ans. The meaning of Polymorphism is something like one name many forms. Polymorphism enables one entity to be used as as general category for different types of actions. The specific action is determined by the exact nature of the situation. The concept of polymorphism can be explained as "one interface, multiple methods".

Q.5. Explain the different forms of Polymorphism.

Ans. From a practical programming viewpoint, polymorphism exists in three distinct forms in Java:

Method overloading

Method overriding through inheritance

Method overriding through the Java interface

Q.6. What are Access Specifiers available in Java?

Ans. ccess specifiers are keywords that determines the type of access to the member of a class. These are:

Public

Protected

Private

Defaults

Q.7. Describe the wrapper classes in Java.

Ans. Wrapper class is wrapper around a primitive data type. An instance of a wrapper class contains, or wraps, a primitive value of the corresponding type.

Following table lists the primitive types and the corresponding wrapper classes:

Primitive Wrapper
boolean java.lang.Boolean
byte java.lang.Byte
char java.lang.Character
double java.lang.Double
float java.lang.Float
int java.lang.Integer
long java.lang.Long
short java.lang.Short
void java.lang.Void

Q.8. Is Iterator a Class or Interface? What is its use?

Ans. Iterator is an interface which is used to step through the elements of a Collection.

Q.9. What is similarities/difference between an Abstract class and Interface?

Ans. **Differences are as follows:** Interfaces provide a form of multiple inheritance. A class can extend only one other class. Interfaces are limited to public methods and constants with no implementation. Abstract classes can have a partial implementation, protected parts, static methods, etc.

A Class may implement several interfaces. But in case of abstract class, a class may extend only one abstract class. Interfaces are slow as it requires extra indirection to to find corresponding method in in the actual class. Abstract classes are fast.

Similarities: Neither Abstract classes or Interface can be instantiated.

Q.10. If a class is located in a package, what do you need to change in the OS environment to be able to use it?

Q.11. You need to add a directory or a jar file that contains the package directories to the CLASSPATH environment variable. Let's say a class Employee belongs to a package com.xyz.hr; and is located in the file c:\dev\com\xyz\hr\Employee.java. In this case, you'd need to add c:\dev to the variable CLASSPATH. If this class contains the method main(), you could test it from a command prompt window as follows:

Ans. c:\>java com.xyz.hr.Employee

Q.12. How many methods in the Serializable interface?

Ans. There is no method in the Serializable interface. The Serializable interface acts as a marker, telling the object serialization tools that your class is serializable.

Q.13. What is the difference between Serializalble and Externalizable interface?

Ans. When you use Serializable interface, your class is serialized automatically by default. But you can override writeObject() and readObject() two methods to control more complex object serailization process. When you use Externalizable interface, you have a complete control over your class's serialization process.

Q.14. What is a transient variable in Java?

Ans. A transient variable is a variable that may not be serialized. If you don't want some field to be serialized, you can mark that field transient or static.

Q.15. What is Java?

Ans. Java is an object-oriented programming language developed initially by James Gosling and colleagues at Sun Microsystems. The language, initially called Oak (named after the oak trees outside Gosling's office), was intended to replace C++, although the feature set better resembles that of Objective C. Java should not be confused with JavaScript, which shares only the name and a similar C-like syntax. Sun Microsystems currently maintains and updates Java regularly.

Q.16. Can you have virtual functions in Java?

Ans. Yes, all functions in Java are virtual by default. This is actually a pseudo trick question because the word "virtual" is not part of the naming convention in Java (as it is in C++, C-sharp and VB.NET), so this would be a foreign concept for someone who has only coded in Java. Virtual functions or virtual methods are functions or methods that will be redefined in derived classes.

Q.17. What is more advisable to create a thread, by implementing a Runnable interface or by extending Thread class?

Ans. Strategically speaking, threads created by implementing Runnable interface are more advisable. If you create a thread by extending a thread class, you cannot extend any other class. If you create a thread by implementing Runnable interface, you save a space for your class to extend another class now or in future.

Q.18. What is NullPointerException and how to handle it?

Q.19. When an object is not initialized, the default value is null. When the following things happen, the NullPointerException is thrown:

- Calling the instance method of a null object.
- Accessing or modifying the field of a null object.
- Taking the length of a null as if it were an array.
- Accessing or modifying the slots of null as if it were an array.
- Throwing null as if it were a Throwable value.

The NullPointerException is a runtime exception. The best practice is to catch such exception even if it is not required by language design.

Q.20. What's the difference between an interface and an abstract class?

Ans. An abstract class may contain code in method bodies, which is not allowed in an interface. With abstract classes, you have to inherit your class from it and Java does not allow multiple inheritance. On the other hand, you can implement multiple interfaces in your class.

Q.21. What is synchronization and why is it important?

Ans. With respect to multithreading, synchronization is the capability to control the access of multiple threads to shared resources. Without synchronization, it is possible for one thread to modify a shared object while another thread is in the process of using or updating that object's value. This often causes dirty data and leads to significant errors.

Q.22. What are three ways in which a thread can enter the waiting state?

Ans. A thread can enter the waiting state by invoking its sleep() method, by blocking on IO, by unsuccessfully attempting to acquire an object's lock, or by invoking an object's wait() method. It can also enter the waiting state by invoking its (deprecated) suspend() method.

Q.23. What would you use to compare two String variables - the operator == or the method equals()?

Ans. I'd use the method equals() to compare the values of the Strings and the == to check if two variables point at the same instance of a String object.

Q.24. What is thread?

Ans. A thread is an independent path of execution in a system.

Q.25. What is multi-threading?

Ans. Multi-threading means various threads that run in a system.

Q.26. How does multi-threading take place on a computer with a single CPU?

Ans. The operating system's task scheduler allocates execution time to multiple tasks. By quickly switching between executing tasks, it creates the impression that tasks execute sequentially.

Q.27. How to create a thread in a program?

Ans. You have two ways to do so. First, making your class "extends" Thread class. Second, making your class "implements" Runnable interface. Put jobs in a run() method and call start() method to start the thread.

Q.28. What invokes a thread's run() method?

Ans. After a thread is started, via its start() method of the Thread class, the JVM invokes the thread's run() method when the thread is initially executed.

Q.29. What is the difference between yielding and sleeping?

Ans. When a task invokes its yield() method, it returns to the ready state. When a task invokes its sleep() method, it returns to the waiting state.

Q.30. What is the difference between Process and Thread?

Ans. A process can contain multiple threads. In most multithreading operating systems, a process gets its own memory address space; a thread doesn't. Threads typically share the heap belonging to their parent process. For instance, a JVM runs in a single process in the host O/S. Threads in the JVM share the heap belonging to that process; that's why several threads may access the same object. Typically, even though they share a common heap, threads have their own stack space. This is how one thread's invocation of a method is kept separate from another's. This is all a gross oversimplification, but it's accurate enough at a high level. Lots of details differ between operating systems. Process vs. Thread A program vs. similar to a sequential program an run on its own vs. Cannot run on its own Unit of allocation vs. Unit of execution Have its own memory space vs. Share with others Each process has one or more threads vs. Each thread belongs to one process Expensive, need to context switch vs. Cheap, can use process memory and may not need to context switch More secure. One process cannot corrupt another process vs. Less secure. A thread can write the memory used by another thread

Q.31. Can an inner class declared inside of a method access local variables of this method?

Ans. It's possible if these variables are final.

Q.32. What is the Vector class?

Ans. The Vector class provides the capability to implement a growable array of objects.

Q.33. What modifiers may be used with an inner class that is a member of an outer class?

Ans. A (non-local) inner class may be declared as public, protected, private, static, final, or abstract.

Q.34. If a method is declared as protected, where may the method be accessed?

Ans. A protected method may only be accessed by classes or interfaces of the same package or by subclasses of the class in which it is declared.

Q.35. What's the main difference between a Vector and an ArrayList?

Ans. Java Vector class is internally synchronized and ArrayList is not.

Q.36. What are wrapped classes?

Ans. Wrapped classes are classes that allow primitive types to be accessed as objects.

Q.37. Does garbage collection guarantee that a program will not run out of memory?

Ans. No, it doesn't. It is possible for programs to use up memory resources faster than they are garbage collected. It is also possible for programs to create objects that are not subject to garbage collection.

Q.38. What is the difference between preemptive scheduling and time slicing?

Ans. Under preemptive scheduling, the highest priority task executes until it enters the waiting or dead states or a higher priority task comes into existence. Under time slicing, a task executes for a predefined slice of time and then reenters the pool of ready tasks. The scheduler then determines which task should execute next, based on priority and other factors.

Q.39. Name Component subclasses that support painting ?

Ans. The Canvas, Frame, Panel, and Applet classes support painting.

Q.40. What is a native method?

Ans. A native method is a method that is implemented in a language other than Java.

Q.41. How can you write a loop indefinitely?

Ans. for(;;)—for loop; while(true)—always true, etc.

Q.42. Can an anonymous class be declared as implementing an interface and extending a class?

Ans. An anonymous class may implement an interface or extend a superclass, but may not be declared to do both.

Q.43. What is the purpose of finalization?

Ans. The purpose of finalization is to give an unreachable object the opportunity to perform any cleanup processing before the object is garbage collected.

Q.44. How does Java handle integer overflows and underflows?

Ans. It uses low order bytes of the result that can fit into the size of the type allowed by the operation.

Q.45. What is the difference between the Boolean & operator and the && operator?

Ans. If an expression involving the Boolean & operator is evaluated, both operands are evaluated. Then the & operator is applied to the operand. When an expression involving the && operator is evaluated, the first operand is evaluated. If the first operand returns a value of true then the second operand is evaluated. The && operator is then applied to the first and second operands. If the first operand evaluates to false, the evaluation of the second operand is skipped.

Operator & has no chance to skip both sides evaluation and && operator does. If asked why, give details as above.

Q.46. How can a subclass call a method or a constructor defined in a superclass?

Ans. Use the following syntax: super.myMethod(); To call a constructor of the superclass, just write super(); in the first line of the subclass's constructor.

Q.47. What is the purpose of the System class?

Ans. The purpose of the System class is to provide access to system resources.

Q48. What is the purpose of the finally clause of a try-catch-finally statement?

Ans. The finally clause is used to provide the capability to execute code no matter whether or not an exception is thrown or caught.

Q.49. What is an abstract method?

Ans. An abstract method is a method whose implementation is deferred to a subclass. Or, a method that has no implementation.

Q.50. What is the difference between interface and abstract class?

Ans. Interface contains methods that must be abstract; abstract class may contain concrete methods. interface contains variables that must be static and final; abstract class may contain non-final and final variables. members in an interface are public by default, abstract class may contain non-public members. interface is used to "implements"; whereas abstract class is used to "extends". interface can be used to achieve multiple inheritance; abstract class can be used as a single inheritance. interface can "extends" another interface, abstract class can "extends" another class and "implements" multiple interfaces. interface is absolutely abstract; abstract class can be invoked if a main() exists. interface is more flexible than abstract class because one class can only "extends" one super class, but "implements" multiple interfaces. If given a choice, use interface instead of abstract class.

Q.51. What is a static method?

Ans. A static method is a method that belongs to the class rather than any object of the class and doesn't apply to an object or even require that any objects of the class have been instantiated.

Q.52. What is a protected method?

Ans. A protected method is a method that can be accessed by any method in its package and inherited by any subclass of its class.

Q.53. What is the difference between a static and a non-static inner class?

Ans. A non-static inner class may have object instances that are associated with instances of the class's outer class. A static inner class does not have any object instances.

Q.54. What classes of exceptions may be caught by a catch clause?

Ans. A catch clause can catch any exception that may be assigned to the Throwable type. This includes the Error and Exception types.

Q.55. What is the difference between throw and throws keywords?

Ans. The throw keyword denotes a statement that causes an exception to be initiated. It takes the Exception object to be thrown as argument. The exception will be caught by an immediately encompassing try-catch construction or propagated further up the calling hierarchy.

The throws keyword is a modifier of a method that designates that exceptions may come out of the method, either by virtue of the method throwing the exception itself or because it fails to catch such exceptions that a method it calls may throw.

Q.56. Does a class inherit the constructors of its super class?

Ans. A class does not inherit constructors from any of its superclasses.

Q.57. Name primitive Java types.

Ans. The primitive types are byte, char, short, int, long, float, double, and Boolean.

Q.58. What is the difference between static and non-static variables?

Ans. A static variable is associated with the class as a whole rather than with specific instances of a class. Non-static variables take on unique values with each object instance.

Q.59. What is the difference between the paint() and repaint() methods?

Ans. The paint() method supports painting via a Graphics object. The repaint() method is used to cause paint() to be invoked by the AWT painting thread.

Q.60. What is the purpose of the File class?

Ans. The File class is used to create objects that provide access to the files and directories of a local file system.

Q.61. What restrictions are placed on method overriding?

Ans. Overridden methods must have the same name, argument list, and return type. The overriding method may not limit the access of the method it overrides. The overriding method may not throw any exceptions that may not be thrown by the overridden method.

Q.62. What is casting?

Ans. There are two types of casting, casting between primitive numeric types and casting between object references. Casting between numeric types is used to convert larger values, such as double values, to smaller values, such as byte values. Casting between object references is used to refer to an object by a compatible class, interface, or array type reference.

Q.63. How are this() and super() used with constructors?

Ans. this() is used to invoke a constructor of the same class. super() is used to invoke a superclass constructor.

Q.64. How is it possible for two String objects with identical values not to be equal under the == operator? How are this() and super() used with constructors?

Ans. The == operator compares two objects to determine if they are the same objects in memory. It is possible for two String objects to have the same value, but located in different areas of memory.

Q.65. How can you force garbage collection?

Ans. You can't force GC, but could request it by calling System.gc(). JVM does not guarantee that GC will be started immediately.

Q.66. What are the Object and Class classes used for?

Ans. The Object class is the highest-level class in the Java class hierarchy. The Class class is used to represent the classes and interfaces that are loaded by a Java program.

Q.67. What is Serialization and deserialization ?

Ans. Serialization is the process of writing the state of an object to a byte stream. Deserialization is the process of restoring these objects.

Q.68. Explain the usage of Java packages.

Ans. This is a way to organize files when a project consists of multiple modules. It also helps resolve naming conflicts when different packages have classes with the same names. Packages access level also allows you to protect data from being used by the non-authorized classes.

Q.69. Is Java a super set of JavaScript?

Ans. No. They are completely different. Some syntax may be similar.

Q.70. How the object oriented approach helps us keep complexity of software development under control?

Ans. We can discuss such issue from the following aspects:

Objects allow procedures to be encapsulated with their data to reduce potential interference.

Inheritance allows well-tested procedures to be reused and enables changes to make once and have effect in all relevant places.

The well-defined separations of interface and implementation allow constraints to be imposed on inheriting classes while still allowing the flexibility of overriding and overloading.

Q.71. What is polymorphism?

Ans. Polymorphism means "having many forms". It allows methods (may be variables) to be written that needn't be concerned about the specifics of the objects they will be applied to. That is, the method can be specified at a higher level of abstraction and can be counted on to work even on objects of un-conceived classes.

Q.72. What is the benefit of subclass?

Ans. **Generally:** The sub class inherits all the public methods and the implementation.

The sub class inherits all the protected methods and their implementation.

The sub class inherits all the default(non-access modifier) methods and their implementation.

The sub class also inherits all the public, protected and default member variables from the super class.

The constructors are not part of this inheritance model.

Q.73. In System.out.println(),what is System,out and println,pls explain?

Ans. System is a predefined final class,out is a PrintStream object acting as a field member and println is a built-in overloaded method in the out object.

Q.74. Can you write a Java class that could be used both as an applet as well as an application?

Ans. Yes. Add a main() method to the applet.

Q.75. What is the output of x > y? a:b = p*q when x=1,y=2,p=3,q=4?

Ans. When this kind of question has been asked, find the problems you think is necessary to ask back before you give an answer. Ask if variables a and b have been declared or initialized. If the answer is yes. You can say that the syntax is wrong. If the statement is rewritten as: x

Q.76. Why Java does not support pointers?

Ans. Because pointers are unsafe. Java uses reference types to hide pointers and programmers feel easier to deal with reference types without pointers. This is why Java and C-sharp shine.

Q.77. Can you declare a class as private?

Ans. Yes, we can declare a private class as an inner class.

What is the difference between shallow copy and deep copy?

Shallow copy shares the same reference with the original object like cloning, whereas the deep copy get a duplicate instance of the original object. If the shallow copy has been changed, the original object will be reflected and vice versa.

Q.78. Why is multiple inheritance not possible in Java?

Ans. It depends on how you understand "inheritance". Java can only "extends" one super class, but can "implements" many interfaces; that doesn't mean the multiple inheritance is not possible. You may use interfaces to make inheritance work for you. Or you may need to work around. For example, if you cannot get a feature from a class because your class has a super class already, you may get that class's feature by declaring it as a member field or getting an instance of that class. So the answer is that multiple inheritance in Java is possible.

Q.79. What's the difference between constructors and other methods?

Ans. Constructors must have the same name as the class and can not return a value. They are only called once while regular methods could be called many times.

Q.80. What are the drawbacks of inheritance?

Ans. Since inheritance inherits everything from the super class and interface, it may make the subclass too clustering and sometimes error-prone when dynamic overriding or dynamic overloading in some situation. In addition, the inheritance may make peers hardly understand your code if they don't know how your super-class acts and add learning curve to the process of development.

Usually, when you want to use a functionality of a class, you may use subclass to inherit such function or use an instance of this class in your class. Which is better, depends on your specification.

Q.81. Is there any other way that you can achieve inheritance in Java?

Ans. There are a couple of ways. As you know, the straight way is to "extends" and/ or "implements". The other way is to get an instance of the class to achieve the inheritance. That means to make the supposed-super-class be a field member. When you use an instance of the class, actually you get every function available from this class, but you may lose the dynamic features of OOP.

Q.82. What is the difference between final, finally and finalize?

Ans. **Short answer:**

final - declares constant

finally - relates with exception handling

finalize - helps in garbage collection

If asked to give details, explain:

final field, final method, final class

try/finally, try/catch/finally

protected void finalize() in Object class

Q.83. The following statement prints true or false, why?

Ans.

```
byte[] a = { 1, 2, 3 };
byte[] b = (byte[]) a.clone();
System.out.println(a == b);
```

The false will be printed out. Because the two arrays have distinctive memory addresses. Starting in Java 1.2, we can use java.util.Arrays.equals(a, b) to compare whether two arrays have the same contents.

Q.84. What is the output from System.out.println("Hello"+null);?

Ans. Hellonull

Q.85. Is null a keyword?

Ans. The null value is not a keyword.

Q.86. Is sizeof a keyword?

Ans. The sizeof operator is not a keyword.

Q.87. What is clipping?

Ans. Clipping is the process of confining paint operations to a limited area or shape.

Q.88. What is the range of the short type?

Ans. The range of the short type is -(2^15) to 2^15 - 1.

Q.89. What is the range of the char type?

Ans. The range of the char type is 0 to 2^16 - 1.

Q.90. What is the argument type of a program's main() method?

Ans. A program's main() method takes an argument of the String[] type.

Q.91. Which Java operator is right associative?

Ans. The = operator is right associative.

Q.92. What is the difference between the String and StringBuffer classes?

Ans. String objects are constants. StringBuffer objects are not.

Q.93. What is the % operator?

Ans. It is referred to as the modulo or remainder operator. It returns the remainder of dividing the first operand by the second operand.

Q.94. Is &&= a valid Java operator?

Ans. No, it is not.

Q.95. Can there be an abstract class with no abstract methods in it?

Ans. yes.

Q.96. Can an Interface be final?

Ans. yes.

Q.97. Can an Interface have an inner class?

Ans. Yes public interface abc { static int i=0; void dd(); class a1 { a1() { int j; System.out.println("in interfia"); }; public static void main(String a1[]) { System.out.println("in interfia"); } } }

Q.98. Can we define private and protected modifiers for variables in interfaces?

Ans. Yes.

Q.99. Why isn't there operator overloading?

Ans. Because C++ has proven by example that operator overloading makes code almost impossible to maintain. In fact there very nearly wasn't even method overloading in Java, but it was thought that this was too useful for some very basic methods like print(). Note that some of the classes like DataOutputStream have unoverloaded methods like writeInt() and writeByte().

Q.100. What is the main difference between Java platform and other platforms?

Ans. The Java platform differs from most other platforms in that it's a software-only platform that runs on top of other hardware-based platforms.

The Java platform has three elements:

Java programming language

The Java Virtual Machine (Java VM)

The Java Application Programming Interface (Java API)

Q.101. What is the Java Virtual Machine?

Ans. The Java Virtual Machine is a software that can be ported onto various hardware-based platforms.

Q.102. Is Java code slower than native code?

Ans. Not really. As a platform-independent environment, the Java platform can be a bit slower than native code. However, smart compilers, well-tuned interpreters, and just-in-time bytecode compilers can bring performance close to that of native code without threatening portability.

Q.103. Can main() method be overloaded?

Ans. Yes. the main() method is a special method for a program entry. You can overload main() method in any ways. But if you change the signature of the main method, the entry point for the program will be gone.

Q.104. What is JIT?

Ans. JIT stands for Just In Time compiler. It compiles java byte code to native code.

Q.105. What are class members and Instance members?

Ans. Any global members(Variables, methods etc.) which are static are called as Class level members and those which are non-static are called as Instance level members.

Q.106. What is final?

Ans. A final class can't be extended ie., final class may not be subclassed. A final method can't be overridden when its class is inherited. You can't change value of a final variable (is a constant).

Q.107. What if the main method is declared as private?

Ans. The program compiles properly but at runtime it will give "Main method not public." message.

Q.108. What if the static modifier is removed from the signature of the main method?

Ans. Program compiles. But at runtime throws an error "NoSuchMethodError".

Q.109. What if I write static public void instead of public static void?

Ans. Program compiles and runs properly.

Q.110. What if I do not provide the String array as the argument to the method?

Ans. Program compiles but throws a runtime error "NoSuchMethodError".

Q.111. What is the first argument of the String array in main method?

Ans. The String array is empty. It does not have any element. This is unlike C/C++ where the first element by default is the program name.

Q.112. If I do not provide any arguments on the command line, then the String array of Main method will be empty or null?

Ans. It is empty. But not null.

Q.113. What is Overriding?

Ans. When a class defines a method using the same name, return type, and arguments as a method in its superclass, the method in the class overrides the method in the superclass.

When the method is invoked for an object of the class, it is the new definition of the method that is called, and not the method definition from superclass. Methods may be overridden to be more public, not more private.

Q.114. Is Empty .java file a valid source file?

Ans. Yes, an empty .java file is a perfectly valid source file.

Q.115. Can a .java file contain more than one java classes?

Ans. Yes, a .java file contain more than one java classes, provided at the most one of them is a public class.

Q.116. Is String a primitive data type in Java?

Ans. No String is not a primitive data type in Java, even though it is one of the most extensively used object. Strings in Java are instances of String class defined in java.lang package.

Q.117. Is main a keyword in Java?

Ans. No, main is not a keyword in Java.

Q.118. Is next a keyword in Java?

Ans. No, next is not a keyword.

Q.119. Is delete a keyword in Java?

Ans. No, delete is not a keyword in Java. Java does not make use of explicit destructors the way C++ does.

❑❑❑

MODEL QUESTION PAPERS

MODEL QUESTION PAPERS

MODEL QUESTION PAPER-I

OBJECT ORIENTED TECHNIQUES WITH JAVA

Time: **3** *Hours* *Total Marks:* **100**

Note: (*i*) *All questions carry equal marks.*
(*ii*) *Attempt all questions.*

1. Attempt *any four* parts of the following **(5 × 4 = 20)**

(*a*) What is object oriented programming? List the various advantages and disadvantages of object oriented programming.

(*b*) Using the object modeling technique, accept a numeric value by the user and calculate its factorial value.

(*c*) Draw an object model off workstation window management system (e.g., X window system).

(*d*) What is generalization? Explain.

(*e*) Draw a class diagram that shows multiple inheritance from over lapping classes.

(*f*) What is "operation-overriding"?

2. Attempt *any four* parts of the following: **(5 × 4 = 20)**

(*a*) Prepare a list of objects that you would expect each of the following systems to handle.

(*i*) A Telephone Answering Machine
(*ii*) A catalog store order entry system.

(*b*) Define the following terms, giving at least one example of each:

(*i*) Inheritance
(*ii*) Data Abstraction

(*c*) Discuss various ways of implementing DFD.

(*d*) What is dynamic modelling? Explain the purpose.

(*e*) What is polymorphism? Give the need of polymorphism, with the help of an example.

(*f*) What is the role of nested state diagram in dynamic modeling?

3. Attempt *any four* parts of the following: **(5 × 4 = 20)**

(*a*) Prepare a dataflow diagram for computing the area and perimeter a right angled triangle. Inputs are base and hypotenuse of the triangle, outputs are area and perimeter.

(*b*) Describe the relation of functional model, object model and dynamic models.

(*c*) Differentiate between SA/SD and OMT modeling.

(*d*) What are the advantage of modeling?

(*e*) What is relationship and difference between OOA and ODD (object oriented Analysis and object oriented Design)?

4. Attempt *any two* of the following: **(10 × 2 = 20)**

(*a*) (*i*) Explain the following code:

result = somecondition ? Value1: Value 2;

(*ii*) Why Java is called 100% pure object oriented language?

(*iii*) What is BYTE code?

(*iv*) What is Java Virtual Machine?

(*v*) What is Thread?

(*b*) (*i*) There are exactly 2.54 centimeter to an inch. Write a program that takes a number of inches from the command line and convert it to centimeters.

(*ii*) Explain the Type casting. And how it is done in Java?

(*c*) (*i*) What is the difference between an object and class?

(*ii*) What do you know about the garbage collector?

5. Attempt *any Two* of the following: **(10 × 2 = 20)**

(*a*) Write down the difference between following:

(*i*) Abstract class and Interface

(*ii*) Applet and Application

(*b*) (*i*) What do you mean by JavaBeans?

(*ii*) What is Java servlet? Discuss the life cycle of a servlet.

(*c*) Write short notes on any two of the following

(*i*) JDBC

(*ii*) Java swing

(*iii*) AWT

(*iv*) E-mail.

❐❐❐

MODEL QUESTION PAPER-II

OBJECT ORIENTED SYSTEM WITH JAVA

Time: **3** *Hours* *Total Marks:* **100**

Note: (*i*) *All questions carry equal marks.*
(*ii*) *Attempts all questions.*

1. Attempt *any four* parts of the following **(5 × 4 = 20)**

(*a*) What do you mean by data hiding?
(*b*) What do you mean by object-based programming?
(*c*) What is OMT methodology?
(*d*) What is the different between operation and methods?
(*e*) What do you mean by "Role names"?
(*f*) Define "ordering"?

2. Attempt *any four* parts of the following **(5 × 4 = 20)**

(*a*) Prepare an object diagram for the tower of Honoi problem. Where Tower have 3 long pegs, and each peg has several disks on it, in a certain order.
(*b*) What is the difference between link and association?
(*c*) Describe events and states. Explain nested states.
(*d*) What is meant by aggregation concurrency.
(*e*) Prepare an object model to describe a directed graph. Use directions as qualifier in your diagram so that it is possible to determine the vertex that is connected to head or to the tail of each edge.
(*f*) Explain Meta data using suitable example.

3. Attempt *any four* parts of the following **(5 × 4 = 20)**

(*a*) What is Swing? How it differs with AWT?
(*b*) Write short note on Java Foundation classes.
(*c*) Explain the state generalization with the help of suitable example.
(*d*) Prepare a DFD for computing the area a circle. Assume that the area calculation is polymorphic function of the class "circle" that inherits other properties from class "shape".
(*e*) Compare the OMT methodology with SA/SD methodology.
(*f*) Which is the best methodology for the software development? Explain.

4. Attempt *any Two* of the following: **(10 × 2 = 20)**

(*a*) Prepare an object diagram for the dining philosopher's problem. There are 5 philosophers and 5 forks around a circle table. Each philosopher access to 2 forks on either side. Each fork is shared by 2 philosophers. Each fork may be either on the table or in use by one philosopher. A philosopher must have 2 forks to eat.

(*b*) How many types of layout manager are used in java programming? Describe each type in short.

(*c*) How can we handle string in java? Write an application program in java to check whether the two input strings are equal.

5. Attempt *any two* of the following: (10 × 2 = 20)

(*a*) Differentiate between following:

(*i*) Aggregation and Association

(*ii*) Generalization and Inheritance

(*b*) What is the role of Garbage collection? Compare it with finalization.

(*c*) Write short notes on any two of following:

(*i*) Exception handling

(*ii*) Multithreading

(*iii*) Video conferencing

(*iv*) Java Swings.

❑❑❑

PREVIOUS YEAR QUESTION PAPER

(SEM. V) ODD SEMESTER THEORY EXAMINATION 2010-11

OBJECT ORIENTED TECHNIQUES

Time: **3** *Hours* *Total Marks:* **100**

Note: *Attempt all questions.*

1. Attempt *any two* parts: **(10 × 2 = 20)**

(*a*) (*i*) What do you mean by encapsulation? How does the object-oriented concept of message passing help to encapsulate the implementation of an object, including its data?

(*ii*) What do you mean by polymorphism? Is this concept only applicable to object oriented systems? Explain.

(*b*) (*i*) What do you mean by activity diagram? What are the two special states shown in an activity diagram? Explain with an example.

(*ii*) Define an abstract class. When is it required to create an abstarct class? Explain it with an example.

(*c*) (*i*) Define object oriented modeling (OOM). Describe various steps involved in OOM process. Explain.

(*ii*) What do you mean by multiple inheritances? Explain it with an example. Can you implement multiple inheritances in Java?

2. Attempt *any two* parts: **(10 × 2 = 20)**

(*a*) Describe an AUTOMOBILE base class in Java. Define its all methods and data structures. Throu8gh inheritance mechanism, create one class namelty CAR. Implement its data structures and important methods. Observe the following while designing the classes:

(*i*) Clear;ly indicate the private and public classes.

(*ii*) Design constructors in each class and explain its purpose.

(*iii*) Identify data structures and methods which can be inherited (mnake suitable assumption where required).

(*b*) What do you mean by a collaboration diagram? Explain various terms and symbols used in a collaboration diagram. How polymorphism is desired using a collaboration diagram? Explain using an example.

(*c*) (*i*) What do you understand by architectural modeling? Explain its variuous concepts and diagrams with suitable example.

(*ii*) Write a short note on use case diagram and time diagram with suitable diagram and their utility in system design.

3. Attempt *any two* parts: (10 × 2 = 20)

(*a*) Describe the following with example:

(*i*) Steps of object oriented design

(*ii*) Generalization and specialization

(*iii*) Modeling association as a class

(*iv*) Physical packaging

(*b*) (*i*) How do you map the object oriented concepts using non-object oriented languages? Explain with an example.

(*ii*) Describe the various features of object oriented languages. Also compare any two object oriented languages.

(*c*) Write short notes on the following:

(*i*) Reusability and robustness

(*ii*) Translating object oriented design into an implementation.

4. Attempt *any two* parts: (10 × 2 = 20)

(*a*) (*i*) Describe the main features of Java. Also discuss the features that make Java different from C++.

(*ii*) What do you mean by multithreading? Does it have an impact in the performance of Java? Explain.

(*b*) (*i*) Write a program in Java to calculate the sum of the digits of a given positive integer number. For example, if the given number is 12345 then the program should display : 15.

(*ii*) Design a class using java to represent a student record having the following attributes and methods:

(*i*) Attributes are as follows: Enroll_no, Name, Father_name, Branch, Year_of_admission, Student_semester, Student_address, Student_status.

(*ii*) The methods are as follows : to assign the initial values to all attributes, to admit a new student, display the list of students admitted in a given year.

(*c*) Write short notes on the following:

(*i*) Enterprise Java Beans

(*ii*) Abstract methods and classes.

5. Attempt *any Two* parts (10 × 2 = 20)

(*a*) What do you understand by ODBC? Why is it required? How it is implemented using Java? Explain with an example.

(*b*) (*i*) Define Applet. Also discuss Applet life cycle using a diagram and explain its various states.

(*ii*) Write a short note on the utility of Java as Internet programming language.

(*c*) Write short notes on the following with suitable examples:

(*i*) Java Servlets

(*ii*) Exception handling techniques.

❐❐❐

INDEX

F...

G

H...

I...

J...

K...

L

M...

U...

V...

W...

Win Prizes

ATTENTION READERS

Win Prizes

We request for your frank assessment regarding various aspects of the book as given below :

OBJECT ORIENTED TECHNIQUES

(WITH JAVA)

By: Lalit Kishore Arora & Dr. Raj Kumar Bhatia

Please Fill up the Shet in Neat and Clean Writing

(i) What topics of your syllabus, important for your examination point of view, are not covered in this book ?

....

....

....

....

(ii) What are the chapters wherein the treatment of the subject matter is not systematic or organised or updated ?

....

....

....

....

(iii) Have you come across misprints/mistakes/factual inaccuracies in the book? Please specify the chapters and the page numbers.

....

....

....

....

(iv) What other books on the same subject have you found/heard better than the present book? Please specify in terms of price and quality

....

....

....

....

(v) Further suggestions and comments for the improvement of this book.

....

....

....

....

(PS: If need be please attach a separate sheet to write your views)

OTHER DETAILS

(i) Who recommended you this book ? (Please tick (✓) in the box near the option relevant to you.

☐ Teacher ☐ Friends ☐ Bookseller

....

....

....

....

(ii) Name of the recommending teacher, his designation and address :

....

....

....

....

(iii) Name and address of the bookseller you purchased the book from :

....

....

....

....

(iv) Name and address of your institution (Please mention the University or Board as the case may be)

....

....

....

....

(v) Your name and complete postal address :

....

....

....

....

(vi) Course you are enrolled in :

....

....

....

....

The best assessment will be awarded each month: The award will be in the form of our publications as decided by the Editorial Board.

Please mail the filled up coupon at your earliest to :

S.K.KATARIA & SONS®
4760-61/23, Ansari Road,
Darya Ganj, New Delhi-110002
☎: 23243489, 23269384
Email : katariabooks@yahoo.com
katson_sanjeev@yahoo.co.in